Mathematical, Physical, Chemical, and Biological Methods for Ice and Water Problems

Mathematical, Physical, Chemical, and Biological Methods for Ice and Water Problems

Guest Editors

Zhijun Li

Fang Li

MDPI

Basel • Beijing • Wuhan • Barcelona • Belgrade • Novi Sad • Cluj • Manchester

Guest Editors
Zhijun Li
State Key Laboratory of Coastal and Offshore Engineering
Dalian University of Technology
Dalian
China

Fang Li
School of Naval Architecture, Ocean and Civil Engineering
Shanghai Jiao Tong University
Shanghai
China

Editorial Office
MDPI AG
Grosspeteranlage 5
4052 Basel, Switzerland

This is a reprint of the Special Issue, published open access by the journal *Water* (ISSN 2073-4441), freely accessible at: https://www.mdpi.com/journal/water/special_issues/70W809Z68S.

For citation purposes, cite each article independently as indicated on the article page online and as indicated below:

Lastname, A.A.; Lastname, B.B. Article Title. *Journal Name* **Year**, *Volume Number*, Page Range.

ISBN 978-3-7258-6922-0 (Hbk)
ISBN 978-3-7258-6923-7 (PDF)
https://doi.org/10.3390/books978-3-7258-6923-7

Contents

About the Editors . . . . **vii**

Preface . . . . **ix**

Zhijun Li and Fang Li
Mathematical, Physical, Chemical and Biological Methods for Ice and Water Problems
Reprinted from: *Water* **2026**, *18*, 414, https://doi.org/10.3390/w18030414 . . . . **1**

Bing Tan, Chunjiang Li, Shengbo Hu, Zhijun Li, Honglan Ji, Yu Deng and Limin Zhang
Review and Prospect of the Uncertainties in Mathematical Models and Methods for Yellow River Ice
Reprinted from: *Water* **2025**, *17*, 1291, https://doi.org/10.3390/w17091291 . . . . **6**

Zhi Liu, Hongwei Han, Yu Li, Enliang Wang and Xingchao Liu
Forecasting the River Ice Break-Up Date in the Upper Reaches of the Heilongjiang River Based on Machine Learning
Reprinted from: *Water* **2025**, *17*, 434, https://doi.org/10.3390/w17030434 . . . . **42**

Yupeng Leng, Chunjiang Li, Peng Lu, Xiang Fu and Shengbo Hu
Analysis of Open-Water Changes and Ice Microstructure Characteristics in Different River Channel Types of the Yellow River in Inner Mongolia Based on Satellite Images and Field Sampling
Reprinted from: *Water* **2025**, *17*, 1898, https://doi.org/10.3390/w17131898 . . . . **65**

Ge Li, Song Gao, Xue Chen, Yan Jiao, Linfeng Wang, Qiaokun Hou, et al.
Designing the Engineering Parameters of the Sea Ice Based on a Refined Grid in the Southern Bohai Sea
Reprinted from: *Water* **2025**, *17*, 2465, https://doi.org/10.3390/w17162465 . . . . **85**

Bingyan Gao, Yang Liu, Peng Lu, Lei Wang and Hui Liao
Advancing Sea Ice Thickness Hindcast with Deep Learning: A WGAN-LSTM Approach
Reprinted from: *Water* **2025**, *17*, 1263, https://doi.org/10.3390/w17091263 . . . . **110**

Bing Tan, Yanming Chang, Chunchun Gao, Ting Wang, Peng Lu, Yingzhe Fan and Qingkai Wang
Statistical Optimization and Analysis on the Spatial Distributions of Ice Ridge Keel in the Northwestern Weddell Sea, Antarctica
Reprinted from: *Water* **2025**, *17*, 1643, https://doi.org/10.3390/w17111643 . . . . **130**

Chaoge Yu and Yukui Tian
Numerical Simulation of Regular Wave and Ice Floe Interaction Using Coupled Eulerian–Lagrangian Method
Reprinted from: *Water* **2025**, *17*, 1879, https://doi.org/10.3390/w17131879 . . . . **146**

Shaopeng Ji, Yubo Liu, Qingkai Wang, Peng Lu and Shunqi Yuan
In Situ Tests on the Flexural Strength and Effective Elastic Modulus of Brackish Ice During Different Ice Periods
Reprinted from: *Water* **2025**, *17*, 3189, https://doi.org/10.3390/w17223189 . . . . **164**

Jiachen Li, Zhichao Wang, Weiping Li, Shuangyi Jing, Caio Graco-Roza and Lauri Arvola
Impact of the Reynolds Numbers on the Velocity of Floating Microplastics in Open Channels
Reprinted from: *Water* **2025**, *17*, 588, https://doi.org/10.3390/w17040588 . . . . **188**

Runling Li, Jianjun Wang, Wei Han, Jianying Gong and Jun Ding
Mathematical Description of the Immunochemical Response of the Manila Clam (*Ruditapes philippinarum*) to Extreme Temperature Reductions
Reprinted from: *Water* **2025**, *17*, 93, https://doi.org/10.3390/w17010093 . . . . **201**

About the Editors

Zhijun Li

Zhijun Li was a Full Professor of Ice Properties at the State Key Laboratory of Coastal and Offshore Engineering, Dalian University of Technology in China before he retired in 2025. Currently, he is Professor of Ice Properties in the "Xinjiang Tianchi Distinguished Professor Program" at College of Urban and Environmental Sciences, Shihezi University in China. He has researched ice physical and mechanical properties of China Bohai sea ice and polar sea ice; non-frozen model ice and its application for ice loads in harbor piles along the Bohai coast; lake ice growth and decay; ecosystem under ice in Chinese lakes; Yellow River ice characteristics and in situ observation technology since 1985.

Fang Li

Fang Li currently serves as an Associate Professor at School of Ocean and Civil Engineering, Shanghai Jiao Tong University. His received his B.Eng. degree at Dalian University of Technology in China, M.Sc. degree NTNU in Norway and Aalto University in Finland, and a Ph.D. from Aalto University in Finland. His research interests include polar ship and offshore engineering, with particular emphasis on sea ice mechanics and ice loads; the performance and safety of polar ships and marine structures; data-driven modeling of ship–ice interaction; and polar navigation strategies and route planning. He has been selected by the National Talent Programme and involved in various scientific projects funded by different sources.

Preface

This Reprint is a collection of research articles from the Special Issue of *Water*, "Mathematical, Physical, Chemical, and Biological Methods for Ice and Water Problems". It addresses new achievements in water–ice-related challenges to navigation, flood control, and ecological stability, etc., by using machine learning algorithms, numerical modeling, high-resolution remote sensing, in situ mechanical testing, and laboratory testing. The main goal is successfully achieved through the joint efforts of authors, anonymous reviewers, and editorial managers. In total, one review article and nine research articles from the Special Issue are included in this Reprint. These articles cover river ice formation and break-up, the interaction between waves and sea ice, and the environmental responses of aquatic systems in science. It presents new theoretical insights and practical guidelines for disaster prevention and water resource management in engineering. This Reprint could be used as a reference for postgraduate students, researchers and engineers in the fields of lake, river and sea ice.

Zhijun Li and Fang Li
Guest Editors

water

MDPI

Editorial

Mathematical, Physical, Chemical and Biological Methods for Ice and Water Problems

Zhijun Li [1,2] and Fang Li [3,*]

1 College of Urban and Environmental Sciences, Shihezi University, Shihezi 832003, China; lizhijun@dlut.edu.cn
2 State Key Laboratory of Coastal and Offshore Engineering, Dalian University of Technology, Dalian 116024, China
3 School of Ocean and Civil Engineering, Shanghai Jiao Tong University, Shanghai 200240, China
* Correspondence: fang.li@sjtu.edu.cn

1. Introduction

High-latitude and cold-region environments feature tightly coupled hydrological, cryospheric, and ecological subsystems, where seasonal freeze–thaw cycles, snow cover, permafrost, and river and lake ice fundamentally shape water flows and ecosystem processes [1,2]. However, the warming climate is disrupting these linkages. The Arctic and subarctic regions are warming at rates well above the global average, leading to accelerated permafrost thaw and altering drainage connectivity [3,4]. Consequently, ice-driven hazards have become more frequent and unpredictable. Ice-jam floods, often more destructive than open-water floods, are posing increasing risks to riverside communities due to changing winter regimes [5,6]. Furthermore, seasonal ice cover on lakes is diminishing globally, and recent pan-Arctic studies indicate that ice-covered lakes are highly sensitive to winter warming; the loss of ice cover greatly amplifies light penetration and thermal shifts, triggering cascading effects on aquatic biodiversity and water quality [7,8].

To tackle these emerging challenges, researchers have achieved significant advancements by utilizing cutting-edge technologies, and integrated numerical modeling and physical experiments have been widely adopted to analyze the mechanical behavior of ice and its interactions with offshore structures, providing insights critical for the design of safety systems in ice-covered waters [9,10]. Concurrently, the application of remote sensing and geospatial analysis has enabled comprehensive monitoring of cryospheric changes, and high-resolution satellite imagery now allows for the precise tracking of river ice evolution and sea ice drift, offering data support for disaster early warning [11,12]. Additionally, artificial intelligence has revolutionized hydrological forecasting, and deep learning algorithms, such as Long Short-Term Memory (LSTM) networks, have shown superior performance compared to traditional statistical approaches in processing nonlinear time series data for the prediction of variables like sea ice thickness and river break-up dates [13,14]. Beyond physics, interdisciplinary studies are also quantifying the transport of emerging contaminants in cold waters and assessing the physiological stress of aquatic organisms under extreme thermal variability, thereby informing integrated ecological management strategies [15].

This Special Issue brings together these and related modern methodologies to improve safety and sustainability in high-latitude environments. Through advanced numerical modeling, physical experiments, and integrated monitoring techniques, the collected studies analyze critical issues ranging from ice mechanics and hydrodynamic processes to biological responses. The findings provide critical theoretical foundations and technical support for ensuring infrastructure resilience, mitigating environmental risks, and optimizing water resource management in high-latitude regions.

2. An Overview of Published Articles

The monitoring, simulation, and risk assessment of hydrological dynamics and ice engineering processes are important in terms of ensuring the safety of infrastructure and the preservation of aquatic environments in cold regions. These complex systems, characterized by the interplay of riverine and sea ice, hydrodynamics, and environmental stressors, pose significant challenges to navigation, flood control, and ecological stability. To address these challenges, cutting-edge methodologies—ranging from machine learning algorithms and numerical modeling to high-resolution remote sensing and in situ mechanical testing—are increasingly being adopted. This Special Issue compiles ten innovative studies that delve into the mechanisms of ice formation and break-up, the interaction between waves and sea ice, and the environmental responses of aquatic systems, offering new theoretical insights and practical guidelines for disaster prevention and water resource management.

Tan et al. (Contribution (5)) provided a comprehensive review of the mathematical models and methods used for studying ice in the Yellow River, highlighting the unique uncertainties associated with the river's mixed ice–water–sediment transport. The study summarizes universal thermodynamic and hydrodynamic equations while addressing specific parameterization schemes that reflect the Yellow River's distinctive characteristics. Their results indicate that, while significant progress has been made in deterministic and stochastic modeling, challenges remain in predicting freeze-up and break-up times and in early warning for ice disasters. The findings point toward future research directions focused on integrating multi-source data to improve the accuracy of ice simulations and disaster forecasting in sediment-laden rivers.

Liu et al. (Contribution (2)) developed a forecasting model for river ice break-up dates in the upper reaches of the Heilongjiang River using advanced machine learning techniques. By comparing various feature selection methods and models, including XGBoost and Random Forest, the study identified that ice reserves and accumulated temperature during the break-up period are the most critical predictors. The results of their study show that the XGBoost model combined with Pearson Correlation Coefficient feature selection achieved the highest prediction accuracy. The findings provide a more effective and scientific approach for forecasting ice-jam floods, thereby enhancing disaster management capabilities in high-latitude river basins.

Leng et al. (Contribution (8)) investigated the spatiotemporal variations in open water and ice microstructure in the Inner Mongolia reaches of the Yellow River using Sentinel-2 satellite imagery and field sampling. The study reveals that open water transformation is distinct across different channel morphologies and is significantly driven by temperature fluctuations. Microstructural analysis indicates that ice crystals predominantly exist as columnar or granular forms, with their distribution and equivalent diameters systematically varying with channel configuration and ice thickness. These findings offer quantitative data on river ice physical properties, which are essential to understanding the heterogeneity of ice cover and its implications for river mechanics.

Li et al. (Contribution (9)) derived refined design parameters for sea ice engineering in the southern Bohai Sea using the NEMO-LIM2 ice–ocean coupling model. By partitioning the region into a high-resolution grid, the study calculated key parameters such as ice thickness, concentration, and strength for return periods ranging from 1 to 100 years. The results demonstrate that ice conditions exhibit a distinct spatial gradient, with the most severe conditions occurring in nearshore zones like Bohai Bay. The derived design strengths—specifically compressive and shear strengths—are comparable to current standards, while

 https://doi.org/10.3390/w18030414

the refined grid captures more detailed spatial variations, providing accurate data support for the anti-ice design of offshore structures.

Gao et al. (Contribution (4)) proposed a novel Wasserstein Generative Adversarial Network–Long Short-Term Memory (WGAN-LSTM) model to hindcast Arctic sea ice thickness (SIT) under conditions of data scarcity. By integrating the data generation capabilities of WGAN with the temporal prediction strengths of LSTM, and utilizing a comprehensive index (DISO) in the loss function, the model significantly outperforms traditional methods. The team's experimental results show that the WGAN-LSTM model improves prediction performance by over 50% compared to standard LSTM models, effectively capturing nonlinear changes in sea ice. The study advances short-term SIT prediction techniques, offering a robust tool for Arctic climate research and navigation safety.

Tan et al. (Contribution (6)) established a statistical optimization model to analyze the spatial distributions of ice ridge keels in the northwestern Weddell Sea, Antarctica. Using data from helicopter-borne electromagnetic induction surveys, the researchers determined an optimal keel cutoff draft of 3.8 m to differentiate keels from level ice. Their results indicate that the Wadhams'80 function and lognormal function effectively describe keel draft and spacing distributions, respectively, across different ridging intensity regimes. These findings provide a solid theoretical foundation for inverting sea ice thickness from surface data, improving the accuracy of remote sensing algorithms in polar regions.

Yu and Tian (Contribution (7)) simulated the interactions between regular waves and ice floes using the Coupled Eulerian–Lagrangian (CEL) method. The study quantitatively analyzed wave propagation and ice floe dynamics, revealing that ice floe motion is synchronized with incident wave periods and that wave height attenuation increases significantly with rising ice concentration. Furthermore, the simulation identified that ice fragmentation predominantly occurs at wave trough phases due to the concentration of flexural stress. The findings clarify energy transfer mechanisms between waves and ice, which is critical to assessing navigation safety in marginal ice zones.

Ji et al. (Contribution (10)) conducted full-scale in situ cantilever beam tests to determine the flexural strength and effective elastic modulus of brackish ice in a generic lake environment. By covering the growth, stable, and melt periods, the tests revealed that the square root of bulk porosity is the optimal predictor for flexural performance, and the results show that both strength and modulus decrease systematically with increasing porosity, being significantly lower during the melt period. This study provides valuable field data on brackish ice mechanics, filling a gap between freshwater and sea ice research and supporting ice load assessments in estuarine environments.

Li et al. (Contribution (3)) explored the transport of microplastics in open channels through flume experiments and, by tracking the trajectories of polyethylene, polypropylene, and polystyrene particles, established relationships among the Reynolds number, particle density, and floating velocity. The results of the study show that the Reynolds number dominates the horizontal migration velocity, while particle density primarily affects dispersion. The findings offer a theoretical framework for predicting the transport and fate of microplastics in water bodies, contributing to pollution control strategies in riverine ecosystems.

Li et al. (Contribution (1)) examined the immunochemical response of the Manila clam to extreme temperature reductions and, by measuring the activity of key immune enzymes under rapid cooling conditions, developed a mathematical model describing the stress and direct temperature responses. The research team's results indicate that enzyme activity fluctuates significantly as a stress response before stabilizing or changing due to low temperatures. The study provides a mathematical description of aquatic organism

https://doi.org/10.3390/w18030414

immunity under thermal stress, which is vital for assessing the ecological impacts of cold weather events on aquaculture.

3. Conclusions

This Special Issue integrates mathematical, physical, chemical, and biological methodologies to address complex challenges in cold region environments. Ranging from deep learning-based sea ice hindcasting to the immunochemical analysis of aquatic organisms, these studies elucidate critical water–ice dynamics and ecological responses. By synthesizing advanced numerical modeling with in situ physical and chemical experiments, this collection establishes a robust scientific framework for enhancing disaster prediction, infrastructure resilience, and environmental preservation in high-latitude waters.

The main conclusions are as follows:

(1) Advanced data-driven models, particularly those utilizing deep learning, significantly outperform traditional methods in forecasting sea ice thickness and river break-up events;

(2) Refined grid systems and porosity-based mechanical indices provide essential design parameters for ensuring infrastructure safety in ice-covered waters;

(3) The synchronization of wave-ice interactions and stress-induced fragmentation governs energy attenuation dynamics and navigation safety in marginal ice zones;

(4) Channel morphology and thermal regimes are the fundamental drivers regulating the spatiotemporal heterogeneity and microstructural evolution of river ice;

(5) Hydrodynamic forces and thermal stress critically influence both the transport mechanisms of pollutants and the physiological resilience of aquatic ecosystems.

Looking ahead, a promising direction is the integration of multi-physics coupling models with artificial intelligence to address the increasing complexity of cold region environments. As global warming alters ice phenology, developing integrated monitoring systems that combine high-resolution remote sensing with in situ mechanical testing will be crucial. Solutions based on these high-precision predictive frameworks, together with comprehensive ecological risk assessments, will be indispensable for sustainable engineering and environmental protection in high-latitude regions.

Author Contributions: Writing—original draft preparation, F.L.; writing—review and editing, Z.L. All authors have read and agreed to the published version of the manuscript.

Funding: The organization of this Special Issue was supported by the National Natural Science Foundation of China (grant numbers 32460290, 52301331).

Conflicts of Interest: The authors declare that there are no conflicts of interest related to the publication.

List of Contributions:

1. Li, R.; Wang, J.; Han, W.; Gong, J.; Ding, J. Mathematical Description of the Immunochemical Response of the Manila Clam (*Ruditapes philippinarum*) to Extreme Temperature Reductions. *Water* **2025**, *17*, 93. https://doi.org/10.3390/w17010093.
2. Liu, Z.; Han, H.; Li, Y.; Wang, E.; Liu, X. Forecasting the River Ice Break-Up Date in the Upper Reaches of the Heilongjiang River Based on Machine Learning. *Water* **2025**, *17*, 434. https://doi.org/10.3390/w17030434.
3. Li, J.; Wang, Z.; Li, W.; Jing, S.; Graco-Roza, C.; Arvola, L. Impact of the Reynolds Numbers on the Velocity of Floating Microplastics in Open Channels. *Water* **2025**, *17*, 588. https://doi.org/10.3390/w17040588.
4. Gao, B.; Liu, Y.; Lu, P.; Wang, L.; Liao, H. Advancing Sea Ice Thickness Hindcast with Deep Learning: A WGAN-LSTM Approach. *Water* **2025**, *17*, 1263. https://doi.org/10.3390/w17091263.

https://doi.org/10.3390/w18030414

5. Tan, B.; Li, C.; Hu, S.; Li, Z.; Ji, H.; Deng, Y.; Zhang, L. Review and Prospect of the Uncertainties in Mathematical Models and Methods for Yellow River Ice. *Water* **2025**, *17*, 1291. https://doi.org/10.3390/w17091291.
6. Tan, B.; Chang, Y.; Gao, C.; Wang, T.; Lu, P.; Fan, Y.; Wang, Q. Statistical Optimization and Analysis on the Spatial Distributions of Ice Ridge Keel in the Northwestern Weddell Sea, Antarctica. *Water* **2025**, *17*, 1643. https://doi.org/10.3390/w17111643.
7. Yu, C.; Tian, Y. Numerical Simulation of Regular Wave and Ice Floe Interaction Using Coupled Eulerian–Lagrangian Method. *Water* **2025**, *17*, 1879. https://doi.org/10.3390/w17131879.
8. Leng, Y.; Li, C.; Lu, P.; Fu, X.; Hu, S. Analysis of Open-Water Changes and Ice Microstructure Characteristics in Different River Channel Types of the Yellow River in Inner Mongolia Based on Satellite Images and Field Sampling. *Water* **2025**, *17*, 1898. https://doi.org/10.3390/w17131898.
9. Li, G.; Gao, S.; Chen, X.; Jiao, Y.; Wang, L.; Hou, Q.; Guo, D.; Zhao, Y.; Ruan, C.; Wang, Q. Designing the Engineering Parameters of the Sea Ice Based on a Refined Grid in the Southern Bohai Sea. *Water* **2025**, *17*, 2465. https://doi.org/10.3390/w17162465.
10. Ji, S.; Liu, Y.; Wang, Q.; Lu, P.; Yuan, S. In Situ Tests on the Flexural Strength and Effective Elastic Modulus of Brackish Ice During Different Ice Periods. *Water* **2025**, *17*, 3189. https://doi.org/10.3390/w17223189.

References

1. Miner, K.R.; Turetsky, M.R.; Malina, E.; Bartsch, A.; Tamminen, J.; McGuire, A.D.; Fix, A.; Sweeney, C.; Elder, C.D.; Miller, C.E. Permafrost carbon emissions in a changing Arctic. *Nat. Rev. Earth Environ.* **2022**, *3*, 55–67. [CrossRef]
2. Hieronymus, M.; Kalén, O. Sea-level rise projections for Sweden based on the new IPCC special report: The ocean and cryosphere in a changing climate. *AMBIO* **2020**, *49*, 1587–1600. [CrossRef] [PubMed]
3. Rantanen, M.; Karpechko, A.Y.; Lipponen, A.; Nordling, K.; Hyvärinen, O.; Ruosteenoja, K.; Vihma, T.; Laaksonen, A. The Arctic has warmed nearly four times faster than the globe since 1979. *Commun. Earth Environ.* **2022**, *3*, 168. [CrossRef]
4. Yang, X.; Pavelsky, T.M.; Allen, G.H. The past and future of global river ice. *Nature* **2020**, *577*, 69–73. [CrossRef] [PubMed]
5. Rokaya, P.; Budhathoki, S.; Lindenschmidt, K.E. Trends in the timing and magnitude of ice-jam floods in Canada. *Sci. Rep.* **2018**, *8*, 5834. [CrossRef] [PubMed]
6. Beltaos, S.; Prowse, T. River-ice hydrology in a shrinking cryosphere. *Hydrol. Process. Int. J.* **2009**, *23*, 122–144. [CrossRef]
7. Sharma, S.; Blagrave, K.; Magnuson, J.J.; O'Reilly, C.M.; Oliver, S.; Batt, R.D.; Magee, M.R.; Straile, D.; Weyhenmeyer, G.A.; Winslow, L.; et al. Widespread loss of lake ice around the Northern Hemisphere in a warming world. *Nat. Clim. Change* **2019**, *9*, 227–231. [CrossRef]
8. Woolway, R.I.; Kraemer, B.M.; Lenters, J.D.; Merchant, C.J.; O'Reilly, C.M.; Sharma, S. Global lake responses to climate change. *Nat. Rev. Earth Environ.* **2020**, *1*, 388–403. [CrossRef]
9. Tuhkuri, J.; Polojärvi, A. A review of discrete element simulation of ice–structure interaction. *Philos. Trans. R. Soc. A Math. Phys. Eng. Sci.* **2018**, *376*, 20170335. [CrossRef] [PubMed]
10. Squire, V.A. Ocean wave interactions with sea ice: A reappraisal. *Annu. Rev. Fluid Mech.* **2020**, *52*, 37–60. [CrossRef]
11. Stroeve, J.; Notz, D. Changing state of Arctic sea ice across all seasons. *Environ. Res. Lett.* **2018**, *13*, 103001. [CrossRef]
12. Reichstein, M.; Camps-Valls, G.; Stevens, B.; Jung, M.; Denzler, J.; Carvalhais, N.; Prabhat, F. Deep learning and process understanding for data-driven Earth system science. *Nature* **2019**, *566*, 195–204. [CrossRef] [PubMed]
13. Andersson, T.R.; Hosking, J.S.; Pérez-Ortiz, M.; Paige, B.; Elliott, A.; Russell, C.; Law, S.; Jones, D.C.; Wilkinson, J.; Phillips, T.; et al. Seasonal Arctic sea ice forecasting with probabilistic deep learning. *Nat. Commun.* **2021**, *12*, 5124. [CrossRef] [PubMed]
14. Ji, H.; Zhang, A.; Gao, R.; Zhang, B.; Xu, J. Application of the break-up date prediction model in the Inner Mongolia Reach of the Yellow River. *Adv. Sci. Technol. Water Res.* **2012**, *32*, 42–45.
15. Bergmann, M.; Collard, F.; Fabres, J.; Gabrielsen, G.W.; Provencher, J.F.; Rochman, C.M.; Sebille, E.V.; Tekman, M.B. Plastic pollution in the Arctic. *Nat. Rev. Earth Environ.* **2022**, *3*, 323–337. [CrossRef]

https://doi.org/10.3390/w18030414

water

MDPI

Review

Review and Prospect of the Uncertainties in Mathematical Models and Methods for Yellow River Ice

Bing Tan [1], Chunjiang Li [2], Shengbo Hu [3], Zhijun Li [3,*], Honglan Ji [4], Yu Deng [5] and Limin Zhang [3]

[1] School of Mathematics and Statistics, Nanyang Normal University, Nanyang 473061, China; tanbing@nynu.edu.cn
[2] School of Energy and Environment, Inner Mongolia University of Science and Technology, Baotou 014010, China; lichunjiang0405@imust.edu.cn
[3] State Key Laboratory of Coastal and Offshore Engineering, Dalian University of Technology, Dalian 116024, China; hushengbo@mail.dlut.edu.cn (S.H.); zhanglimin@dlut.edu.cn (L.Z.)
[4] College of Water Conservancy and Civil Engineering, Inner Mongolia Agricultural University, Hohhot 010018, China; honglanji@imau.edu.cn
[5] Yellow River Institute of Hydraulic Research, Zhengzhou 450003, China; dengyu@hky.yrcc.gov.cn
* Correspondence: lizhijun@dlut.edu.cn

Abstract: Mathematical models and methods serve as fundamental tools for studying ice-related phenomena in the Yellow River. River ice is driven and constrained by hydrometeorological and geographical conditions, creating a complex system. Regarding the Yellow River, there are some uncertainties that manifest in unique features in this context, including ice–water–sediment mixed transport processes and the distribution of sediment both within the ice and on its surface. These distinctive characteristics are considered to different degrees across different scales. Mathematical models for Yellow River ice developed over the past few decades not only encompass models for the large-scale deterministic evolution of river ice formation and melting, but also uncertainty parameter schemes for deterministic mathematical models reflecting the Yellow River's particular ice-related characteristics. Moreover, there are modern mathematical results quantitatively describing these characteristics with uncertainty, allowing for a better understanding of the unique ice phenomena in the Yellow River. This review summarizes (a) universal equations established according to thermodynamic and hydrodynamic principles in river ice mathematical models, as well as (b) uncertainty sources caused by the river's characteristics, ice properties, and hydrometeorological conditions, embedded in parametric schemes reflecting the Yellow River's ice. The intractable uncertainty-related problems in space–sky–ground telemetric image segmentation and the current status of mathematical processing methods are reviewed. In particular, the current status and difficulties faced by various mathematical models in terms of predicting the freeze-up and break-up times, the formation of ice jams and dams, and the early warning of ice disasters are presented. This review discusses the prospects related to the uncertainties in research results regarding the simulation and prediction of Yellow River ice while also exploring potential future trends in research related to mathematical methods for uncertain problems.

Keywords: uncertainty; mathematical models; the Yellow River ice; data analysis; review and prospect

1. Introduction

The formation of river ice is a process controlled by thermal, dynamical, and morphological factors. Considering its temperature-sensitive nature and proximity to human habitats, river ice has exhibited significant changes in the context of global warming, reflecting the difficulties that humans face in adapting to its impacts. Prowse et al. [1] classified these issues into ecological and socio-economic effects and adaptations, including six dimensions: lentic and lotic water ecosystems, river delta ecosystems, cold region infrastructure, transportation, traditional lifestyles, and hydropower generation.

In both the natural and engineering sciences, mathematical tools are essential for comprehending and quantitatively describing the essence of research subjects. The complexity of a research subject determines the mathematical methods required for its analysis. Indeed, the application degree of mathematics serves as an indicator of a scientific field's maturity. Zheng [2], citing reports from other scholars, has stated that the maturity of any scientific discipline is largely dependent on the successful integration of mathematical tools.

As mankind's understanding of the change rule of river ice advances, there is an increasing demand for more precise assessment and prediction methodologies. Mathematical models and methods play a crucial role in this field, enabling researchers to simulate ice evolution processes under diverse hydrometeorological conditions and transform qualitative observations into quantifiable mathematical representations. The laws and characteristics of ice formation and melting process can be accurately expressed through analytical or numerical calculations [3,4]. While certain river ice processes remain mathematically unmodeled, various mathematical tools facilitate deeper understanding of the complex phenomena and laws governing its evolution. These tools aid in elucidating the behavioral characteristics [5] and predicting the development of ice [6].

The complexity of natural sciences necessitates the development of mathematical models that reflect essential characteristics while emphasizing the primary cause of the problem, thereby facilitating mathematical derivation. A judiciously simplified mathematical model must ensure that a computed result with acceptable error margins is obtained while satisfying the necessary preconditions of the chosen mathematical methodology. Multiple mathematical models can be formulated for a natural problem, with the optimal model emerging through iterative testing and comparative analysis during the research process.

Mathematical models undergo continuous refinement through iterative testing and application, evolving from problem-specific models into universal models describing the same natural phenomenon [7]. In river ice research, two primary mathematical models predominate: deterministic models [3,4] and stochastic models [6,8]. Deterministic models employ various mathematical equations to characterize regular ice phenomena governed by fixed causes, while stochastic models utilize traditional and contemporary mathematical techniques to represent probabilistic ice phenomena. The research object is treated as a stochastic process, which follows statistical laws while encompassing multiple possible outcomes. Notably, the deterministic and stochastic aspects of river ice phenomena are not completely separate. Certain existing deterministic models introduce multiple stochastic factors, resulting in the formation of mathematical models such as statistical equations. The inherent randomness can also be expressed using specified parameter ranges, manifesting as bifurcated and chaotic stochastic behavior, which are established through corresponding mathematical parameterization schemes [9].

Various mathematical models of ice encompass multi-factor and multi-process coupled equations. Contemporary model development is carried out by expanding upon the foundational ideas of one-dimensional river ice mathematical models. Numerous parametric schemes persist, which aim to capture the uncertainties in the fundamental process equations of these models. As examples of one-dimensional river ice mathematical models, this review delves into the mathematical models and methodologies pertinent to four aspects of river ice: ice forming and melting, river ice applications, space–air–ground/ice integrated investigation, and ice disaster warning. Consequently, the study encompasses deterministic models, statistical equations for certain parameters within some deterministic models, and the latest advancements in stochastic modeling. In addition to scrutinizing mathematical models and methodologies, this review succinctly delineates the essential characteristics of the Yellow River's ice. It also sheds light on existing research gaps, challenges, and future research directions, with the aim of improving mathematical models and methods for the study of ice phenomena.

2. Mathematical Equations and Methods for the Processes of River Ice Formation and Melting

2.1. General Processes of River Ice Formation and Melting

The evolution of river ice formation and melting is primarily controlled by three types of factors: thermal, dynamical, and morphological factors. Thermal factors include heat exchange at air, ice, water, and riverbed interfaces, driven by hydrometeorological conditions including solar radiation, convective heat exchange, evaporation/condensation latent heat, precipitation/snowfall, and riverbed thermal flux. Dynamic factors comprise meteorological and hydraulic elements that influence the transport of frazil ice and ice floes, primarily flow rate, flow velocity, water level, and wind. Morphological factors include the horizontal and vertical boundary constraints affecting the movement of water, wind, and ice, mainly including channel alignment, geographic location, and topography. Different stages of the ice formation process in the Yellow River have been documented in detail [10–12].

During the river ice period, the evolution of river ice formation and melting according to the time scale can be categorized into following periods: supercooled water, ice runs, ice cover formation, freeze-up thermal growth, and break-up [13]. Ice crystal formation initiates when the water's temperature decreases to the freezing point (around 0 °C) or below due to cold air. In this process, two primary ice formation patterns emerge. Ice crystals accumulate along riverbanks where flow velocities are low, forming incipient border ice. Under sustained negative temperatures, this incipient border ice increases in thickness and extends toward the channel's center, eventually forming a stable ice cover connected to the riverbank. Continuous heat loss in the river's central region increases the volume of frazil particles developed by crystal nuclei. When the buoyant forces acting on these particles exceed the vertical downward mixing forces induced by turbulent flow, a dense layer of frazil particles accumulates at the water's surface. The frazil particles aggregate into dense, flocculent frazil particle clusters due to continued heat release. These frazil particle clusters transform into ice disk-shaped frazil particles when the water inside the pores of the clusters freezes, resulting in increased bond strength and thickness. As these ice disk-shaped frazil particles are transported downstream, their volume and strength increase continuously, while their edges become upturned through repeated collisions, forming distinctive ice pans. These ice pans typically accumulate at downstream river bends, channel constrictions, and hydraulic structures, forming initial ice covers through mechanical obstruction. These ice covers subsequently impede the movement of continuous floating ice blocks, leading to upstream ice accumulation and ice cover development. The rate of upstream progression of the ice cover's leading edge is governed by two primary factors: the incoming ice volume and the thickness of the newly formed ice cover, which is determined by hydrodynamic conditions at the ice cover's leading edge. Under low-velocity conditions, a single-layer ice cover forms with thickness approximating that of ice pans, which is called juxtaposed freeze-up ice. Conversely, high-velocity conditions promote the squeezing, overlapping, and accumulation of floating ice blocks, resulting in thicker composite ice covers. Once the broken ice freezes, these formations develop into deposition freeze-up ice [11]. When flow velocities at the ice cover's leading edge exceed a critical threshold, floating ice blocks and frazil particles are forced beneath the existing ice cover rather than accumulating upstream. These submerged ice blocks either accumulate beneath the existing ice cover or continue downstream to unfrozen river sections [12].

During this period, thermal exchange processes at the interface can be categorized into two cases: air–water heat exchange persists on the open water surface, while air–ice heat exchange exists on the ice-covered surface. The presence of ice cover transforms the river's hydrodynamic pattern from an open flow to a pressurized flow under the ice cover, consequently reducing its water conveyance capacity. However, ice formation and melting do not occur only at the bottom of ice covers during this period.

The onset of positive temperatures, coupled with increased solar radiation at mid-latitudes, initiates the river's break-up period. The melting process typically begins with border ice, which detaches from riverbanks due to accelerated warming at the bank. As air temperatures continue to rise, hydraulic and wind forces fragment the melting ice covers into discrete floating ice blocks. The subsequent behavior of this ice varies, according to local hydrodynamic conditions. In reaches with weak hydrodynamic forces, ice blocks primarily drift with the water flow while undergoing thermal melt, reducing the probability of ice jam or ice dam formation [14,15]. Conversely, in reaches with strong hydrodynamic forces, fast-moving ice accumulates and extrudes at locations where the river channel is obstructed, significantly increasing the probability of ice jam or ice dam development [11,14]. In 1986, the IAHR Working Group on River Ice Hydraulics defined an ice jam as a stationary accumulation of fragmented ice or frazil which restricts flow that occurs in ice runs during ice formation periods, whereas an ice dam is radically different from other jams in the ice decay and run period [14–16].

It can be concluded that the evolution process of river ice represents a complex system full of uncertainties. There are many spatiotemporal variations in the evolution of river ice, particularly in the Yellow River, due to its distinctive geographical characteristics. Mathematical equations and methods serve as crucial tools in river ice research. Based on the understanding of river ice, natural processes have been simplified into various mathematical equations. Through numerical analysis of the intrinsic connections and constraints of air–ice–water–riverbed systems, researchers can develop comprehensive insights into complex phenomena and establish practical solutions for scientific and engineering challenges. In the study of natural phenomena, two mathematical approaches are primarily employed: mechanistic analysis and statistical analysis [17]. Mechanistic analysis utilizes classical mathematical tools to analyze the causal relationships relating to phenomena, while statistical analysis involves the application of classical mathematical tools to determine statistical laws from extensive observational data through parameterizations of certain deterministic models. The setting up of more modern stochastic models using extensive observational data is also an important research method. Finally, the phenomenon of interest is described using some mathematical relationship or pattern. To date, the main mathematical models used in river ice research include analytical models, numerical models, and multivariate statistical models. Analytical models are limited to simplified scenarios in which ice can be simplified to a homogeneous and continuous medium, the water flow under the ice is stable, and the river channel is regular. For more complex situations, which are beyond the scope of analytical models, numerical models employing corresponding partial differential equations can provide numerical solutions. As river ice is a multivariate system, multivariate statistical models have also been used in an attempt to solve associated problems. In recent years, new mathematical methods and techniques have emerged [18–20].

There are certain differences between the deterministic models of Yellow River ice with physical basis developed by various researchers, reflecting their respective research objectives and problems. These models have been derived from the conditions, phenomena, information, and data on the evolution, formation, and melting of river ice obtained through field investigations and/or laboratory experiments.

2.2. Key Characteristics of the Yellow River Ice and Its Investigations

The evolution of river ice is influenced by a multitude of factors, with the most significant influences arising from air temperature conditions, hydraulic conditions, and the geomorphological traits of the river channel. Foundational river ice models are constructed based upon these factors, then subsequently honed to provide precise representations of specific rivers, accounting for their distinctive features. The Yellow River—distinguished by its unique geographical, hydrological, and meteorological conditions—presents both similarities and notable differences in its ice evolution process when compared with rivers elsewhere in the world.

In addition to the typical characteristics of freezing rivers in cold regions, Wang et al. [13] have summarized the main features of the Yellow River ice and applied the Canadian river ice model River1D [21] to simulate the ice formation process in the Yellow River. The Yellow River, which originates from the Qinghai–Tibet Plateau and flows into the Bohai Sea, stretches across a total length of 5464 km. The section of the river from the Qinghai–Tibet Plateau to Ningxia is primarily mountainous. After that, it carries a large amount of suspended yellow sand from the Loess Plateau, which imparts the river with its characteristic yellow hue and makes it the river with the highest sediment concentration among the world's major rivers. Since 1974, the construction of the Liujiaxia, Longyangxia, Wanjiazhai, and Xiaolangdi reservoirs has significantly mitigated ice flood disasters in the Yellow River Basin [22]. However, the northernmost section of the Inner Mongolia Reach still frequently suffers from ice flood disasters [19,23]. Existing research on ice disasters in the Yellow River has mainly focused on the Inner Mongolia Reach, which spans over 823 km (from 37°35′ N to 41°50′ N and 106°10′ E to 112°50′ E) and is primarily characterized as a plain river.

The Inner Mongolia Plateau and its surrounding mountains significantly influence the movement and intensity of cold air, with cold waves in the Yellow River Basin predominantly originating from the north, northwest, west, and east. Among these, the northern cold waves lead to the strongest temperature drops, and the northwest cold waves are the most recurrent. Due to variations in cold air paths each winter and late spring, the ice-on and ice-off locations in the Inner Mongolia Reach can vary, with initial ice often occurring between Sanhuhekou and Baotou. Ice formation in the Inner Mongolia Reach typically starts downstream and moves upstream, while ice-off starts upstream and moves downstream. The average ice run date before ice freeze-up is from 17 November to 1 December, the average ice freeze-up date is between 1 December and 4 January, the average ice break-up date is 4 March, and the ice-off date is 27 March. The average ice period in the Inner Mongolia Reach is 117 days, with the longest in recent history being 150 days in 1969/1970 and the shortest being 76 days in 1989/1990 [24]. At the top of the large inverted U-shaped bend in the Inner Mongolia Reach, the water flows from west to east with a total drop of only 162.5 m, and it is frequently frozen. The riverbed gradually changes from narrow and deep to wide and shallow from upstream to downstream, with a gradient of only 0.09‰ to 0.11‰ from Zhaojunfen to Toudaoguai. The widest point (at Dukoutang) is 400 m to 1200 m, and the narrowest (at Zhaojunfen) is 200 m to 600 m. Along its course, shallow areas, bends, and branches are common, often causing ice blocks to become stuck and form ice jams or dams. The section from Guaishang to Wanjiazhai flows from north to south with a large gradient and high velocity, and it generally did not freeze before the construction of the Wanjiazhai Reservoir. After its construction, the water surface gradient and velocity decreased at the backwater end, reducing the area with ice transport capacity, thus extending the area of increasing ice jams and dams upstream. As such, previously unfrozen sections have become frozen. In recent years, due to global climate warming and human activities, the ice conditions in the Inner Mongolia section of the Yellow River have exhibited new characteristics [25]. Particularly, frequent extreme temperature fluctuations in winter and strong mid-latitude solar radiation have led to overall higher air temperatures and larger air temperature differences. Since the 1990s, influenced by rising temperatures, reduced water volume, and winter irrigation, the dates of ice-on in the Inner Mongolia Reach have been delayed, while the date of ice-off has advanced. The number of stable freeze-up days has decreased, and the ice cover thickness has become thinner. The river sometimes experiences two freezing and melting cycles. Generally, ice coverage exceeds 80% of the total reach area [24]. Approximately 20% of the river surface typically remains ice-free, allowing for continued frazil particle formation. These frazil particles either consolidate beneath open water surfaces to form new ice covers at the downstream side of the open water zone/pond, according to recent field observations, or are transported beneath existing ice cover, where they accumulate and freeze to form frazil ice or continue downstream to open water zones/ponds. For an assessment of the future ice condition trend under global climate warming, Zhou et al. [26] used eight global climate models from CMIP6 to simulate the ice conditions in the Inner Mongolia Reach of the Yellow River from 2015 to 2100 and predicted that, by 2070, the Yellow River might not experience complete freezing. In the near term, the duration of ice runs before river freeze-up and after break-up will continue to increase, while the period of stable ice cover will shorten.

The likelihood of ice jams and dams increases significantly in areas with gentle gradients, sharp bends, and water conservancy structures that narrow the river channel. Moreover, the construction of over 20 reservoirs in the upper reaches of the Yellow River has significantly altered the thermal, dynamic, and hydraulic conditions due to the discharged water volume. The regulated water volume and heat not only alter the river's natural properties but also change the sediment and thermal conditions due to reservoir discharge, resulting in downstream sections remaining unfrozen due to higher water temperatures (Figure 1a). During reservoir regulation, elevated water levels worsen ice jam disasters during ice freeze-up but reduce the occurrence of ice dams during ice break-up. This directly affects the length of the frozen river downstream of the reservoir, as well as the dates of ice runs during freezing and melting. The several "n–shaped" bends and constructed bridges in the Inner Mongolia Reach of the Yellow River constrain the flow of ice. Throughout the Inner Mongolia Reach, juxtaposed ice cover, hummocky ice cover, and open water zones/ponds coexist, leading to a reduction in the length of frozen river sections in the Inner Mongolia Reach (Figure 1b). Once a stable ice cover forms, the increased roughness beneath the ice leads to greater water storage in the river channel, resulting in the coupled flow of ice, water, and sediment in the Inner Mongolia section. During ice break-up, a larger water discharge volume can accelerate the process and lead to a "violent ice break-up", while a lower water discharge volume typically results in a "tranquil ice break-up". These scenarios are closely linked to the mechanical properties of the ice cover under hydrodynamic forces [27]. Sediment is not only present in the river water but also within the ice, making it a distinctive feature of Yellow River ice [28]. From a composite material perspective, the sediment influences the density, thermal, electrical, optical, and mechanical properties of the ice. Combined with the river's mid-latitude location, where ice temperatures are relatively higher and solar radiation is stronger compared with high-latitude regions, the Yellow River ice can be viewed as a four-phase composite material comprising air, ice, water, and sediment. Additionally, sediment inclusions are often found in hummocky ice, and the non-frozen water content is higher in ice during spring. Although snowfall is typically not an important factor, dust particles tend to settle on ice surfaces during spring sandstorms in the Inner Mongolia Reach of the Yellow River, being located in one of the arid and semi-arid regions in China.

(**a**) The ice conditions in February 2025

(**b**) Photo of ice covers and open water ponds near bridge piles

Figure 1. Distributions of ice covers and image of open water zones/ponds of ice cover in the Inner Mongolia Reach of the Yellow River.

Since the 1950s, extensive research on the Yellow River ice has been conducted by scholars, spanning various aspects. Early studies included field observations of its conditions [29,30] and investigations into the characteristics of ice in sediment-laden rivers [31]. In recent years, significant progress has been achieved in Yellow River ice research, due to enhanced field methods and the development of new observational instruments. Beyond comprehensive ice surveys using air–space–ground/ice integrated methods (see Figure 2), studies have been conducted on the crystal structure, sediment content [32], mechanical properties [33], and dielectric constant [34] of this river ice. Particularly, the sediment in the Inner Mongolia reaches of the Yellow River—whether on the riverbed or banks—is influenced not only by the physical and mechanical properties of the sediment but also by hydrodynamic forces, especially those under the ice cover [35,36]. Recent systematic winter surveys of under-ice velocity profiles at the Shensifenzi Bend in Inner Mongolia have revealed some flow field patterns [37,38] and the mechanical properties of riverbank sediment [39]. However, theoretical concepts alone are insufficient for river ice models; extensive field data, including high-precision river topography, real-time hydrological, and meteorological data, are required for such modeling. At present, research efforts regarding the Yellow River—particularly riverbank or ice surface surveys—are still sparse and scattered across certain sections and stations, with most river sections lacking sufficient measured data to fully support the validation and refinement of deterministic and stochastic mathematical models.

Figure 2. Photo of comprehensive space–sky–ground integrated ice investigation.

3. Uncertainties in Deterministic Mathematical Models for the Evolution of River Ice Formation and Melting

3.1. Development Status of Evolutionary Models of River Ice Formation and Melting

When analyzing mathematical models of river ice in terms of their mathematical universality, the greatest differences in ice formation and melting processes globally stem from differences in river characteristics and water flow parameters. A comprehensive mathematical model of river ice that can completely describe the process of river icing, freezing, and thawing should incorporate three sub-models: hydraulic, thermal, and thermodynamic models [17,40]. These sub-models reflect interconnected relationships, wherein hydraulic conditions influence thermal exchange and freezing processes, thermal conditions govern freezing processes, and freezing processes influence both hydraulic and thermal conditions.

Shen et al. [41] made a significant advancement by combining the one-dimensional non-steady Saint-Venant equation with equilibrium equations for river ice motion, mass conservation, and floating ice concentration. This foundation for river ice dynamics modeling allows for the effective simulation of both ice movement and ice jam development processes. Building upon this work, Lal and Shen [42] developed a one-dimensional non-steady flow river ice model (RICE). This model incorporates simulations of frazil ice, floating ice, border ice, and ice jam processes while accounting for water temperature variations, thermal growth processes, and ice flow interactions. Subsequently, Shen et al. [43] improved the RICE model to create the RICEN model, which incorporates additional modules for complex river networks, anchor ice, and frazil ice transport based on ice transport rates. The RICEN model was successfully implemented to simulate the formation of ice jams in the backwater area of the Sanshenggong Water Control Structure in the Inner Mongolia Reach, which was in good agreement with the actual observed backwater height [44]. Then, this one-dimensional river ice model was further upgraded to the CRISSP1D and RICE–E models, which provide comprehensive simulation capabilities for the entire lifecycle of river ice formation, development, and evolution [45–47]. Pan et al. [48] extended the CRISSP1D model by incorporating an anchor ice flood wave module that accounts for the growth and release of anchor ice. This enhancement enabled analysis of the riverbed's topographic elevation, cross-sectional flow changes, and variations in comprehensive riverbed roughness caused by anchor ice formation. Notably, their research identified that changes in bed roughness due to anchor ice significantly influence the water level and flow fluctuations.

To address the effects of complex terrain and riverbanks, Shen et al. [49] developed a two-dimensional river ice dynamics model (DynaRICE), extending their previous one-dimensional approach. This model integrates a Eulerian-based finite element method to solve two-dimensional shallow water equations with a Lagrangian-based meshless smooth particle method for computation of the motion of river ice. Building upon this foundation, Liu et al. [50] introduced CRISSP2D, a comprehensive two-dimensional mathematical modeling system that is capable of simulating river ice formation and melting. CRISSP2D can simulate multiple processes, including complex topographical flow dynamics; temperature fluctuations influenced by solar radiation and wind patterns; various ice formation processes (frazil ice, anchor ice, border ice, and floating ice); transport and subsidence processes of floating ice under ice cover; the formation, development, and release processes of ice jams or ice dams; thermal growth and melting of the ice cover; and the process of mechanical ice cover break-up. Knack and Shen [51] coupled the two-dimensional river ice module with the numerical sand module to develop CRISSP2D into a river ice–sand thermodynamic model. This enhancement enabled the simulation of bed load, suspended load sediment transport, and bed deformation processes in ice-affected northern rivers. Pan and Shen [52] subsequently evolved CRISSP2D into RICES2D, considering sediment transport, riverbed morphological changes, and bank erosion under ice conditions, thereby establishing a framework for studying water–ice–sand coupling mechanisms. Notably, efforts in the development of both two- and three-dimensional mathematical river ice models continue to build upon the initial one-dimensional mathematical river ice models [4,42,43,53]. Based on the plane body-fitted coordinate transformation and the vertical σ coordinate transformation, a three-dimensional river ice model was established for the Yellow River, and the finite volume method was used to discretize the control equation of the k–ε turbulence model. To avoid fluctuation of the pressure field, the velocity equation at the interface of the control body was derived through momentum interpolation, and the accumulation of ice was numerically simulated [54,55].

The theoretical foundation for mathematical river ice modeling traces back to the 1960s, when Pariset and Hausser applied the basic equations of hydraulics to ice jam analysis. This pioneering work sparked extensive scholarly contributions to ice jam theory and mathematical models of river ice [4,14,56,57]. Based on the static equilibrium theory, Flato and Gerard [58] developed a non-equilibrium ice jam model (ICEJAM) under a one-dimensional steady flow. Beltaos [59] created the wide river ice jam model (RIVJAM), which considers the effect of water seepage in ice jams.

Following on from the above ice models, static equilibrium ice jam theory was subsequently integrated into commercial non-steady one-dimensional models such as HEC–RAS [60]. There are also a number of other comprehensive river ice models, such as RIVICE, MIKE11–ICE [61], and River1D [21]. While these models can effectively predict ice thickness and water levels under extreme conditions, they are limited by their inability to simulate dynamic ice movement processes or predict ice jam formation locations.

3.2. Uncertainties in Parametric Schemes for One-Dimensional River Ice Mathematical Models

Xu [40] provided a comprehensive summary of one-dimensional river ice models, identifying twenty-five distinct mathematical formulations. Beyond the continuity equation and the set of momentum equations in hydraulic models, there were six basic mathematical equations. The others included mathematical models of the water temperature distribution, frazil ice concentration, floating ice transport, ice jams or ice dams, and thermal ice formation and melting processes. While one-dimensional river ice mathematical models enable simulation of the complete evolution of ice (icing, freezing, and thawing) and support the analysis of specific ice processes through their sub-models, the accuracy of their predictions often deviates from observed conditions. These discrepancies can be primarily attributed to the selection of parametric schemes and parameter values within the models.

3.2.1. Uncertainties Caused by Ice Bottom Roughness in One-Dimensional River Ice Mathematical Models

The composite roughness coefficient plays a crucial role in ice–water dynamics, which is related to the riverbed's roughness, ice bottom's roughness, and freeze-up conditions. This coefficient exhibits both spatial and temporal variability. Guo et al. [62] investigated the uncertainties associated with this parameter and identified multiple contributing factors within the composite roughness calculation, as expressed in Equation (1):

$$n_c = \left(\frac{{n_b}^{3/2} + f\frac{B}{P}{n_i}^{3/2}}{1 + f\frac{B}{P}} \right)^{2/3} \tag{1}$$

where $f = (B - B_0)/B$ is the percentage of ice coverage width to the total river width, B is the total river width, B_0 is the width of the open water surface, n_c is the composite roughness, n_b is the riverbed roughness, n_i is the ice bottom roughness, and P is the wetted perimeter of the channel (for wide and shallow channels, $B/P \approx 1$).

Variations in the ice bottom's roughness are influenced by multiple factors, including the specific types of floating ice, ice layer characteristics, meteorological conditions, and their change rates. During the initial ice cover formation period, the ice layer primarily comprises floating ice and frazil ice, characterized by low ice thickness and high ice bottom roughness. As the ice layer thickens and undergoes continuous water flow exposure, its bottom surface progressively smoothens, resulting in decreased roughness. The ice layer's bottom surface roughness exhibits temporal dependency, decreasing with prolonged ice bottom flow scour until reaching an equilibrium value. Therefore, the temporal evolution of the ice layer bottom's roughness can be expressed mathematically through Equation (2) [42].

In reality, it is difficult to accurately determine the ice bottom's roughness from ice cover formation through to the break-up period.

$$n_i = n_{i,e} + (n_{i,i} - n_{i,e})e^{-kt_{ic}} \quad (2)$$

where $n_{i,e}$ is the ice layer's bottom surface roughness at the beginning of the ice break-up period (it is believed that this value is close to 0.008~0.015 when the bottom surface of the ice layer is smooth); k is the attenuation coefficient, which varies according to winter conditions (flat winter, warm winter, cold winter) and freezing degree, with the range of this parameter being 0.02~0.03 when the ice layer has no open water surface [63]; t_{ic} is the number of freezing days; and $n_{i,i}$ is the initial bottom surface roughness of the ice layer. This initial ice bottom roughness, which exhibits considerable variation range, is dependent on the ice layer's characteristics and the morphology of the river channel.

Numerous observational studies have investigated the ice layer bottom surface's roughness and obtained comprehensive roughness coefficients [63], resulting in value tables. However, significant variability exists in the measured roughness values, with the initial ice roughness ranging from 0.015 to 0.050. The standards published by the Ministry of Water Resources of the People's Republic of China [64] specify a range of 0.015 to 0.100. In Equation (2), the change rate $\partial n_i/\partial n_{i,i}$ of n_i relative to $n_{i,i}$ is larger than the change rate $\partial n_i/\partial n_{i,e}$ of n_i relative to $n_{i,e}$ during the initial ice cover formation period. Uncertainties in the value of $n_{i,i}$ significantly influence the bottom surface roughness of ice layer, consequently affecting variations in composite roughness.

Moreover, the Yellow River's riverbed contains substantial sediment, necessitating the incorporation of ice–water–sediment coupled transport equations [12]. Consequently, the riverbed roughness exhibits temporal and spatial variability which is more complex than that of ice bottom roughness, thereby amplifying the uncertainties in composite roughness calculations.

3.2.2. Uncertainties in the Physical and Thermal Properties of Ice in One-Dimensional River Ice Mathematical Models

Understanding of the ice growth mode has evolved to cover a range of complex processes. However, one-dimensional river ice mathematical models continue to rely on simplified computational approaches. These approaches are primarily founded on one-dimensional ice thermal growth principles under static water conditions, making them more applicable to lake and reservoir ice formation [65]. According to the actual situation of river ice, a given model may undergo subsequent modifications to develop a more suitable river ice growth model.

Stefan (1891) pioneered the development of an analytical model for calculating the formation of ice under static water conditions [66]. This model operates under the assumption that the ice surface temperature is equal to the ambient air temperature. Therefore, one only needs to consider the heat balance at the bottom of ice layer while assuming a linear distribution of ice temperature.

$$\frac{dh_i}{dt} = -\frac{k_i}{h_i \rho_i L_i}\left(T_{sfc} - T_f\right) \quad (3)$$

where h_i is the thickness of the ice; t is the time; k_i, ρ_i, and L_i are the thermal conductivity (W/m·K), density (g/cm^3), and latent heat of freezing (J/g) of ice, respectively; T_{sfc} is the surface temperature of ice (°C); and T_f is the ice bottom temperature (which is also called the freezing temperature).

Assuming that t_f is the terminal moment of the computation, both sides of Stefan's formula are integrated with respect to time $t = [0, t_f]$.

$$\frac{1}{2}h_i^2 = \frac{k_i}{\rho_i L_i}\int_0^{t_f}\left(T_f - T_{sfc}\right)dt \tag{4}$$

$$FDD = \int_0^{t_f}\left(T_f - T_{sfc}\right)dt \tag{5}$$

When the integration time step dt = 1 d, FDD is the cumulative freezing-degree days (°C·d). In this case, the ice thickness h_i is calculated as shown in Equations (6) and (7).

$$h_i = \sqrt{a^2 FDD} \tag{6}$$

$$a = \sqrt{2k_i/\rho_i L_i} \tag{7}$$

As river ice contains frazil particles, Shen et al. [67] have proposed an improved method for determination of cumulative freezing degree-days to simulate winter ice thickness variations. Although the formula is simple, the method requires a variable coefficient. This coefficient takes different values for the ice layer growth and melting periods; thus, there is uncertainty in the model parameters. In the RICE model, the thermal growth and melting of the ice layer can be described by Equation (8) [42].

$$\rho_i L_i \frac{dh_i}{dt} = k_a\left(T_{sfc} - T_a\right) + k_{wi}\left(T_f - T_w\right) \tag{8}$$

where h_f is the thickness of frazil ice in the lower part of ice layer, h_i is the thickness of thermally grown ice, T_{sfc} is the ice surface temperature, T_f is the freezing temperature, T_w is the water temperature, k_{wi} is the heat exchange coefficient at the ice–water interface, and k_a is the linear heat exchange coefficient at the air–ice interface. When a frazil deposit exists on the underside of the cover, Equation (8) should be replaced by Equations (9) and (10):

$$e_f \rho_i L_i \frac{dh_i}{dt} = k_a\left(T_{sfc} - T_a\right) \tag{9}$$

$$\left(1 - e_f\right)\rho_i L_i \frac{dh_i}{dt} = k_{wi}\left(T_f - T_w\right) \tag{10}$$

where e_f denotes the porosity of the frazil accumulation [4], which reflects the portion of water in frazil.

As can be seen from Equations (8)–(10), porosity, the heat exchange coefficient at the ice–water interface, and that at the air–ice interface have significant influences on the variation in ice thickness. Other factors, including air temperature and flow rate, are measured values. The value of k_{wi} has been expressed in reference [42] as a nonlinear function of flow velocity and water depth. The value of k_a is influenced by various factors, such as latitude, humidity, and meteorological conditions. Shen et al. [67] took the typical value of k_a as 20 W/(m^2·°C) in their calculations, while reference [68] took k_a as 21.87 W/(m^2·°C) when calculating the ice thickness in different sections of the Yellow River. However, setting k_a as 10 W/(m^2·°C) when calculating the thickness of Yellow River ice has also been reported [69].

In the one-dimensional river ice mathematical model summarized by Xu [40], the ice thickness formula was further simplified based on Stefan's equation [66]:

$$h_i = A'(FDD)^{\zeta_{is}} \tag{11}$$

where h_i is the ice thickness (cm), A' is an empirical coefficient, ζ_{is} is an empirical index, and FDD is the cumulative freezing degree-days (°C·d), counted from the day when the temperature turns negative steadily. The two empirical coefficients in Equation (11) are specified in the specification of the Ministry of Water Resources of the People's Republic of China [70]. For Northeast China, ζ_{is} = 0.50~0.56 and A' = 2.0~2.3; and for North China, ζ_{is} = 0.50~0.56 and A' = 2.6~3.0. Therefore, all of the parameters in Equation (6) are empirical values that can be determined geographically. While ice formation and melting involve multiple processes, the thermal growth process of ice, which is mainly controlled by cumulative freezing degree-days, predominates within the entirety of ice processes. A detailed analysis of the Yellow River ice can be found in the referenced study [71].

The roots of these uncertainties are essentially caused by uncertainties in the physical and thermal properties of ice. Both simple and complex ice growth models require parameters such as the latent heat of freezing (L_i), ice density (ρ_i), ice thermal conductivity (k_i), freezing temperature (T_f), heat exchange coefficient at the ice–water interface (k_{wi}), and linear heat exchange coefficient at the air–ice interface (k_a). As the Yellow River ice is natural ice, it generally consists of pure ice, gas, unfrozen water, and sediment. Therefore, all the parameters of ice characteristics involved in the one-dimensional river ice mathematical model deviate from reality.

According to the continuous high-frequency measured vertical ice temperature profile data, the thermal conductivity, specific heat, and density are treated as a unified coefficient $k_i/(c_i \cdot \rho_i)$, termed the thermal conductivity coefficient. The effective thermal conductivity coefficients of reservoir ice are obtained by mathematically inverting the measured data, which are obtained as a linearly distributed section of ice temperature profiles below the ice surface and above the ice bottom [72]. These coefficients vary with the composition of ice, particularly with respect to the content of impurities. Such variations introduce uncertainties in the thermal conductivity of natural ice, subsequently affecting the accuracy of ice simulation calculations during the freeze-up period. The inversion results of these ice thermal properties have been summarized by Li et al. [73].

3.2.3. Uncertainties in the Linear Heat Exchange Coefficients at the Air–Ice and Ice–Water Interfaces in One-Dimensional River Ice Mathematical Models

The accurate determination of heat exchange coefficients at both the ice–water and air–ice interfaces remains a significant challenge in ice growth models [62]. Shen et al. (1995) have addressed this challenge by developing a parameterization scheme for river ice models, introducing a linear approximation formula to calculate heat exchange at air–water/ice interfaces [43,74]:

$$\phi = -\phi_s + \beta + k_a\left(T_{sfc} - T_a\right) \tag{12}$$

where ϕ is the net heat flux lost from the river surface (W/m^2), ϕ_s is the heat flux from solar radiation (W/m^2), β is the local climate and morphology-related empirical heat flux (W/m^2), k_a is the heat exchange coefficient at the air–water/ice interface (W/(m^2·°C)), T_{sfc} is the water/ice surface temperature (°C), and T_a is the air temperature (°C). For long-term, large-scale numerical simulations of river ice for rivers with lengths exceeding hundreds of kilometers providing forecasts over two weeks in advance, Equation (12) can be further simplified to Equation (13) [43,74]:

$$\phi = \alpha'\left(T_{sfc} - T_a\right) \tag{13}$$

where α' is the comprehensive heat exchange coefficient (W/(m^2·°C)), which takes into account solar radiation and other heat losses.

While parametric schemes exist for radiation, convection, and conduction in the heat balance equation at the air–water/ice interface on clear days, these schemes lack comprehensive consideration of cloudiness and latitude effects. The complexity of ice surface albedo parameterization in the heat exchange at the air–ice interface varies with the scale of numerical simulation [65]. Large-scale models employ the simplest approach, merely assigning albedo values to specific ice types. Regional climate models incorporate surface temperature effects, while detailed ice physics studies necessitate the consideration of multiple factors, including the solar altitude angle, atmospheric properties, visible and near-infrared band variations, surface temperature, snow thickness, and ice thickness [75]. These varying albedo parameterization schemes yield significantly different simulation outcomes [76].

In lake ice model analysis, Yang [65] selected Shine's parameterization scheme for incident shortwave radiation [77] in conjunction with Bennett's suggested reflection coefficients for shortwave radiation from cloud cover [78]. This scheme treated the effect of cloud cover as a linear relationship. Incident longwave radiation was chosen from Efimova's [79] parameterization scheme for incident longwave radiation and Jacobs's parameterization scheme for cloud cover in relation to longwave radiation, which also treated the effect of cloud cover as a linear relationship [80].

Cao et al. [81] obtained a linear relationship between solar radiation flux and cloud cover for the semi-arid region of Western China:

$$\phi_{c1} = -4.29C_1 + 758.41 \tag{14}$$

where ϕ_{c1} is solar radiation in the presence of clouds (W/m^2) and C_1 is the cloud cover, which is divided into 10 levels (where $C_1 = 0$ means a clear sky without clouds and $C_1 = 10$ means full coverage of clouds).

This statistical formula demonstrated significant dispersion from observed values, resulting in low correlation coefficients. To solve the river ice problem in China, Wang et al. [13] applied the methodology developed by Ashton and Kennedy [82], utilizing Equation (15) to estimate solar radiation under both clear and cloudy conditions.

$$\phi_{c1} = \phi_0\left(1 - 0.0065C_1^2\right) \tag{15}$$

where ϕ denotes the solar radiation in the absence of clouds.

Comparative analysis between Wang et al.'s [13] parameterization scheme and the data measured by Cao [81] revealed limitations regarding this equation's accuracy. Li et al. [83] addressed these limitations by re-analyzing the same dataset using a logistic function approach. Their analysis demonstrated that solar radiation exhibited a gradual decrease with increasing cloud cover, from 550 W/m^2 when cloud cover was below 50%, followed by a pronounced decline at higher cloud coverage, ultimately reaching 170 W/m^2 under complete cloud cover. This approach achieved a correlation coefficient of 0.8845 at the 0.01 significance level. This logistic approach provided superior representation of the variable rates of solar radiation decreasing with increasing cloud cover when compared with previous linear and parabolic parameterization schemes, effectively addressing the tendency to overestimate radiation at low cloud coverage and underestimate it at high coverage.

$$\phi_{c1} = \frac{425}{1 + 2.5e^{(-18.8+0.2C_1)}} + 125 \quad (16)$$

The fitting results from Equation (16) aligned well with observed solar radiation values in the Wuliangsuhai region along the Yellow River, which ranged from 160 W/m^2 to 770 W/m^2. As noted by Wang et al. [13], the absence of comprehensive cloud cover data poses a significant challenge in establishing parametric relationships between local solar radiation and cloud coverage. Nevertheless, logistic models show promise for future applications as cloud observation data becomes more readily available and valued. Wang et al. [13] further demonstrated that the albedo of the water surface is correlated with both the water's turbidity and suspended sediment content. Similarly, Li [84] established that the albedo of ice varies with the quantity of sand deposited on its surface. In practice, the river ice albedo differs from place to place. Traditional mathematical models often neglect these albedo effects in solar radiation calculations, resulting in inaccurate heat exchange estimations. To address this limitation, researchers have implemented varying albedo values to reflect the changes in solar reflectance for different temporal periods and river ice conditions, deriving coefficients that express solar reflectance through model inversion. The coefficient is classified into six types of reflectance in ice simulations, including those associated with the early part of the ice run period R_{b1}, ice cover formation period R_{b2}, ice freeze-up period R_{b3}, early period of ice break-up R_{b4}, ice break-up period R_{b5}, and snow-covered ice R_{b6} [13]. While these improvements have reduced uncertainties in solar radiation parameterization, some degree of uncertainty still remains, as has been discussed by Yang et al. [85].

Furthermore, whatever river ice mathematical model is used, meteorological information is required as a driving condition. As far as existing weather forecast information is concerned, in the context of ice forecasting and warning systems for the Yellow River, the reliability of forecasted ice conditions exhibits a temporal gradient: predictions within a three-day window demonstrate high credibility and forecasts extending from four to seven days show increasing uncertainty, while projections beyond seven days (up to fourteen days) serve merely as indicative estimates.

Under identical meteorological conditions, heat exchange at the ice–water interface on the ice bottom is significantly lower than that at the air–water/ice interface on the surface, with their ratio ranging from 0.205 to 0.970. Xu [40] incorporated this ratio to replace the heat flux from the water body under the ice in a one-dimensional river ice model. The parameterization of heat flux at the ice–water interface on the ice bottom during static water freezing remains a subject of international debate, with various studies reporting different results across multiple lakes [86]. In river ice, it is more difficult to determine the flux from water under the ice. Bai et al. [87] have performed mathematical inversion calculations of in-ice thermal conductivity coefficients using long-term continuous monitoring data of vertical ice temperature profiles from lakes and reservoirs, assuming constant ice–water heat flux during data analysis. Cao [88] proposed novel perspectives on temperature and radiative transfer within ice based on long-term lake ice observations, although these findings remained qualitative rather than quantitative. For the ice of the Yellow River's Shisifenzi bend, Li [84] employed mathematical inversion techniques to determine the short-term heat flux at the ice–water interface, assuming constant thermal conductivity within the ice for engineering applications. It is obvious that the heat flux at the ice–water interface remains an uncertainty problem when the conjoint analysis of air–ice interface and the ice–water interface is conducted asynchronously. Yang et al. [85] also reported similar conclusions.

4. Uncertainties in Other Deterministic Models of Yellow River Ice with Physical Basis

4.1. Uncertainties of Mechanical Properties in Yellow River Ice Layer Bearing Capacity Models

The Yellow River embankment is protected through the deployment of gabion mattresses on ice, with the efficacy of this technique being primarily dependent on the bearing capacity of the ice. International researchers have developed various mechanistic models for the ice layer's bearing capacity, founded on plate and shell mechanical principles, which have subsequently been adopted by Chinese researchers. While these models predominantly address transient loads on single-point circular contact surfaces [89], some account for load distributions of varying geometries. The ice bearing capacity can be derived as a function of flexure strength, elastic modulus, and ice thickness, regardless of the mechanical theory employed.

The internal temperature of ice significantly influences both its flexure strength and elastic modulus. When gabion mattresses are deployed on the ice surface, increased intra-ice temperatures can lead to diminished strength of the ice layer. An expression for the ice layer's bearing capacity incorporating ice thickness, ice temperature, and decay time after sinking gabion mattresses can be theoretically derived through the combination of physical indicators relating to the Yellow River, such as those associated with ice crystals, density, sediment content [32], and the flexure strength and elastic modulus of the ice layer [33]. There remain uncertain parameters in the physical and thermodynamic models of ice layers, as well as uncertainties in the experimentally determined parameters reflecting the mechanical properties of ice.

Li et al. [90] employed the HIGHTSI model [91] to calculate the temperature of each layer of ice, enabling subsequent calculations of equivalent flexure strength and elastic modulus for corresponding layers. Then, the long-term loaded flexure strength and elastic modulus of ice layer were established through the power function form of the long-term strength of ice, including a time variable. In this way, the long-term flexure strength and elastic modulus of ice could be calculated.

Li et al. [90] analyzed ice deformation measurements around gabion mattresses to identify the moment of ice destabilization, subsequently determining the attenuation coefficient regarding the long-term strength of ice to be 0.08 through iterative calculations. While this represents the first quantitative estimate for long-term ice bearing capacity assessment, its applicability to ice conditions in other rivers remains uncertain.

4.2. Uncertainties in Dielectric Constants for Single-Point Radar Thickness Measurement Applications in the Yellow River Ice Layer

The advantages of radar-based ice thickness detection have been shown in Yellow River ice surveys. The physical basis for detecting ice thickness relies on the speed of electromagnetic wave propagation within ice, which is determined by the ice's dielectric constant. The propagation speed of radar waves in air is 0.30 m/ns, and the dielectric constants vary significantly across different media: 3.17 for pure ice, 81 for pure water, and 1 for air [92]. Natural ice exhibits dielectric constants of 3.15 ± 0.05 [93], while ice containing air shows values of 3.18 ± 0.002 [94] and free water exhibits a dielectric constant of 80.37 [95]. At present, Yellow River ice thickness measurements generally employ the pure ice dielectric constant of 3.17. However, the Yellow River ice typically contains air bubbles and solid impurities [28,32], necessitating consideration of their effects on radar wave speed for improved measurement accuracy. Theoretically, freshwater ice can be characterized as a three-phase composite of pure ice, air bubbles, and unfrozen water. At lower temperatures, when unfrozen water around crystals solidifies, it becomes a two-phase pure ice–air bubble composite. The Yellow River ice presents additional complexity due to its non-uniform sediment distribution [32]. According to the sediment content within the ice, a four-phase pure ice–air bubble–unfrozen water–sediment composite can be considered. In such composite materials, phase components are randomly distributed in size and shape. When these components are smaller than the radar wavelength, scattering effects become negligible. At this time, the composite's dielectric constant model considers only the volume fraction of each phase component, disregarding their shapes and sizes. Theoretically, the presence of unfrozen water decreases radar wave speeds, while sediment content increases them. The dielectric constants in the range of 2.5–4.5 reported by Fu et al. [28] for Yellow River ice reflect the combined influence of both unfrozen water and sediment content.

During winter of 2020–2021, thirteen ice thickness measurements were conducted on juxtaposed freeze-up ice beneath a fixed radar installation, as well as ice sampling at one point. Measurements included ice temperature at sampling time, along with subsequent analysis of ice crystal structure and density profiles. The measured data were used to analyze the variation in radar wave speed in both juxtaposed freeze-up ice and deposition freeze-up ice formations in the Yellow River. The investigation examined three key relationships: the dependence of radar wave speed (or dielectric constant) on unfrozen water content within the ice, the influence of ice temperature on unfrozen water content, and the combined effects of air temperature, radiation, and ice thickness on ice temperature. These findings led to the development of a parameterization scheme for radar wave speed correction based on air temperature and ice thickness. The average dielectric coefficient and radar wave speed of the granular or columnar ice were calculated by summing the radar wave travel times through different ice layers. The results revealed average dielectric constants of 3.1605 for granular ice, 3.1607 for granular ice with sediment, and 3.1586 for columnar ice. These values are marginally lower than the pure ice constant of 3.17, attributable to the presence of air bubbles in the Yellow River ice.

Under the principle of mass conservation in Yellow River ice, the volumes of unfrozen water and ice fluctuate with temperature variations. Minor changes in the proportion of unfrozen water can significantly affect the dielectric coefficient. Although the variation range of unfrozen water content within ice is small and its relationship with the dielectric constant is relatively straightforward, the unfrozen water content within ice is nonlinearly related to the ice's temperature which, in turn, is nonlinearly related to external hydrometeorological environmental factors. Li et al. [34] validated the physical basis and feasibility of their parameterization approach through a field study. They conducted measurements at 13 drill sites during winter, collecting data on ice thickness and meteorological parameters including air temperature, radiation, wind speed, and cloud cover. The researchers integrated these measurements with a one-dimensional thermodynamic ice model to compute ice temperature profiles and thicknesses [65,90]. Subsequently, they established a continuous record of thermodynamic ice formation and melting processes using fixed radar observations during the winter of 2020–2021. The collected data enabled the researchers to characterize the relationship between the air temperature and radar wave propagation speed in both granular and columnar ice structures. This relationship served as the foundation for temperature-based corrections of radar wave speed. The radar wave speed within the ice exhibited a nonlinear distribution with the ratio of two phases, falling between the speeds observed in completely frozen Yellow River ice and liquid water. To maintain the accuracy of the simulation results, the researchers employed a logistic statistical model. The system was treated as a three-phase composite (ice–air–sediment) at low temperatures and as a two-phase composite (water–sediment) at high temperatures. Ice profiles were successfully expressed using these four parameters [34].

5. Uncertainties in Mathematical Models for Yellow River Ice Investigations and Disaster Warning

River ice phenomena present complex variability in response to hydrometeorological conditions, and field investigations remain indispensable in understanding these phenomena. The analysis of survey data—particularly for complex phenomena—necessitates the application of stochastic mathematical models; however, it is also an undeniable fact that field investigations and data processing are difficult.

5.1. Uncertainties in Mathematical Image Processing Techniques for Geometric Characterization of Parameters of Floating Ice Surfaces in Space–Air–Ground/Ice Integrated Investigations

During freeze-up periods, integrated space–air–ground/ice monitoring systems enable the progression from qualitative to quantitative description of river ice phenomena. These systems facilitate the acquisition of comprehensive data, including the locations where rivers freeze and open; the concentration, morphology, and flow velocities of ice; and the formation of ice jams or dams. Contemporary monitoring techniques have evolved to capture multiple parameters across temporal and spatial dimensions throughout the entire process. However, the diversity of monitoring methodologies has led to variations in the methods for extraction and the accuracy of floating ice parameters, resulting in different mathematical formulations. Furthermore, the varying focuses and objectives of researchers can yield distinct mathematical expressions, even when analyzing data obtained using identical monitoring techniques.

In general, space–air–ground/ice monitoring systems typically comprise three tiers: satellites in space [96,97], unmanned aerial vehicles (UAVs) in the sky [98,99], and equipment on the ground [100,101]. Satellite monitoring offers extensive spatial coverage that is unrestricted by geographical or political boundaries. However, its effectiveness is constrained by revisit intervals spanning several days, and satellites offering shorter revisit periods often tend to be either costly or low-resolution, limiting their real-time observation capabilities. While UAV systems provide rapid deployment and minimal terrain-based interference, their operational duration remains limited. Ground-based monitoring installations enable continuous data collection but face significant siting constraints, particularly in mountainous terrain.

During the ice freeze-up period, detection sensors generate graphical representations that capture various characteristics of ice, including freeze-up locations, the morphological features of the freeze-up ice, and the location, shape, and size of open water surfaces. Image analysis primarily employs mathematical techniques focused on image segmentation and quantitative data analysis. While river ice image analysis methodologies build upon established sea ice analysis techniques [102], they require additional refinement to account for topographical influences. For image segmentation, three approaches are predominantly utilized: traditional thresholding methods, machine learning algorithms, and neural network-based techniques. To obtain accurate geometric characterization parameters of floating ice, a variety of mathematical techniques need to be continuously developed, improved, and validated. However, existing mathematical approaches for ice image segmentation remain in developmental stages, with significant uncertainties in their application and reliability.

In recent decades, image processing techniques for ice analysis have evolved significantly. Early approaches primarily utilized classical histogram threshold segmentation methods and clustering algorithms [103]. This foundation was subsequently expanded to include more sophisticated methodologies, such as watershed algorithms [104], wavelet transforms [105], active contour models [106], gradient vector flow (GVF) [107], snake models [108], Markov random fields, spectral clustering, and neural networks [109]. The Otsu algorithm and k-means clustering algorithm remain fundamental techniques, providing a theoretical and technical framework for traditional histogram-based separation of background and foreground in grayscale images. Accurate ice concentration and velocity measurements fundamentally depend on precise morphological characterization. However, the complexity of floating ice imagery often precludes its successful analysis using a single mathematical approach. When ice blocks exhibit small contrast with the surrounding water, automated computer processes struggle to effectively delineate the ice's boundaries or extract its surface morphology and melt pool characteristics. These challenges are further compounded by the presence of tightly connected and overlapping ice edges [110], which introduce additional segmentation complexities.

Watershed segmentation techniques exhibit over-segmentation when applied to ice surfaces with non-uniform illumination and ice blocks, which vary in size and shape [111]. While an enhanced watershed method incorporating neighboring region merging algorithms can improve edge detection performance [102], its reliance on binary images results in information loss at overlapping ice block boundaries. The marker-controlled watershed (MCW) technique [112], which integrates nonlinear support vector machine (SVM) analysis with traditional watershed algorithms, was developed to avoid over-segmentation issues. However, this approach remains inadequate for the analysis of irregular ice blocks with ambiguous boundaries and uneven illumination conditions. The gradient vector flow (GVF) method [107,113] attempts to address the complex shapes of ice through a multi-step process: separating seemingly connected ice blocks, highlighting the ice's morphology through enhancement algorithms, identifying individual ice formations, and finally recognizing various complex shapes of floating ice blocks. Nevertheless, this methodology proves insufficient for distinguishing tightly packed ice blocks under conditions of uneven illumination and indistinct boundaries.

Recent advancements in machine learning (ML) have facilitated promising applications in the ice image processing field, including deep machine learning methods for river ice segmentation [114]. While ML models have gained widespread adoption across various domains, their application in ice recognition remains an emerging field requiring further development. In summary, traditional threshold methods—which rely on grayscale values for image segmentation—possess the advantages of simple computation and high accuracy [111]. However, these methods are susceptible to noise and brightness, making it challenging to determine appropriate threshold values. Although traditional machine learning techniques perform adequately in simple scenes, they exhibit limitations when processing complex scenes, often requiring human intervention and potentially compromising segmentation accuracy. In contrast, deep learning approaches—specifically, convolutional neural network-based semantic segmentation—offer superior nonlinear fitting and learning capabilities when compared with traditional methods. These advanced approaches show considerable promise for future developments in ice image processing [115].

Liu et al. [116] have developed a river ice identification method utilizing Landsat-7 ETM and images to distinguish river ice from water in remote sensing data of the Yellow River. This foundational work has paved the way for more sophisticated ice type classification approaches incorporating advanced image processing techniques and machine learning algorithms. Building upon this research, Huang et al. established a supervised classification method for river ice based on PolSAR data [117] and an unsupervised decision tree-based river ice classification method [118].

5.2. Uncertainties in Mathematical Techniques for Space–Air–Ground/Ice Integrated Investigation of River Ice Thickness

If there are technical difficulties in the space–sky monitoring of geometric parameters of floating ice, it is even more difficult to measure the thickness of ice along the vertical axis. Nevertheless, research in this domain continues to advance. Recent comprehensive reviews by Chinese researchers have documented the progress made in terms of river ice remote sensing monitoring techniques [119]. In the specific area of ice thickness detection, Yang [120] studied ice thickness inversion methodologies for the upstream section of the Wanjiazhai reservoir in the Yellow River. Their study employed varying albedo for ice thickness inversion, with the Lebedev model serving as a validation tool to demonstrate the effectiveness of the methodology.

Prior to the advent of unmanned aerial vehicles (UAVs), helicopters served as the primary monitoring platform for sea ice observations [121]. Contemporary river ice monitoring has shifted toward cost-effective UAV systems, which have demonstrated significant advantages in terms of river ice detection [98,122]. The predominant ice thickness measurement techniques employed by UAVs involve ground-penetrating radar (GPR) systems. Initially, GPR measurements were conducted by towing radar equipment across ice surfaces, with data interpretation being carried out with respect to established physical and mathematical principles developed in other fields. The complexity of ice analysis necessitates diverse mathematical approaches due to several key factors. Natural ice, being a multiphase anisotropic composite material, requires signal processing techniques similar to those used in geotechnical applications [123]. The presence of large air bubbles within ice structures calls for mathematical methods analogous to those used in determining the dielectric coefficients of cavity-containing concrete [124]. Additionally, the thin nature of ice layers demands mathematical techniques developed for target recognition in shallow radar images [125]. The Yellow River ice presents additional complexities beyond these characteristics, including its crystalline properties and the presence of sediment. Consequently, comprehensive applications incorporate not only the aforementioned mathematical techniques but also the influence of various factors on the radar propagation velocity, including unfrozen water content, sediment content, and ice crystal morphology, determined through field test data [34]. This mathematical principle is based on propagation formulas for composite anisotropic materials in different directions. The uncertainties in ice thickness measurements stem not from mathematical formulations but from variations in phase component proportions and their spatial distribution patterns within the ice. These represent the uncertainty sources for radar detection of ice thickness at a single point at different times and thus do not represent the uncertainties in towed radar [28] or airborne radar detection of ice thickness at different spatial scales at the same time. Furthermore, spatial analyses of ice thickness must account for additional uncertainties arising from the non-uniform distribution of ice thickness and radar resolution limitations.

5.3. Uncertainties in Prediction Models for River Freeze-Up and Break-Up Dates

The evolution of freeze-up date prediction methods has progressed through several distinct phases. Initial methods relied primarily on indicator methods, where freeze-up date predictions were based on changes in selected relevant indicators according to observation data. The 1960s marked the development of empirical correlation methods grounded in physical principles, leading to the establishment of numerous predictive formulas that played a great role in ice date prediction for preventive purposes. The 1980s witnessed the emergence of mathematical methods for the prediction of freeze-up and break-up dates; this development process has been documented in detail by Ji [126].

Contemporary freeze-up and break-up date prediction methods for the Yellow River predominantly employ thermal and hydraulic factor correlation methods and mathematical models [127], in alignment with international methods. Empirical formulas are established for prediction based on various parameters including water storage, ice thickness, and cumulative positive temperatures when the air temperature turns positive. Among empirical models, neural networks developed first, with machine learning applications currently being in exploratory phases. Ice forecasting systems have been expanded through the implementation of artificial intelligence-based approaches. Chen and Ji [9] successfully applied fuzzy optimal selection back propagation (BP) neural networks to predict freeze-up and break-up dates in the Ningxia–Inner Mongolia Reach of the Yellow River. However, classical BP neural networks, based on gradient descent methods, exhibited limitations including slow network approximation and susceptibility to local minima, which impeded their learning capabilities. To address these constraints, the Levenberg–Marquardt algorithm was implemented to enhance the performance of the BP neural network in terms of ice dam prediction. This algorithm allows for performance optimization by replacing the mean square error with the squared error to minimize the sum of squared errors, combining the advantages of both Newton's method and gradient descent approaches. The Levenberg–Marquardt algorithm has demonstrated superior convergence rates when compared with gradient descent methods [23]. Guo et al. [128] developed a comprehensive ice prediction database for the Ningxia–Inner Mongolia Reach of the Yellow River, implementing a geographic information system (GIS)-based decision support system for ice forecasting. Since its implementation in 2004, this system has been shown to possess a long forecast period, high prediction accuracy, and stable operation.

5.4. Uncertainties in Prediction Models for Ice Jams or Dams

Ice jam and dam formation in the Yellow River has been highly regarded in the relevant literature. Sun et al. [10] conducted systematic field observations investigating the formation mechanisms of ice jams resulting from frazil ice accumulation beneath ice layers, which yielded empirical formulas for riverbed roughness and ice jam water levels. The formation and evolution of ice dams during ice break-up periods emerge from complex interactions among hydrological, meteorological, hydraulic, thermal, and dynamic forces. The complexity of these influencing factors has limited the number of mathematical models that are capable of accurately predicting and simulating ice dams during break-up periods. Since the 1990s, the China Yellow River Conservancy Commission has performed decades-long systematic observations of the Yellow River's ice conditions, establishing several practical thermodynamics-based methods for forecasting freeze-up dates, break-up dates, and ice jams or dams [68,69,129]. While recent advances in river ice numerical modeling have shown promise, several fundamental problems remain unresolved.

The inherent complexity of ice jams or dams exceeds the descriptive capabilities of deterministic mathematical models, while non-deterministic stochastic models can only reflect a part of the actual phenomena. For ice jams or dams that are difficult to express via deterministic and stochastic expressions, Ji [126] applied mathematical fuzzy set theory to characterize their uncertainties. This approach conceptualized the uncertainties of ice jams or dams through the dual aspects of randomness and fuzziness, leading to the development of fuzzy optimal selection neural network prediction models. Additionally, various mathematical methods for image segmentation, aside from fuzzy theory, have been extensively applied to ice jam and dam analysis [130], further highlighting the uncertainties and challenges in predicting ice jams or dams.

Artificial neural networks (ANNs) have the characteristics of strong approximation ability for complex nonlinear mapping relations, good robustness and fault tolerance for information processing, and strong adaptability to fuzzy and incomplete information. These attributes make ANNs particularly suitable for ice dam prediction as it is a complex nonlinear problem influenced by multiple factors that resist precise correlation analysis. Different neural network-based methods for different types of river ice are still being explored, including clustering algorithms [131], genetic algorithms [132], and swarm intelligence algorithms [133]. Notable advances in this field include Wang et al.'s development of a neural network prediction model for ice jam water levels and ice thickness in curved channels, utilizing data derived from laboratory curved channel section experiments [134].

5.5. Uncertainties in Warning Models of Ice Disasters in the Yellow River

Ice flood risks in the Yellow River mainly include riverbank inundation caused by backwater, backward flow in dike-through drainage ditches, overflow, and levee-breach ice flood, which may cause damage to engineering facilities [135]. Gao et al. [136] and Song et al. [137] sorted the research status of river ice's basic theory, prototype observation, and regime forecasting while also analyzing recent flood characteristics and reservoir management practices along the Yellow River. Guo et al. [138] investigated the factors influencing ice floods and the characteristics of disaster evolution in the Ningxia–Inner Mongolia Reach of the Yellow River, utilizing historical ice regime measurements and disaster statistics. Yuan et al. [139,140] and Tian [130] have systematically studied the one- to two-dimensional coupled model and a two-dimensional hydrodynamic model spanning from river channels to floodplains. Their research analyzed ice flood disaster mechanisms in the Ningxia–Inner Mongolia Reach of the Yellow River, while a series of quantitative technical methods were proposed, such as ice jam hazard diagnosis, flood embankment evaluation, and ice flood risk assessment.

The main factors leading to ice flood disasters include the thermal environment, dynamic factors, river boundary conditions, and human activities. Human activities mainly include reservoir regulation, embankment and river training projects, and the joint regulation of ice flood–diversion zones. The joint dispatch of the Longyangxia Reservoir and Liujiaxia Reservoir in the upper reaches of the Yellow River, the river regulation project in the Ningxia–Inner Mongolia reaches of the Yellow River, and the emergency regulation of ice flood–diversion zones are important measures for the defense and emergency response to ice flood disasters. Through comprehensive literature reviews and analytical investigations, researchers have identified four primary factors affecting ice flood disasters relating to the Yellow River: air temperature, water level, floating ice concentration, and ice thickness. These parameters serve as warning indicators, with quantified early warning thresholds establishing a warning system comprising blue, yellow, orange, and red alert levels [141].

The presence of human factors introduces significant uncertainties into ice flood risk assessment methodologies. Luo et al. [8], Luo [142], and Luo et al. [143] have proposed a three-parameter interval gray number evaluation method to handle fuzzy multiple indicators for assessing ice jam disaster risk in the Ningxia–Inner Mongolia reaches of the Yellow River. Subsequently, they introduced assessment methods incorporating gray information combined with flood disaster risk models for the Yellow River, as well as a VIKOR extension method, and applied these approaches to evaluate ice flood disaster risks across different regions of the Yellow River's Inner Mongolia section. Wu et al. [144] have developed a comprehensive ice disaster risk evaluation model integrating projection pursuit, fuzzy clustering, and accelerating genetic algorithms to assess the ice disaster risk in the Ningxia–Inner Mongolia reaches of the Yellow River from 1991 to 2010. Li [145] constructed a grey decision model for ice jam and dam risk assessment in the Inner Mongolia section of the Yellow River, while Wu [146] applied a mutation evaluation method to evaluate the risk of ice flood disasters on the Yellow River. A loss assessment system for ice flood disasters was established to assess the post-disaster loss of Yellow River ice flood disasters. Further methodological advances include Wang et al.'s [147] mutation theory-based ice flood risk assessment for the Bayannur region of Inner Mongolia, as well as Hu's [148] refinement of the mutation evaluation method to provide highly comprehensive evaluation values. The risk evaluation of flooding disasters in the Inner Mongolia section of the Yellow River was carried out according to this improved method. However, these approaches collectively fail to account for uncertainties in the subjective perceptions of decision-makers.

6. Research Gaps, Challenges, and Future Directions for Mathematical Models of Yellow River Ice

The natural phenomena associated with river ice conditions in cold regions worldwide share both similarities and unique characteristics. In contrast to river ice in North America, Northern Europe, Russia, and other regions, the Yellow River's ice conditions resist full explication through mathematical models. There are numerous common challenges in mathematical modeling and methodologies, as well as unique gaps due to differences in geographical and physical characteristics. For the Yellow River ice, the scientific obstacles faced at present are multifaceted, arising from the complexity and randomness of the natural phenomena associated with the formation of ice. The technical difficulties in detecting these phenomena, especially during the long and segmented ice run periods, pose high risks for ice surveys, and there is a need to improve mathematical models and methods through the efforts of both mathematical and natural scientists. While the common challenges posed by river ice can be addressed through advancements in international ice research, the unique problems relating to the Yellow River must be tackled by Chinese scholars. Here, we focus on the uncertainties and current development status of mathematical models and methods for the Yellow River's ice conditions, proposing common issues to advocate for international collaboration and emphasizing the principle of independent development to address unique problems. Some preliminary and partial ideas are presented here, inviting academic peers to discuss and provide constructive feedback to propel advancements in this research domain.

6.1. Major Research Gaps in the Improvement of Mathematical Models and Methods for Yellow River Ice

(1). Parts of the Yellow River are located in arid and semi-arid regions, where the deposition of spring dust on the ice surface directly affects its solar albedo. Although there is international research on the impact of black carbon dust on the albedo of ice, studies of dust deposition on ice surfaces are rare. There is also limited international research on dust deposition on glacier surfaces in arid and semi-arid regions, though there have been explorations of the impacts of gravel on the albedo of glacier surfaces. Addressing this gap requires worldwide efforts from scholars.

(2). The joint influence of mid-latitude temperature and radiation on the physical, thermal, electrical, and mechanical properties of ice covers is a common issue for freezing water bodies in mid-latitude cold regions. Improving coupled air–ice–water models controlled by hydrometeorological factors requires mutual learning among scholars globally. However, the presence of sediment in the Yellow River ice—particularly in frazil ice—is a unique phenomenon. Although investigations into the physical characteristics of the Yellow River ice (including ice crystals and sediment within the ice) have been initiated, and it has been recognized that the presence of sediment within the ice is related to frazil ice [32], these single-point ice sample surveys have not yet determined the proportion of different ice types in the Inner Mongolia Reach of the Yellow River. This means that it is not possible to assess how much sediment is contained within the ice at different locations which, in turn, affects the evaluation of the physical, thermal, electrical, optical, and mechanical properties of the ice in specific sections of the Yellow River and impacts the refinement of associated models. In the future, Chinese scholars should seek to develop a four-phase (air, ice, water, sediment) composite model for Yellow River ice, which will be valuable for the evaluation of its physical, thermal, electrical, and mechanical properties.

(3). The sediment in the Inner Mongolia section of the Yellow River is distinctive. Although ice jam/dam models incorporate ice–water–sediment coupling, the mathematical description of ice block shapes and stacking porosity in the Yellow River ice dams, controlled by factors such as ice thickness, temperature, hydraulic conditions, and transport distance, remains a challenge. Chinese scholars need to overcome the difficulties of random ice dam locations and their short durations of occurrence through extensive research, thus developing models for ice block mechanics, ice dam formation, and breach patterns.

(4). The integration of space–air–ground surveys with artificial intelligence and traditional ice surface and ground surveys is a hot topic in international research. Advanced imaging techniques for segmenting ice flow block images are affected by ice block compression, stacking, and the distribution of light and shadow in the river channel image, posing a common challenge for scholars worldwide. However, developing models for ice block morphology parameters that reflect the joint influence of multiple factors such as the variable and multi-point ice break-up locations, uneven ice thickness, temperature fluctuations, accumulated temperature alternations, and ice block transport distances in the Yellow River requires further efforts by Chinese scholars.

(5). The Yellow River has many wandering channels characterized by varying water depths and significant spatiotemporal changes in hydraulic conditions. While models for river ice in straight and curved channels and the feedback mechanisms of sediment transport on riverbed and ice bottom roughness have been explored internationally, the branching channels of the Yellow River—especially the topographic changes caused by sediment-induced water depth variations—require further research by Chinese scholars.

(6). The stable ice coverage in the Inner Mongolia Reach of the Yellow River is about 80%. The formation, development, and disappearance processes of open water areas and their impact on ice jams/dams and ice break-up methods have not been well studied. This is another gap to be addressed by Chinese scholars.

(7). Quantifying the impacts of extreme climate events (e.g., warm winters) on the ice conditions of the Yellow River, as well as developing a coupled climate–hydrology–ice model and integrating it into ice disaster warning systems for the Yellow River require further efforts and improvement. While research has revealed that, under the eight global climate models of CMIP6, the Yellow River may not undergo complete freezing by 2070 [26], it is evident that the duration of ice runs before freeze-up and after break-up will persistently increase, while the period of stable ice freeze-up will diminish in the future. Particularly in high-risk areas for ice jams and dams, such as sections with gentle slopes, sharp bends, and bridges and/or water conservancy structures, it is imperative for Chinese scholars to amalgamate artificial intelligence and big data technologies, introduce regional economic assessments, and develop more astute warning systems and disaster loss assessment frameworks. The establishment of a minute-level ice dam risk warning model is fundamental to address the shortcomings of traditional models.

6.2. Some Research Challenges of Potential Mathematical Models and Methods for Yellow River Ice

Aside from the mathematical developments relating to the Yellow River ice in China, there are some mathematical models and methods that can be used for ice-related problems in general, some of which have potential for application to Yellow River ice problems. In particular, the following three methods can be considered in future studies.

(1). Logistic regression models with upper and lower limits provide essential limits for ice-related physical phenomena, such as the effect of cloud cover on solar irradiance. Shortwave radiation in the air–ice heat balance in one-dimensional mathematical river ice models is typically considered only for clear days. Although the Yellow River region experiences predominantly sunny conditions during winter, cloudy days are not uncommon. Some regions utilize solar irradiance measurements from clear days as upper limits and cloudy day measurements as lower limits (with the cloud cover parameter set as 0 and 10, respectively) [83]. While linear relationships between solar irradiance and cloud cover have been established using available data [13], quadratic relationships fail to accurately represent conditions within the considered range of cloud cover values (0 and 10). This highlights a critical limitation: statistical relationships derived only within the range of measured data may not align with theoretical logic when data under limiting conditions are lacking.

In other parameterized statistics regarding the Yellow River ice, logistic regression models have been successfully applied to correlate intra-ice radar wave speed with air temperature [34]. It is recommended that this statistical relationship be used between thermal diffusion coefficients and broad ice temperature ranges [149].

(2). Wavelet analysis has proven effective in analyzing periodic ice problems, such as ice temperature variations [150] and ice forces on structures [151]. In studies of daily variations in Yellow River and lake border ice conditions [152], wavelet analysis combined with local meteorological data provided a robust mathematical framework for understanding meteorological driving forces. Bai et al. [153] have analyzed the daily process of dissolved oxygen in Finland's Valkea-Kotinen Lake using wavelet analysis. This lake is at a high latitude such that it experiences short days and long nights for long periods in winter. While the influence of meteorological factors on this lake appears less pronounced than in the Yellow River, the success of wavelet analysis suggests promising applications for characterizing daily ice variations in the Yellow River. In 2024, a paper on the water channel ice was published, which used a wavelet-based method for the Chinese north-to-south water division channel [154]. Similar mid-latitude ice is influenced not only by distinct daily variations in temperature but also by distinct daily variations in radiation. Consequently, wavelet analysis shows potential for examining ice phenomena governed by fundamental physical properties such as temperature and water content within the ice.

The winter sunshine duration varies with latitude and Julian day across regions. The integration of wavelet analysis with geophysical concepts enables the precise calculation of sunrise, sunset, and local noon times across different latitudes and Julian days [83]. This approach also addresses the mathematical challenge of asymmetric waveform midpoints resulting from seasonal variations in the day–night duration across geographical locations. It should also be noted that the problem of asymmetric midpoint position in waveforms caused by variations in winter day–night duration at different places also has a mathematical solution.

(3). Fractal geometry is characterized by the property of self-similarity, which can be absolute or statistical. Fractal dimensions can describe irregular and fragmented geometrical forms, making it particularly suitable for characterizing the complex forms of natural materials. While natural river and lake ice exhibit diverse morphological characteristics, their shapes fundamentally reflect break-up and drift processes driven by external forces, which are ultimately a reflection of the ice's microstructure as a crystalline material. The growth environment's complexity leads to varied ice crystal morphologies; however, all such morphologies maintain characteristics derived from ice's fundamental hexagonal crystal system, which imparts both distinctive and similar features across ice crystals. This internal crystalline structure fundamentally determines the initial shape and dimensions of the ice. Preliminary research has documented these characteristics in natural lake ice [155] and river ice broken up in different ways [156]. Therefore, fractal theory is a powerful mathematical tool to study ice blocks and cracks, which needs to be further developed by more scholars.

6.3. Future Research Directions for the Development of Mathematical Methods in Yellow River Ice

(1). The modeling of ice–water–sediment interactions in the Yellow River necessitates further refinement. While existing models already encompass hydrodynamics, thermodynamics, and sediment transport, the unique topography of the Inner Mongolia Reach of the Yellow River introduces certain complexities. Factors such as the gentle slope gradient, variations in channel width (from wide to narrow and vice versa), bend curvature, water depth, tributaries, hidden shoals, the physical and mechanical properties of sediment, the content and distribution of sediment within the ice, and dust deposition from sandstorms on the ice surface all influence the river's water behavior and heat exchange processes. Thus, employing the mathematical principles of fractal geometry and clustering, the Inner Mongolia Reach of the Yellow River can be divided into units of different types. These units can then be further classified into more refined categories, based on hydrodynamic, thermodynamic, and sediment behavior parameters, which are specific to different terrains. This approach would enable the development of a multidimensional model that integrates hydrodynamics, thermodynamics, and morphological changes in the river channel, systematically reflecting the river's flow velocity, temperature distribution, and morphological evolution. Subsequently, leveraging the dominant hydrological and meteorological factors in current models, the evolution of flow velocity, sediment transport, and temperature distribution under varying flow rates and temperature conditions could be intuitively simulated. This would incorporate changes in river morphology along the Yellow River (e.g., sediment suspension, rolling, and deposition; channel erosion and siltation), hydrodynamic changes (e.g., velocity distribution and flow lines), and regional thermal variations (e.g., the water temperature distribution, driven by hydrological and meteorological factors) into mathematical models and/or methods. This would yield an enhanced multidimensional model of the Yellow River ice, overcoming the present challenge of sparse data at certain cross-sections, as a few cross-sections are insufficient to represent the entire Inner Mongolia section of the river.

(2). While deterministic models have been developed to study ice jams and dams, significant uncertainties persist regarding their spatial and temporal occurrence. Ice jams and dams result from complex interactions between hydrometeorological dynamic and thermal processes, ice behaviors, and river conditions. On the one hand, thermodynamic factors control the thickness and strength of ice, while on the other hand, dynamic factors control the fracture, movement, and accumulation of frazil ice or ice layers. The stable formation of ice jams and dams represents an equilibrium between the shear resistance of bonding among frazil ice or ice blocks and external hydrometeorological forces within the river channel's constraints. Although the physical mechanisms of these processes are relatively clear, definitive conclusions regarding the size and shape of ice blocks during disintegration and transport remain elusive, particularly under varying ice thicknesses [156]. While the size and shape of frazil ice and ice blocks significantly influence the internal porosity and overall cohesion of accumulations, shear mechanics-based constitutive relations for destabilization processes remain underdeveloped. While studies focused on the Yellow River's ice have referenced critical parameters from Ettema and Urroz [157] and Beltaos [158], these applications lack validation through direct observations of the Yellow River. Similarly, numerical simulations of frazil ice distribution patterns under ice layers during early freeze-up periods rely heavily on certain assumptions [159]. In order to realize the formation and melting processes of ice jams and dams, modern monitoring tools are first needed to obtain the shape and size of ice blocks. The size distribution of frazil ice or ice blocks may be described according to block shape parameters, such as the roundness, flatness, elongation, and relative width for calculation of the shape index; normalized shape factor; morphological complexity index; comprehensive morphological index; and standardized morphological index [160]. Methods for determination of the rolling resistance of granular materials may have potential value [161]. Meanwhile, the accumulation shape of ice jams and dams can be obtained through the use of monitoring tools, with the shape and complexity of ice blocks potentially being described according to their fractal dimensions. Future research on ice jams and dams can draw lessons from research on the accumulation shapes of polar sea ice ridges [162], which progress from individual investigations to mathematical indicators at macro scales. Resolving the problem of ice dam-related damage requires an integrated approach combining mathematical descriptive expressions of ice ridges, block shear constitutive relations, hydrometeorological dynamics-based calculation methods, and river channel mathematical characterizations.

(3). The fundamental process of heat balance on ice surfaces is heat exchange, leading to the formation and melting of ice by means of thermal convection, thermal radiation, and thermal conduction. In the context of the Yellow River, two distinctive characteristics significantly influence these processes: the presence of impurities within the ice structure [28,32] and the presence of dust particles on ice surfaces, deposited during spring sandstorms. These characteristics alter the ice's thermal properties in two critical ways. First, surface dust accumulation modifies the ice's color, significantly affecting the albedo of its surface. Second, internal impurities alter the ice's thermal conductivity, necessitating the application of composite media thermal conductivity models rather than consideration of pure ice's simple thermal constants. This demands both the introduction of established composite thermal conductivity models and the development of advanced mathematical stochastic models to address these challenges [84].

(4). Frozen rivers experience varying weather conditions during winter, being sometimes sunny and sometimes cloudy. In addition to considering the variation in solar radiation intensity and sunlight duration with latitude on clear days, the impact of cloud cover on radiation intensity requires further investigation. Li et al. [83] have demonstrated that linear regression models inadequately represent the relationship between cloud cover and radiation intensity, instead proposing a logistic model for statistical analysis of this relationship. While this statistical approach aligns with both physical laws and mathematical principles in parameterizing river ice models, further comprehensive research is necessary to validate and refine these methods.

(5). Advances in modern monitoring technologies have overcome safety limitations in traditional ice monitoring projects, enabling comprehensive space–air–ground/ice integrated ice monitoring approaches. This technological evolution allows for intelligent, multi-element monitoring of ice conditions throughout its entire lifecycle. However, the validation and refinement of deterministic and stochastic mathematical models for river ice require not only space–air–ground/ice integrated ice condition survey data but also extensive field data, including high-precision river topography, real-time hydrological data, and meteorological data. To date, research and survey efforts on the Yellow River—particularly those focused on riverbanks or ice surfaces—remain sparse and scattered across certain cross-sections and stations. There is a need to enhance space–ground/ice monitoring data. Additionally, the use of different monitoring techniques, each targeting different elements, necessitates different mathematical methods and models for data processing. In this context, the development of mathematical methodologies for data fusion across various technological platforms becomes crucial. As manual monitoring methods transition to intelligent monitoring systems, the processed data must not only be transmitted to users but also enable the system to intelligently determine and decide on subsequent monitoring methods, relevant monitoring elements, and preventive and mitigation measures. Such advanced capabilities require robust mathematical methods including image recognition algorithms, fuzzy clustering algorithms, and neural network algorithms.

(6). The Yellow River region plays a vital role in China's economic development, and the prevention and mitigation of ice disasters in the Yellow River represents a combination of natural and socio-economic sciences. This necessitates the implementation of a comprehensive scientific management program from risk identification to risk assessment and risk assessment to risk control. In the past, when quantitative assessment indicators were identified, there was a lack of quantification of uncertainties in decision-makers' subjective cognitions. This limitation has created information gaps in the assessment of ice flood disaster risk, that is, uncertainty. Uncertainty measurement theory offers a solution to address incomplete objective information while minimizing human subjective bias in scenarios characterized by limited data [163]. Set pair analysis provides a systematic theoretical approach to deal with certainty and uncertainty, unifying the complex relationships between objective things through the principles of identity, difference, and opposition. By coupling uncertainty measurement theory with set pair analysis, the ice flood-prone areas of the Yellow River can be systematically analyzed in sections. This integrated approach is expected to enable the establishment of localized ice flood disaster risk assessment models, introducing novel methodologies for studying ice flood disaster risk throughout the Yellow River.

7. Conclusions

This review comprehensively addressed the inherent uncertainties in mathematical models for ice processes in rivers, with a particular focus on the Yellow River. It emphasized the sources, impacts, and methods for potential resolution of these uncertainties from the perspectives of physical and mathematical theories, the natural conditions of the Yellow River, the capabilities of space–air–ground/ice integrated observation technology, and ice disaster warning systems. The key findings are summarized as follows:

1. The sediment content and unfrozen water within the Yellow River ice play pivotal roles in influencing its density, thermal conductivity, specific heat, dielectric properties, mechanical properties, and more. These factors drive the spatial and temporal variations in the basic properties of the Yellow River ice, thereby introducing significant uncertainties. These uncertainties notably affect the applicability of mathematical models and the accuracy and universality of mathematical methods. They also highlight challenges in terms of numerical simulations of ice thermodynamics and thickness inversion using radar. Future developments must concentrate on advanced models tailored to the four-phase composite medium of air–ice–water–sediment within the Yellow River.

2. To address uncertainties in deterministic models related to the Yellow River's ice conditions—such as ice bottom roughness, heat exchange coefficients at the air–ice/water interfaces, effects of cloud cover on solar radiation, and the basic properties of ice—the development and integration of advanced mathematical tools, high-resolution monitoring systems, and interdisciplinary collaborative approaches is necessary. Short-term focuses should prioritize validating theoretical presumptions through the use of field data while enhancing the predictive accuracy of the developed models.
3. The Yellow River, situated in the mid-latitudes, exhibits distinct regional characteristics regarding the formation and melting of ice. Its freezing and thawing locations are intricately linked to the paths of cold air. The rate of air temperature decline determines the time from drifting ice to stable ice cover, while the rate of air temperature increase determines the time from ice melting to the disappearance of drifting ice. In addition, climate change has increased the frequency of extreme sudden and drastic cooling and warming events. The interactions between the rate of air temperature change, hydrodynamic forces, and ice morphology determine the total amount of drifting ice and the resistance between ice blocks, as well as between ice blocks and riverbanks, with respect to the local Froude number. These factors influence the likelihood of ice jams or dams, ultimately determining the probability of "tranquil" or "violent" river break-ups. Despite the use of long-term observational data from the Yellow River, empirical correlations, and the use of machine learning approaches (e.g., BP neural networks) to enhance the prediction of ice jams/dams, challenges persist in capturing the nonlinear interactions occurring under climate change.
4. Advancements in modern monitoring technologies have facilitated multi-scale data acquisition through the utilization of space–air–ground integrated monitoring systems. Nevertheless, challenges persist regarding the development of effective intelligent segmentation techniques, primarily due to the presence of overlapping ice edges and uneven lighting in ice images. Additionally, differences in spatial and temporal scales pose new demands for data assimilation and fusion techniques in terms of ice condition monitoring. Future efforts should focus on strengthening the integration of multi-source data obtained from space-, air-, and ground-based sensor platforms, while utilizing AI-driven decision systems for real-time ice condition monitoring and risk management.

Author Contributions: Conceptualization, B.T. and Z.L.; validation, C.L., S.H. and L.Z.; investigation, C.L., H.J. and Y.D.; data curation, S.H. and L.Z.; writing—original draft preparation, B.T. and Z.L.; writing—review and editing, S.H., C.L., H.J. and Y.D.; project administration, Z.L.; funding acquisition, Z.L. All authors have read and agreed to the published version of the manuscript.

Funding: This research was supported by joint funds of the National Natural Science Foundation of China (U23A2012), Key Science Foundation for Universities of Henan Province (24A110008), Senior Foreign Expert Introduction Project of Henan Province (HNGD2025034), and Cultivation Project of the National Natural Science Foundation of China.

Data Availability Statement: Not applicable.

Conflicts of Interest: The authors declare no conflicts of interest.

Nomenclature

The following nomenclature is used in this manuscript.

Symbols

B	The total river width (m)
B_0	The width of the open water surface (m)

P	The wetted perimeter of the channel (m)
n_c	The composite roughness
n_b	The riverbed roughness
n_i	The ice bottom roughness
$n_{i,e}$	The ice layer's bottom surface roughness at the end of freezing period
$n_{i,i}$	The initial bottom surface roughness of the ice layer
k	The attenuation coefficient for roughness with time
t	The time (s, min, hr, d)
t_{ic}	The number of freezing days
h_i	The thickness of thermally grown ice (cm)
h_0	The initial ice thickness (cm)
h_f	The thickness of frazil ice in the lower part of the ice layer (cm)
T_a	The air temperature (°C)
T_{sfc}	The ice surface temperature (°C)
T_f	The ice freezing temperature (°C)
T_w	The water temperature (°C)
ρ_i	Ice density (g/cm^3)
L_i	Latent heat of freezing (J/g)
k_i	The ice thermal conductivity (W/m·K)
FDD	The cumulative freezing degree-days (°C·d)
FDD_0	The initial cumulative freezing degree-days (°C·d)
e_f	The porosity of the frazil ice layer
k_a	The linear heat exchange coefficient at the air–ice interface (W/(m^2·°C))
k_{wi}	The heat exchange coefficient at the ice–water interface (W/(m^2·°C))
α'	The comprehensive heat exchange coefficient (W/(m^2·°C))
ϕ	The net heat flux lost from the river surface (W/m^2)
ϕ_0	The solar radiation without clouds (W/m^2)
ϕ_s	The heat flux from solar radiation (W/m^2)
ϕ_{c1}	The solar radiation in the presence of clouds (W/m^2)
C_1	The cloud cover (0–10; 0 indicates clear sky without clouds, 10 is full coverage of clouds)
β	The climate-related empirical parameter (W/m^2)
A'	The empirical coefficient from reference [70]
ζ_{is}	The empirical index from reference [70]
Acronyms	
CRISSP1D	Comprehensive River Ice Simulation System—1D
CRISSP2D	Comprehensive River Ice Simulation System—2D
DynaRICE	Dynamic River Ice Model
HEC-RAS	Hydrologic Engineering Center's River Analysis System
HIGHTSI	High-resolution Thermodynamic Sea Ice Model
ICEJAM	Ice Jam Model
IAHR	International Association for Hydraulic Research
RICE	River Ice Continuum Model
RICEN	River Ice Model with Enhanced Numeric
RICE-E	River Ice Model—Enhanced
RICES2D	River Ice Simulation Model—2D
RIVJAM	River Jam Model
MIKE11-ICE	MIKE 11 Ice Module
Landsat-7 ETM	Landsat-7 Enhanced Thematic Mapper
GIS	Geographic information system
GPR	Ground-penetrating radar
PolSAR	Polarimetric synthetic aperture radar
BP	Back propagation (neural network)
UAV	Unmanned aerial vehicle

ANN	Artificial neural network
GVF	Gradient vector flow
MCW	Marker-controlled watershed
ML	Machine learning
VIKOR	VlseKriterijumska Optimizacija Kompromisno Resenje

References

1. Prowse, T.; Alfredsen, K.; Beltaos, S.; Bonsal, B.R.; Bowden, W.B.; Duguay, C.R.; Korhola, A.; McNamara, J.; Vincent, W.F.; Vuglinsky, V.; et al. Effects of changes in arctic lake and river ice. *Ambio* **2011**, *40*, 63–74. [CrossRef]
2. Zheng, J. Research in the mathematical sciences in view of some new areas of applied mathematics. *Sci. Sci. Manag. Sci. Technol.* **1994**, *15*, 22–24. (In Chinese)
3. Leppäranta, M. A review of analytical models of sea-ice growth. *Atmos. Ocean* **1993**, *31*, 123–138. [CrossRef]
4. Shen, H.T. Mathematical modeling of river ice processes. *Cold Reg. Sci. Technol.* **2010**, *62*, 3–13. [CrossRef]
5. Beltaos, S. Threshold between mechanical and thermal breakup of river ice cover. *Cold Reg. Sci. Technol.* **2003**, *37*, 1–13. [CrossRef]
6. Massie, D.D.; White, K.D.; Daly, S.F.; McDonald, R. Predicting ice jams with neural networks. In Proceedings of the International Conference on Offshore Mechanics and Arctic Engineering, Oslo, Norway, 23–28 June 2002; Volume 36134, pp. 669–673.
7. Xu, G.H.; Li, X.Y.; Shi, C.H. The complexity-stability relationship: Progress in mathematical models. *Biodivers. Sci.* **2019**, *27*, 1364–1378. (In Chinese)
8. Luo, D.; Mao, W.X.; Sun, H.F. Risk assessment and analysis of ice disaster in Ning–Meng reach of Yellow River based on a two-phased intelligent model under grey information environment. *Nat. Hazards* **2017**, *88*, 591–610. [CrossRef]
9. Chen, S.Y.; Ji, H.L. Fuzzy optimization neural network approach for ice forecast in the Inner Mongolia reach of the Yellow River. *Hydrol. Sci. J.* **2005**, *50*, 319–330.
10. Sun, Z.C.; Yao, K.Z.; Zhou, M.; Fu, Y.W. Analysis of cause for the ice run flooding in January of 1982 at Hequ section of the Yellow River. *Yellow River* **1990**, *1*, 19–22. (In Chinese)
11. Ke, S.J. A Study of Freeze-Up Forecasting Mathematical Model in the Inner Mongolia Reach of the Yellow River. Master's Thesis, Hefei University of Technology, Hefei, China, 2002. (In Chinese)
12. Luo, H.C. Characteristics of Ice-Flow-Sediment Transport and Its Coupling Mechanism of Bed Evolution at Sharp Bend in Inner Mongolia Section of the Yellow River. Ph.D. Thesis, Inner Mongolia Agricultural University, Hohhot, China, 2021. (In Chinese)
13. Wang, T.; Guo, X.L.; Liu, J.F.; Chen, Y.Z.; She, Y.T.; Pan, J.J. Ice process simulation on hydraulic characteristics in the Yellow River. *J. Hydraul. Eng.* **2024**, *150*, 05024001. [CrossRef]
14. Beltaos, S. Progress in the study and management of river ice jams. *Cold Reg. Sci. Technol.* **2008**, *51*, 2–19. [CrossRef]
15. IAHR. River ice jams: A state of the art report. In *Proceedings of the 8th International Association for Hydraulics Research, Symposium on Ice; Working Group on Ice*; Institute of Hydraulic Research University of Iowa: Iowa, IA, USA, 1986; Volume 3.
16. Wang, T.; Liu, Z.P.; Guo, X.L.; Fu, H.; Liu, W.B. Prediction of breakup ice jam with Artificial Neural Networks. *J. Hydraul. Eng.* **2017**, *48*, 1355–1362. (In Chinese)
17. Pan, J.J.; Shen, H.T.; Guo, X.L.; Wang, T. A two-dimensional flow-ice-sediment coupled model for rivers I: Theory and methods. *J. Hydraul. Eng.* **2021**, *52*, 700–711. (In Chinese)
18. Hu, J.B.; Liu, L.; Huang, Z.P.; You, Y.; Rao, S.Q. Ice Breakup Date Forecast with Hybrid Artificial Neural Networks. In Proceedings of the 4th International Conference on Natural Computation, Jinan, China, 18–20 October 2008.
19. Fu, C.; Popescu, I.; Wang, C.; Myneet, A.E.; Zhang, F. Challenges in modelling river flow and ice regime on the Ningxia–Inner Mongolia reach of the Yellow River, China. *Hydrol. Earth Syst. Sci.* **2014**, *18*, 1225–1237. [CrossRef]
20. Rokaya, P.; Morales-Marin, L.; Lindenschmidt, K.E. A physically-based modelling framework for operational forecasting of river ice breakup. *Adv. Water Resour.* **2020**, *139*, 103554. [CrossRef]
21. Blackburn, J.; She, Y. A comprehensive public-domain river ice process model and its application to a complex natural river. *Cold Reg. Sci. Technol.* **2019**, *163*, 44–58. [CrossRef]
22. Bai, T.; Wei, J.; Chang, F.; Yang, W.; Huang, Q. Optimize multi-objective transformation rules of water-sediment regulation for cascade reservoirs in the upper Yellow River of China. *J. Hydrol.* **2019**, *577*, 123987. [CrossRef]
23. Wang, T.; Yang, K.L.; Guo, Y.X. Application of artificial neural networks to forecasting ice conditions of the Yellow River in the Inner Mongolia reach. *J. Hydrol. Eng.* **2008**, *13*, 811–816.
24. Yao, H.M.; Qin, F.X.; Shen, G.C.; Dong, X.N. Ice regime characteristics in the Ningxia-Inner Mongolia reach of Yellow River. *Adv. Water Sci.* **2007**, *18*, 893–899. (In Chinese)
25. Xie, J.; Xu, Y.-P.; Wang, Y.; Gu, H.; Wang, F.; Pan, S. Influences of climatic variability and human activities on terrestrial water storage variations across the Yellow River basin in the recent decade. *J. Hydrol.* **2019**, *579*, 124218. [CrossRef]

26. Zhou, Z.Y.; Wang, T.; Sun, Y.F.; Chen, Y.Z.; Lu, J.Z. Analysis of the evolution of ice characteristics in the Yellow River under future climate scenarios. *J. Hydraul. Eng.* **2024**, *55*, 355–366. (In Chinese)
27. Wang, J.; Shi, F.Y.; Chen, P.P. Stress criterion for mechanical ice-cover breakup based on elastic-plastic theory. *J. Hydrol. Eng.* **2013**, *44*, 344–348. (In Chinese)
28. Fu, H.; Liu, Z.P.; Guo, X.L.; Cui, H.T. Double-frequency ground penetrating radar for measurement of ice thickness and water depth in rivers and canals: Development, verification and application. *Cold Reg. Sci. Technol.* **2018**, *154*, 85–94. [CrossRef]
29. Cai, L. *Characteristics of River Ice Regime in China*; Yellow River Conservancy Press: Zhengzhou, China, 2008. (In Chinese)
30. Sui, J.; Karney, B.W.; Sun, Z.; Wang, D. Field investigation of frazil jam evolution: A Case Study. *J. Hydraul. Eng.* **2002**, *128*, 781–787. [CrossRef]
31. Wang, J.; Sui, J.; Zhang, H.; Chen, P.; Hirshfield, F. Mechanisms of ice accumulation in a river bend—An experimental study. *Int. J. Sediment Res.* **2012**, *27*, 521–537. [CrossRef]
32. Zhang, Y.D.; Li, Z.J.; Xiu, Y.R.; Li, C.J.; Zhang, B.S.; Deng, Y. Microstructural characteristics of frazil particles and the physical properties of frazil ice in the Yellow River, China. *Crystals* **2021**, *11*, 617. [CrossRef]
33. Wang, J.K.; Cao, X.W.; Wang, Q.K.; Yan, L.H.; Li, Z.J. Experimental relationship between flexural strength, elastic modulus of ice cover and equivalent ice temperature. *South North Water Transf. Water Sci. Technol.* **2016**, *14*, 75–80. (In Chinese)
34. Li, Z.J.; Li, C.J.; Yang, Y.; Zhang, B.S.; Deng, Y.; Li, G.H. Physical mechanism and parameterization for correcting radar wave velocity in Yellow River ice with air temperature and ice thickness. *Remote Sens.* **2023**, *15*, 1121. [CrossRef]
35. Wang, J.; Fu, H.; Yi, M.K.; Sun, X.Y.; Chen, P.P. The simulation of flow velocity profile under ice cover. *J. Glaciol. Geocryol.* **2009**, *31*, 705–710. (In Chinese)
36. Wang, J. A study on the relations about balance ice jam thickness with its flow conditions and ice discharge. *J. Lanzhou Univ. Nat. Sci.* **2002**, *38*, 117–121. (In Chinese)
37. Wang, X.; Luo, H.; Ji, H.; Li, B. Experimental study on characteristics of subglacial circulation for Shisifenzi bend flow in Yellow River. *Adv. Sci. Tech. Water Res.* **2023**, *43*, 31–36+51. (In Chinese)
38. Luo, H.; Ji, H.; Chen, Z.; Liu, B.; Xue, Z.; Li, Z. An analytical study for predicting incipient motion velocity of sediments under ice cover. *Sci. Rep.* **2025**, *15*, 1912. [CrossRef]
39. Yang, Z.; Ji, H.L.; Li, H.; Mou, X.Y.; Mao, Y.X.; Song, H.Z. Investigating influence factors and developing regression model for sandy silt strength in Inner Mongolia section of Yellow River under freeze-thaw cycles. *J. Arid Land Res. Environ.* **2024**, *38*, 137–147. (In Chinese)
40. Xu, G.B. One-dimensional mathematical models of river ice evolving process. *J. Water Resour. Water Eng.* **2011**, *22*, 78–83. (In Chinese)
41. Shen, H.T.; Shen, H.; Tsai, S.M. Dynamic transport of river ice. *J. Hydraul. Res.* **1990**, *28*, 659–671. [CrossRef]
42. Lal, A.W.; Shen, H.T. Mathematical model for river ice processes. *J. Hydraul. Eng.* **1991**, *117*, 851–867. [CrossRef]
43. Shen, H.T.; Wang, D.S.; Lal, A.W. Numerical simulation of river ice processes. *J. Cold Reg. Eng.* **1995**, *9*, 107–118. [CrossRef]
44. Shen, H.T. *River Ice Research*; Yellow River Water Resources Press: Zhengzhou, China, 2010. (In Chinese)
45. Shen, H.T. *CRISSP1D Programmer's Manual*; CEATI Report, No. T012700–0401; Department of Civil Engineering, Clarkson University: Potsdam, NY, USA, 2005.
46. Chen, F.H.; Shen, H.T.; Jayasundara, N.C. A one-dimensional comprehensive river ice model. In Proceedings of the 18th IAHR International Symposium on Ice, Sapporo, Japan, 28 August–1 September 2006.
47. Kandamby, A.; Jayasundara, N.; Shen, H.T.; Deyhle, C. A numerical river ice model for Elbe River. In Proceedings of the 20th IAHR International Symposium on Ice, Lahti, Finland, 14–18 June 2010.
48. Uzuner, M.S.; Kennedy, J.F. Theoretical model of river ice jams. *J. Hydraul. Div.* **1976**, *102*, 1365–1383. [CrossRef]
49. Shen, H.T.; Su, J.S.; Liu, L.W. SPH simulation of river ice dynamics. *J. Comput. Phys.* **2000**, *165*, 752–770. [CrossRef]
50. Liu, L.W.; Li, H.; Shen, H.T. A two-dimensional comprehensive river ice model. In Proceedings of the 18th IAHR International Symposium on Ice, Sapporo, Japan, 28 August–1 September 2006.
51. Knack, I.M.; Shen, H.T. A numerical model for sediment transport and bed change with river ice. *J. Hydraul. Res.* **2018**, *56*, 844–856. [CrossRef]
52. Pan, J.J.; Shen, H.T. Modeling ice cover effect on river channel bank stability. *Environ. Fluid Mech.* **2022**, *22*, 1121–1133. [CrossRef]
53. Ma, X.Y.; Fukushima, Y. A numerical model of the river freezing process and its application to the Lena River. *Hydrol. Process.* **2002**, *16*, 2131–2140. [CrossRef]
54. Wang, J.; Chen, P.P.; Sui, J.Y. Numerical simulation of ice jams in natural channels. *J. Hydraul. Eng.* **2011**, *42*, 1117–1121. (In Chinese)
55. Chen, P.P.; Wang, J. Study of the treatment technology of irregular boundaries for numerical simulating river ice. *J. Glaciol. Geocryol.* **2012**, *34*, 1353–1357. (In Chinese)
56. Pariset, E.; Hauser, R. Formation and evolution of ice covers on rivers. *Trans. Eng. Inst. Can.* **1961**, *5*, 41–49.
57. Tatinclaux, J.C.; Gogus, M. Asymmetric plane flow with application to ice jams. *J. Hydraul. Eng.* **1983**, *109*, 1540–1554. [CrossRef]

58. Flato, G.M. Calculation of Ice Jam Profiles. Master's Thesis, University of Alberta, Edmonton, AB, Canada, 1988.
59. Beltaos, S. Numerical computation of river ice jams. *Can. J. Civ. Eng.* **1993**, *20*, 88–99. [CrossRef]
60. Daly, S.F.; Vuyovich, C.M. Modeling river ice with HEC-RAS. In Proceedings of the 12th Workshop on the Hydraulics of Ice Covered Rivers, Edmonton, AB, Canada, 19–20 June 2003.
61. Timalsina, N.P.; Charmasson, J.; Alfredsen, K.T. Simulation of the ice regime in a Norwegian regulated river. *Cold Reg. Sci. Technol.* **2013**, *94*, 61–73. [CrossRef]
62. Guo, X.L.; Yang, K.L.; Fu, H.; Wang, T.; Guo, Y.X. Simulation and analysis of ice processes in an artificial open channel. *J. Hydrodyn.* **2013**, *25*, 542–549. [CrossRef]
63. Yang, K.L.; Wang, T.; Guo, X.L.; Fu, H.; Guo, Y.X. Safety regulations of water conveyance in the middle route of south-to-north water diversion project in ice period. *South North Water Transf. Water Sci. Technol.* **2011**, *9*, 1–4. (In Chinese)
64. Ministry of Water Resources of the People's Republic of China. *SL428-2008 Specification for Ice Flood Computation;* China Water&Power Press: Beijing, China, 2008. (In Chinese)
65. Yang, Y. Physical Parameter Identification and High Resolution Modeling in Ice and Snow Thermodynamic Processes. Ph.D. Thesis, Dalian University of Technology, Dalian, China, 2012. (In Chinese)
66. Stefan, J. Über die Theorie der Eisbildung, insbesondere über die Eisbildung im Polarmeere. *Ann. Phys.* **1891**, *278*, 269–286. (In German) [CrossRef]
67. Shen, H.T.; Yapa, P.D. A unified degree-day method for river ice cover thickness simulation. *Can. J. Civ. Eng.* **1985**, *12*, 54–62. [CrossRef]
68. Zhao, S.; Shen, H.T.; Shi, X.; Li, C.; Li, C.; Zhao, S. Wintertime surface heat exchange for the Inner Mongolia Reach of the Yellow River. *J. Am. Water Resour. Assoc.* **2020**, *56*, 348–356. [CrossRef]
69. Ke, S.J.; Lu, G.Q.; Ren, Z.Y. Study on mechanism of ice jam formation in Bayangaole section of Yellow River. *J. Hydraul. Eng.* **2000**, *7*, 66–69. (In Chinese)
70. Ministry of Water Resources of the People's Republic of China. *SL278-2002 Regulation for Hydrology Computation of Water Resource and Hydropower Projects;* China Water&Power Press: Beijing, China, 2002. (In Chinese)
71. Wang, J.; Hou, Z.X. Simulation of ice cover thickness in Inner Mongolia Section of Yellow River. *South North Water Transf. Water Sci. Technol.* **2021**, *19*, 921–929. (In Chinese)
72. Bai, Y.L.; Xu, H.; Shi, L.Q. Research on the optimal identification of thermal diffusivity of fresh ice in reservoirs of cold regions. In Proceedings of the 21st IAHR International Symposium on Ice, Ice Research for a Sustainable Environment, Dalian, China, 11–15 June 2012.
73. Li, Z.J.; Fu, X.; Shi, L.Q.; Huang, W.F.; Li, C.J. Recent advances and challenges in the inverse identification of thermal diffusivity of natural ice in China. *Water* **2023**, *15*, 1041. [CrossRef]
74. Shen, H.T. River ice processes. In *Advances in Water Resources Management;* Springer: Cham, Switzerland, 2016; pp. 483–530.
75. Briegleb, B.; Bitz, C.; Hunke, E.; Lipscomb, W.; Holland, M.; Schramm, J.; Moritz, R.E. Scientific description of the sea ice component in the Community Climate System Model. *Version* **2004**, *3*, 70.
76. Curry, J.; Schramm, J.; Perovich, D.; Pinto, J. Applications of SHEBA/FIRE data to evaluation of snow/ice albedo parameterizations. *J. Geophys. Res. Atmos.* **2001**, *106*, 15345–15355. [CrossRef]
77. Shine, K. Parametrization of the shortwave flux over high albedo surfaces as a function of cloud thickness and surface albedo. *Q. J. R. Meteorol. Soc.* **1984**, *110*, 747–764. [CrossRef]
78. Bennett, T.J., Jr. A Coupled Atmosphere-Sea Ice Model Study of the Role of Sea Ice in Climatic Predictability. Ph.D. Thesis, University of Illinois at Urbana-Champaign, Urbana-Champaign, IL, USA, 1982.
79. Efimova, N. On methods of calculating monthly values of net longwave radiation. *Meterol. Gidrol.* **1961**, *10*, 28–33.
80. Jacobs, J.D. Radiation climate of broughton island. In *Energy Budget Studies in Relation to Fast-Ice Breakup Processes in Davis Strait;* Barry, R.G., Jacobs, J.D., Eds.; University of Colorado, Institute of Arctic and Alpine Research: Boulder, CO, USA, 1978; pp. 105–120.
81. Cao, Y.Q.; Zhang, W.; Yao, J.Y.; Wang, W. Variation of cloud fraction and its relationship with solar radiation over semi-arid region. *J. Arid Meteorol.* **2015**, *33*, 684–693. (In Chinese)
82. Ashton, G.D.; Kennedy, J.F. Ripples on underside of river ice covers. *J. Hydraul. Div.* **1972**, *98*, 1603–1624. [CrossRef]
83. Li, Z.J.; Wang, Q.K.; Tang, M.G.; Lu, P.; Li, G.Y.; Leppäranta, M.; Huotari, J.; Arvola, L.; Shi, L.J. Diurnal cycle model of lake ice surface albedo: A case study of Wuliangsuhai Lake. *Remote Sens.* **2021**, *13*, 3334. [CrossRef]
84. Li, C.J. Comprehensive Applied Study of the Basic Properties of Freshwater Ice in the Yellow River Field Test. Doctor's Thesis, Dalian University of Technology, Dalian, China, 2024. (In Chinese)
85. Yang, J.; She, Y.; Loewen, M. Assessing the uncertainties in modeling water temperatures during river cooling and freeze-up periods. *Cold Reg. Sci. Technol.* **2023**, *210*, 103840. [CrossRef]

86. Aslamov, I.A.; Kozlov, V.V.; Kirillin, G.B.; Mizandrontsev, I.B.; Kucher, K.M.; Makarov, M.M.; Gornov, A.Y.; Granin, N.G. Ice–water heat exchange during ice growth in Lake Baikal. *J. Great Lakes Res.* **2014**, *40*, 599–607. [CrossRef]
87. Bai, Y.L.; Li, Z.J.; Feng, E.M.; Han, M.; Lu, P. Identification and optimization for effective thermal diffusivity of fresh ice. *Chin. J. Comput. Mech.* **2007**, *24*, 187–191. (In Chinese)
88. Cao, X.W. Observation and Simulation Research on Growth and Decay Processes of Ice Cover on Lake Wuliangsuhai. Ph.D. Thesis, Dalian University of Technology, Dalian, China, 2021. (In Chinese)
89. Masterson, D.M. State of the art of ice bearing capacity and ice construction. *Cold Reg. Sci. Technol.* **2009**, *58*, 99–112. [CrossRef]
90. Li, C.J.; Li, Z.J.; Yang, Y.; Wang, Q.K.; Zhang, B.S.; Deng, Y. Theory and application of ice thermodynamics and mechanics for the natural sinking of gabion mattresses on a floating ice cover. *Cold Reg. Sci. Technol.* **2023**, *213*, 103925. [CrossRef]
91. Launiainen, J.; Cheng, B. Modelling of ice thermodynamics in natural water bodies. *Cold Reg. Sci. Technol.* **1998**, *27*, 153–178. [CrossRef]
92. Arcone, S.A.; Delaney, A.J. Airborne river-ice thickness profiling with helicopter-borne UHF short-pulse radar. *J. Glaciol.* **1987**, *33*, 330–340. [CrossRef]
93. Evans, S. Dielectric properties of ice and snow–a review. *J. Glaciol.* **1965**, *5*, 773–792. [CrossRef]
94. Koh, G. Complex dielectric constant of ice at 1.8 GHz. *Cold Reg. Sci. Technol.* **1997**, *25*, 119–121. [CrossRef]
95. Lunt, I.A.; Hubbard, S.S.; Rubin, Y. Soil moisture content estimation using ground-penetrating radar reflection data. *J. Hydrol.* **2005**, *307*, 254–269. [CrossRef]
96. Chu, T.; Lindenschmidt, K.E. Integration of space-borne and air-borne data in monitoring river ice processes in the Slave River, Canada. *Remote Sens. Environ.* **2016**, *181*, 65–81. [CrossRef]
97. Romanov, P. Global multisensor automated satellite-based snow and ice mapping system (GMASI) for cryosphere monitoring. *Remote Sens. Environ.* **2017**, *196*, 42–55. [CrossRef]
98. Liu, H.; Ji, H.L.; Luo, H.C.; Gao, G.M.; Zhang, B.S.; Mou, X.Y. Application of UAV radar in survey of ice thickness and ice structure. *J. Drain. Irrig. Mach. Eng.* **2021**, *39*, 630–636. (In Chinese)
99. Zhang, X.; Zhou, Y.; Jin, J.; Wang, Y.; Fan, M.; Wang, N.; Zhang, Y. ICENETv2: A Fine-grained river ice semantic segmentation network based on UAV images. *Remote Sens.* **2021**, *13*, 633. [CrossRef]
100. Bourgault, D. Shore-based photogrammetry of river ice. *Can. J. Civ. Eng.* **2008**, *35*, 80–86. [CrossRef]
101. Ansari, S.; Rennie, C.; Seidou, O.; Malenchak, J.; Zare, S. Automated monitoring of river ice processes using shore-based imagery. *Cold Reg. Sci. Technol.* **2017**, *142*, 1–16. [CrossRef]
102. Zhang, Q.; Skjetne, R.; Su, B. Automatic image segmentation for boundary detection of apparently connected sea-ice floes. In Proceedings of the 22nd International Conference on Port and Ocean Engineering Under Arctic Conditions, Espoo, Finland, 9–13 June 2013.
103. Ostu, N. A threshold selection method from gray-level histograms. *IEEE Trans. SMC* **1979**, *9*, 62–66.
104. Hu, S.B. Research on Ice Inversion Identification and Migration Process Based on UAV Low-Altitude Remote Sensing. Master's Thesis, Northeast Agricultural University, Harbin, China, 2022. (In Chinese)
105. Yang, X.J.; Chen, P. SAR image denoising algorithm based on Bayes wavelet shrinkage and fast guided filter. *J. Adv. Comput. Intell. Intell. Inf.* **2019**, *23*, 107–113. [CrossRef]
106. Ahmadi, S.; Zoej, M.J.V.; Ebadi, H.; Moghaddam, H.A.; Mohammadzadeh, A. Automatic urban building boundary extraction from high resolution aerial images using an innovative model of active contours. *Int. J. Appl. Earth Obs. Geoinf.* **2010**, *12*, 150–157. [CrossRef]
107. Zhang, X.Q.; Dong, G.G.; Xiong, B.L.; Kuang, G.Y. Refined segmentation of ship target in SAR images based on GVF snake with elliptical constraint. *Remote Sens. Lett.* **2017**, *8*, 791–800. [CrossRef]
108. Lakshmi, K.G.A.; Surling, S.N.N.; Sheeba, O. A novel approach for the removal of artifacts in EEG signals. In Proceedings of the 2017 International Conference on Wireless Communications, Signal Processing and Networking (WiSPNET), Chennai, India, 22–24 March 2017.
109. Singh, A.; Kalke, H.; Loewen, M.; Ray, N. River ice segmentation with deep learning. *IEEE Trans. Geosci. Remote Sens.* **2020**, *58*, 7570–7579. [CrossRef]
110. Zhang, Q.; Skjetne, R.; Metrikin, I.; Løset, S. Image processing for ice floe analyses in broken-ice model testing. *Cold Reg. Sci. Technol.* **2015**, *111*, 27–38. [CrossRef]
111. Akter, S.; Imtiaz, S.; Islam, M.; Ahmed, S.; Zaman, H.; Gash, R. Image based ice-field characterization and load prediction in managed ice field. *Cold Reg. Sci. Technol.* **2025**, *231*, 104381. [CrossRef]
112. Parmiggiani, F.; Moctezuma-Flores, M.; Wadhams, P.; Aulicino, G. Image processing for pancake ice detection and size distribution computation. *Int. J. Remote Sens.* **2019**, *40*, 3368–3383. [CrossRef]
113. Haugen, J.; Imsland, L.; Løset, S.; Skjetne, R. Ice observer system for ice management operations. In Proceedings of the 21st International Ocean and Polar Engineering Conference, Maui, HI, USA, 19–24 June 2011.

114. Ansari, S.; Rennie, C.D.; Clark, S.P.; Seidou, O. IceMaskNet: River ice detection and characterization using deep learning algorithms applied to aerial photography. *Cold Reg. Sci. Technol.* **2021**, *189*, 103324. [CrossRef]
115. Zhang, X.W.; Yue, Y.Z.; Han, L.; Li, F.; Yuan, X.Z.; Fan, M.H.; Zhang, Y.N. River ice monitoring and change detection with multi-spectral and SAR images: Application over yellow river. *Multimed. Tools Appl.* **2021**, *80*, 28989–29004. [CrossRef]
116. Liu, L.M.; Xu, Q.; Yang, S.Q. Identification of river ice on the Yellow River using LANDSAT images. In Proceedings of the 18th International Conference on Geoinformatics, Beijing, China, 18–20 June 2010.
117. Huang, P.P.; Shi, Q.; Xu, W.; Tan, W.X. Research on Yellow River ice supervised classification method based on polarimetric SAR data. In Proceedings of the IEEE International Conference on Signal, Information and Data Processing, Chongqing, China, 11–13 December 2019.
118. Huang, P.P.; Shi, Q.; Tan, W.X.; Xu, W. Yellow River ice decision tree classification method based on polarimetric SAR data. In Proceedings of the IEEE International Geoscience and Remote Sensing Symposium, Yokohama, Japan, 28 July–2 August 2019.
119. Li, H.J.; Li, H.Y.; Wang, J.; Hao, X.H. Advances in remote sensing of river ice. *Adv. Earth Sci.* **2020**, *35*, 1041–1051. (In Chinese)
120. Yang, G. Inversion of Ice Thickness and Its Variation Characteristics in Wanjiazhai Reservoir Area Based on Remote Sensing Data. Master's Thesis, Inner Mongolia Agricultural University, Hohhot, China, 2019. (In Chinese)
121. Lalumiere, L.; Prinsenberg, S.; Peterson, I. Observing pack ice properties with a helicopter-borne video-laser-GPS sensor. In Proceedings of the 10th International Ocean and Polar Engineering Conference, Seattle, DC, USA, 28 May–2 June 2000.
122. Zhang, B.S.; Gao, G.M. Experimental study on the monitoring technology of ice flow on Ningxia-Inner Mongolia section of the Yellow River. *J. Eng. HeiLongjiang Univ.* **2009**, *36*, 90–95. (In Chinese)
123. Fang, H. Studying the Dispersion from Heterogeneous Underground Materials in Ground Penetration Radar. Ph.D. Thesis, China University of Geosciences (Beijing), Beijing, China, 2005. (In Chinese)
124. Zhang, G.J.; Yu, C.; Huang, S.; Shu, Z.L.; Liu, B.X.; Wu, H.K. Mathematical problems in engineering detection of concrete structures with cavity defects by ground penetrating radar and meshless simulation based on improved weight function. *Sci. Technol. Innov.* **2022**, *33*, 122–126. (In Chinese)
125. Li, W. ANN-Based Sub-Surface Objects Identification and Inversion of Dielectric Properties by Means of GPR. Master's Thesis, Nanchang University, Nanchang, China, 2012. (In Chinese)
126. Ji, H.L. Factor Analysis for Ice Flood and Model Research for Freeze-Up Time and Break-Up Time in the Inner Mongolia Reach of the Yellow River. Ph.D. Thesis, Inner Mongolia Agricultural University, Hohhot, China, 2002. (In Chinese)
127. Sun, Y.F.; Wang, T.; Zhou, Z.Y. Ice prediction and identification of influence parameters affecting the initial freeze-up of the Inner Mongolia reach of the Yellow River. *J. China Inst. Water Resour. Hydropower Res.* **2024**, *22*, 149–158. (In Chinese)
128. Guo, Y.X.; Wang, T.; Yang, K.L.; Huo, S.Q.; Rao, S.Q. Design and development of decision support system for ice regime forecast of Ningxia-Inner Mongolia reach of Yellow River. *Water Resour. Hydropower Eng.* **2005**, *36*, 70–72. (In Chinese)
129. Chen, Z.T.; Ke, S.J. River ice modelling in lower reaches of the Yellow River and its utility in sanmen gorge reservoir regulation. *J. Glaciol. Geocryol.* **1994**, *16*, 211–217. (In Chinese)
130. Tian, F.C. Driving Mechanism and Risk Diagnosis and Assessment Methods of Ice Flood Disasters in the Ningxia-Inner Mongolia Reach of Yellow River. Ph.D. Thesis, Tianjin University, Tianjin, China, 2020. (In Chinese)
131. Guo, X.L.; Wang, T.; Fu, H.; Guo, Y.X.; Li, J.Z. Ice-jam forecasting during river breakup based on neural network theory. *J. Cold Reg. Eng.* **2018**, *32*, 04018010. [CrossRef]
132. Wang, S.J. Research of the Ice Condition Forecasting Based on GA Artificial Neural Network. Master's Thesis, Tianjin University, Tianjin, China, 2008. (In Chinese)
133. Bu, S. Application of BP Neural Network Based on Swarm Intelligence Methods for Prediction of the Stage and Thickness for Ice Jam. Master's Thesis, Hefei University of Technology, Hefei, China, 2009. (In Chinese)
134. Wang, J.; Yi, M.K.; Fu, H.; Yin, Y.J.; Gao, Y.X. Application of artificial neural network to predict the increase in stage due to ice jam in a bend. *J. Glaciol. Geocryol.* **2006**, *28*, 782–786. (In Chinese)
135. Hu, L.; Zhang, Q.; Wang, G.; Singh, V.P.; Wu, W.H.; Fan, K.K.; Shen, Z.X. Flood disaster risk and socioeconomic in the Yellow River Basin, China. *J. Hydrol. Reg. Stud.* **2022**, *44*, 101272. [CrossRef]
136. Gao, G.M.; Yu, G.Q.; Wang, Z.L.; Li, S.X. Advances in break-up date forecasting model research in the Ningxia-Inner Mongolia reach of the Yellow River. In Proceedings of the 21th IAHR International Symposium on Ice, Dalian, China, 11–15 June 2012.
137. Song, W.H.; Liu, H.Z.; Cui, P.; Xu, M.Y. Analysis of the characteristics of medium and small floods and reservoir operation in the upper reaches of the Yellow River in recent years. *Yellow River* **2020**, *42*, 34–39. (In Chinese)
138. Guo, L.B.; Zhou, Y.H.; Tian, F.C.; Yuan, X.M.; Wang, L.N. Influence factors and evolution characteristics of ice flood disaster in the Ningxia-Inner Mongolia section of the Yellow River. *Yellow River* **2020**, *42*, 22–26. (In Chinese)
139. Yuan, X.M.; Cao, L.G.; Jia, S.J.; Tian, F.C. Coupled modeling of bed deformation and stability analysis on a typical vulnerable zone on the Ningxia reach of the Yellow River during flood season. *J. Flood Risk Manag.* **2021**, *14*, e12704. [CrossRef]

140. Yuan, X.M.; Zhang, X.Y.; Tian, F.C. Research and application of an intelligent networking model for flood forecasting in the arid mountainous basins. *J. Flood Risk Manag.* **2020**, *13*, e12638. [CrossRef]
141. Wang, Z.M.; Ren, Y.F.; Yang, D. Research on the quantification of disaster warning index and warning index system of ice flood. *Yellow River* **2021**, *43*, 45–50. (In Chinese)
142. Luo, D. Risk evaluation of ice-jam disasters using gray systems theory: The case of Ningxia-Inner Mongolia reaches of the Yellow River. *Nat. Hazards* **2014**, *71*, 1419–1431. [CrossRef]
143. Luo, D.; Jia, H.D. Risk assessment of ice disaster in the Yellow River based on extended VIKOR method. *J. N. China Univ. Water Resour. Electr. Power* **2017**, *38*, 52–57. (In Chinese)
144. Wu, C.G.; Wei, Y.M.; Jin, J.L.; Huang, Q.; Zhou, Y.L.; Liu, L. Comprehensive evaluation of ice disaster risk of the Ningxia–Inner Mongolia reach in the upper Yellow River. *Nat. Hazards* **2015**, *75*, 179–197. [CrossRef]
145. Li, S. Research on the Grey Decision-Making Model and Its Application to the Yellow River Ice Disaster Risk Management. Master's Thesis, North China University of Water Resources and Electric Power, Zhengzhou, China, 2016. (In Chinese)
146. Wu, L. Risk Assessment and Disaster Assessment of Ice Flood Based on Catastrophe Theory. Master's Thesis, Inner Mongolia Agricultural University, Hohhot, China, 2019. (In Chinese)
147. Wang, T.J.; Ji, H.L.; Mou, X.Y.; Luo, H.C.; Zhang, B.S. Risk assessment of ice flood disaster in Bayan Nur of Inner Mongolia based on catastrophe theory. *Water Resour. Power* **2022**, *40*, 84–87. (In Chinese)
148. Hu, Y.L. Risk Evaluation of Ice Flood Disaster of the Ningxia–Inner Mongolia Reach in the Yellow River. Master's Thesis, Zhengzhou University, Zhengzhou, China, 2022. (In Chinese)
149. Li, Z.J.; Fu, X.; Shi, L.Q.; Huang, W.F.; Li, C.J. Inversely identified natural ice thermal diffusivity by using measured vertical ice temperature profiles: Recent advancement and considerations. *J. Glaciol. Geocryol.* **2023**, *45*, 599–611. (In Chinese)
150. Shamshiri, R.; Nahavandchi, H.; Joodaki, G. Seasonal variation analysis of Greenland ice mass time-series. *Acta Geod. Geophys.* **2018**, *53*, 1–14. [CrossRef]
151. Bjerkås, M. Wavelet transforms and ice actions on structures. *Cold Reg. Sci. Technol.* **2006**, *44*, 159–169. [CrossRef]
152. Deng, Y. Time scale variation of the Yellow River ice in the Inner Mongolia section using wavelet analysis. *Appl. Mech. Mater.* **2012**, *220*, 2184–2187. [CrossRef]
153. Bai, Q.X.; Li, R.L.; Li, Z.J.; Leppäranta, M.; Arvola, L.; Li, M. Time-series analyses of water temperature and dissolved oxygen concentration in Lake Valkea-Kotinen (Finland) during ice season. *Ecol. Inf.* **2016**, *36*, 181–189. [CrossRef]
154. Cheng, T.; Wang, J.; Sui, J.; Song, F.; Fu, H.; Wang, T.; Guo, X. Simulation and prediction of water temperature in a water transfer channel during winter periods using a new approach based on the wavelet noise reduction-deep learning method. *J. Hydrol. Hydromech.* **2024**, *72*, 49–63. [CrossRef]
155. Yu, H.; Yu, F.J.; Liu, W.; Lin, H.; Chang, T. Information retrieve and preliminary study of reservoir ice—Surface cracks from digital photos. *Water Conserv. Sci. Technol. Econ.* **2010**, *16*, 1311–1313. (In Chinese)
156. Zhang, H. Analysis of the Difference in Geometrical Characteristics Between Blasted and Natural Ice Blocks in the Yellow River. Master's Thesis, Dalian University of Technology, Dalian, China, 2024. (In Chinese)
157. Ettema, R.; Urroz, G.E. On internal friction and cohesion in unconsolidated ice rubble. *Cold Reg. Sci. Technol.* **1989**, *16*, 237–247. [CrossRef]
158. Beltaos, S. *River Ice Jams*; Water Resources Publication: Highlands Ranch, CO, USA, 1995.
159. Chen, P.P. Experimental and Numerical Simulation Study on the Hydraulic Characteristics of the Ice Jams Influenced by the Bridge Piers. Ph.D. Thesis, Hefei University of Technology, Hefei, China, 2023. (In Chinese)
160. Hiçyilmaz, C.; Ulusoy, U.; Yekeler, M. Effects of the shape properties of talc and quartz particles on the wettability based separation processes. *Appl. Surf. Sci.* **2004**, *233*, 204–212. [CrossRef]
161. Sun, H.; Gao, Z.; Wang, D. Equivalent rolling resistance of particle shape in stable packing of granular material. *J. Water Res. Archit. Eng.* **2022**, *20*, 26–33. (In Chinese)
162. Tan, B.; Wang, T.; Wang, L.; Feng, E.M.; Lu, P.; Li, Z.J. Clustering and analyzing the pressure ridge morphology in northwestern Weddell Sea of Antarctic via an improved k-means clustering algorithm. *Pac. J. Optim.* **2016**, *12*, 623–633.
163. Li, J.X.; Wang, C.M.; Wang, G.C. Landslide risk assessment based on combination weighting-unascertained measure theory. *Rock Soil Mech.* **2013**, *34*, 468–474. (In Chinese)

water

MDPI

Article

Forecasting the River Ice Break-Up Date in the Upper Reaches of the Heilongjiang River Based on Machine Learning

Zhi Liu [1], Hongwei Han [1,2,*], Yu Li [1,*], Enliang Wang [1] and Xingchao Liu [1]

[1] School of Water Conservancy and Civil Engineering, Northeast Agricultural University, Harbin 150030, China; 17853819179@163.com (Z.L.); hljwel@126.com (E.W.); dnlxc@neau.edu.cn (X.L.)

[2] Heilongjiang Provincial Key Laboratory of Water Resources and Water Conservancy Engineering in Cold Region, Northeast Agricultural University, Harbin 150030, China

* Correspondence: hanhongwei@neau.edu.cn (H.H.); 18846771809@163.com (Y.L.)

Abstract: Ice-jam floods (IJFs) are a significant hydrological phenomenon in the upper reaches of the Heilongjiang River, posing substantial threats to public safety and property. This study employed various feature selection techniques, including the Pearson correlation coefficient (PCC), Grey Relational Analysis (GRA), mutual information (MI), and stepwise regression (SR), to identify key predictors of river ice break-up dates. Based on this, we constructed various machine learning models, including Extreme Gradient Boosting (XGBoost), Backpropagation Neural Network (BPNN), Random Forest (RF), and Support Vector Regression (SVR). The results indicate that the ice reserves in the Oupu to Heihe section have the most significant impact on the ice break-up date in the Heihe section. Additionally, the accumulated temperature during the break-up period and average temperature before river ice break-up are identified as features closely related to the river's opening in all four feature selection methods. The choice of feature selection method notably impacts the performance of the machine learning models in predicting the river ice break-up dates. Among the models tested, XGBoost with PCC-based feature selection achieved the highest accuracy ($RMSE = 2.074$, $MAE = 1.571$, $R^2 = 0.784$, $NSE = 0.756$, $TSS = 0.950$). This study provides a more accurate and effective method for predicting river ice break-up dates, offering a scientific basis for preventing and managing IJF disasters.

Keywords: break-up date; feature selection; machine learning; ice reserves; Heilongjiang River

1. Introduction

Ice-jam floods (IJFs) are a hydrological phenomenon caused by the obstruction of water flow due to the accumulation and blockage of ice, which is mainly manifested through the rupture, movement, and evolution of river ice due to sudden changes in temperature and other environmental factors [1–3]. In recent years, IJFs have become increasingly frequent in many high-latitude and high-altitude regions due to global warming [4,5]. At the same time, owing to rapid population growth and accelerated socio-economic development, higher requirements have been placed on the prevention of IJF risks [6–8]. Accurately predicting river ice break-up dates can aid in identifying and assessing potential IJF disasters in advance, optimizing water resource management, and assisting ecological protection departments in implementing appropriate measures to safeguard aquatic habitats and maintain ecological balance. Consequently, developing precise tools to predict river ice break-up dates and evaluate the impact of various factors on their variability has become a complex and critical focus of IJF research.

At present, two primary types of models are utilized to predict river ice break-up dates: numerical simulation methods and statistical methods [9]. Numerical simulation methods employ physical models that account for the mechanics and thermodynamics of ice layers, predicting ice break-up by simulating the rupture process. Common numerical simulation models, such as RIVICE, DYNARICE, ICEJAM, and RIVJAM [10–13], offer detailed dynamic predictions. However, in practice, acquiring real-time and accurate data from field observations is frequently challenging due to adverse environmental conditions and the complexity of ice-flooding regions [14,15]. This scarcity of data constrains the accuracy and reliability of the models, diminishing their effectiveness. Furthermore, a significant limitation of these traditional hydrological models is the complexity associated with parameter calibration. Accurate parameter calibration is crucial for reliable predictions; however, this process often requires substantial time and expertise. Statistical methodology refers to methods for collecting, collating, analyzing, and interpreting statistical data and for drawing certain conclusions about the issues they reflect. Statistical methods do not require a large number of boundary initial conditions or a variety of complex technical parameters and can, therefore, be adapted to many ambiguous background problems. Lei et al. [16] correlated historical hydrometeorological data with the river ice break-up dates and established empirical correlation diagrams between the factors with higher coefficients and the date of river opening for forecasting. The accuracy rate of their forecasting was 72%, demonstrating the effectiveness of statistical methods in addressing complex environmental problems with limited data and uncertainties. However, statistical methods have their own limitations. They often rely on assumptions about data distribution, such as normality or independence, which may not always hold true for real-world data. Furthermore, the accuracy of statistical conclusions is contingent on data quality, and issues such as measurement errors, missing values, or noise can undermine the validity of results. In addition, statistical methods typically focus on correlation rather than causation, and without careful experimental design or control for confounding variables, they may lead to misleading conclusions.

In recent years, the development of Artificial Intelligence (AI) has opened new paths and demonstrated great potential for ice forecasting [17]. By leveraging its ability to process large datasets and analyze complex systems, AI is expected to improve the accuracy of river ice break-up forecasting and enhance model interpretability. Smith et al. [18] used Bayesian logistic regression to estimate the frequency of IJFs while considering uncertainty in the historical record, thereby making the flood frequency analysis more accurate. Liu et al. [19] combined a Long Short-Term Memory (LSTM) network with a CEEMDAN-decomposed time series to predict river ice break-up. Shevnina and Solov'eva [20] investigated the correlation between the autumn accumulation temperature and ice break-up dates, achieving a probability of correct prediction between 67% and 86%. The variability of time-series data has increased due to more unpredictable weather patterns and a higher frequency of extreme events caused by global warming. However, multi-feature forecasting can provide more accurate results by combining and analyzing multiple features and adapting to changes in different data sources and forecasting needs [21–24]. As a data-driven approach, choosing appropriate predictors is the key to successful prediction [25,26]. However, the mechanism of IJFs is complex and influenced by multiple factors, with intricate nonlinear relationships among them, making feature engineering particularly important [27]. Effective feature engineering enables machine learning (ML) models to better capture nonlinear relationships, eliminate redundant features, and minimize the impact of irrelevant inputs on performance [28–30]. Ji et al. [31] utilized the correlation coefficient method for feature selection and applied a fuzzy optimization neural network to predict the river closure at the estuary of the Three Lakes section of the Yellow River, achieving a prediction accuracy of 80%. Li et al. [32] studied the method of determining the forecasting factors in the process of ice forecasting by using AI technology and used Back Propagation Neural Network (BPNN) to forecast river ice break-up in the Mohe section of the Heilongjiang River with a prediction accuracy rate of between 50% and 75%.

Although IJFs have been extensively studied by many scholars [33–35], their suddenness, wide impact and erratic characteristics still limit the accuracy of existing forecasting models [36–38]. Most studies have employed a single feature selection method or a single model for forecasting, with limited consideration of integrating multiple feature selection methods and ML models. In ice forecasting, complex and nonlinear relationships often exist between the forecast target and influencing factors, and a single method may fail to capture all features contributing to the model. Furthermore, different ML models with different assumptions, ways of handling data, and varying adaptability to specific problems may also affect the final prediction. Therefore, in this study, multiple feature selection techniques (PCC, GRA, MI, and SR) were used to screen the contributing factors to reduce the bias caused by a single method, and multiple ML models (XGBoost, BPNN, RF, and SVR) were constructed on this basis. The aim was to elucidate the changing characteristics of river ice break-up dates in the upper Heilongjiang River, along with their influencing factors, and to maximize prediction accuracy. Enhancing disaster prevention capabilities and exploring the complementary advantages of diverse methods can offer valuable insights for improving ML models.

2. Material and Methods

2.1. Overview of the Study Area

The main course of the Heilongjiang River starts at the confluence of the Shilka and Argun Rivers and extends to Khabarovsk, Russia, with a total length of 2821 km. Winter lasts for about 8 months, with an ice-covered period averaging over 150 days. Due to its unique river channel shape, flow rate, and hydrometeorological conditions, IJFs occur frequently. From 1940 to 2023, there were 34 years of severe IJF disasters, which greatly impacted the production and lives of the local people [39,40].

Heihe is located in the mid-temperate zone (Figure 1) [41], where the river typically freezes in November and begins thawing in late April or early May. The annual average temperature ranges from −1.3 °C to 0.4 °C, with extreme minimum temperatures dropping as low as −45 °C. The annual precipitation averages around 500 mm, mostly concentrated during the summer months. Heihe is located directly across from Russia's Amur Region. It plays a "leading" role in early warning and forecasting for flood control, flood prevention, and ice jam disaster mitigation in the Heilongjiang River. Historically, IJFs in the Heilongjiang River have caused significant harm to the people along its banks, and the Heihe section alone has experienced ice dams five times, resulting in severe damage.

2.2. Data Acquisition and Processing

River ice break-up dates, water levels, and ice thicknesses for the Heihe section were obtained from the Heilongjiang Hydrology and Water Resources Center. Daily atmospheric reanalysis data, including temperature, wind speed, snow depth, cloudiness, and precipitation, were sourced from the ERA5-Land dataset provided by the European Centre for Medium-Range Weather Forecasts (ECMWF), with a horizontal resolution of 0.1° × 0.1°. This dataset is available at https://cds.climate.copernicus.eu, accessed on 1 June 2023. River channel data were acquired from the Global River Widths from Landsat Database, which can be accessed on Zenodo (https://doi.org/10.1126/science.aat0636). Data analysis and processing were carried out using Python 3.8, Origin 2022, and Excel 2021.

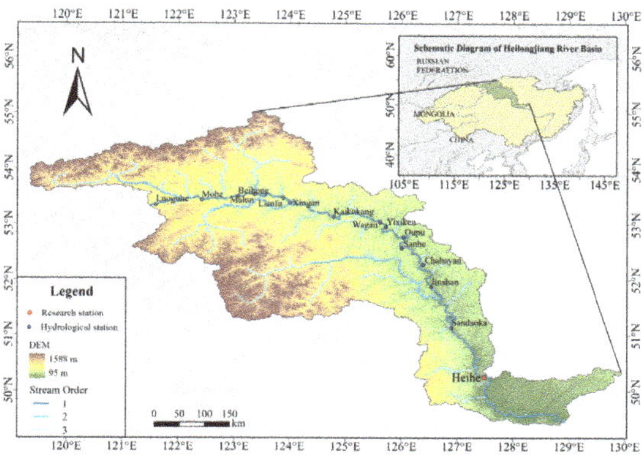

Figure 1. The geography of the study area.

2.3. Calculation of Ice Reserves

During winter, as the temperature drops, a significant amount of water is stored in rivers as solid ice, referred to as ice reserves. Ice reserves serve as an indicator of the amount of ice accumulated in rivers during the winter. In order to obtain an accurate picture of the changes in ice reserves and their impact on the river ice break-up, first, two-dimensional calculations were performed using DEM data with 12.5 m resolution. The area of the river that was frozen was determined based on the frozen water level recorded at each hydrological station during the previous year. Second, the measured maximum ice thickness at the hydrological stations was used to calibrate the ice thickness growth coefficient of the freezing degree day method in order to deduce the river ice thickness and ice reserves during the open river period. The formulas used to calculate ice reserves and ice thickness are provided below [42]:

$$h_m = \alpha\sqrt{FDD}, \quad (1)$$

$$\Delta W_{ice} = \int_0^L h_m dA, \quad (2)$$

where h_m is the maximum ice thickness, m; α is the ice thickness growth factor, cm/(°C·d)$^{-0.5}$; FDD is the freezing degree days, °C·d; and A is the frozen area, m^2. Equation (2) can be expressed as follows:

$$W_{ice} = \sum_{i=1}^{n} \Delta W_{ice} \quad (3)$$

2.4. Mann-Kendall Test

The Mann–Kendall test is a non-parametric method used to analyze trend characteristics and change points in data series. When using the Mann–Kendall algorithm to test broad trends in time series data, the trends are categorized as no apparent trend, increasing trend, or decreasing trend. The test compares the relative magnitude of the sample data rather than the data values themselves, so the underlying data does not need to follow a particular distribution or require the trend of change to be linear. The overall statistic S is calculated by comparing each pair of data points and is then normalized to a Z-value to assess whether a significant trend exists in the time-series data. This method is widely used in the analysis of hydrological and meteorological datasets [43].

2.5. Feature Selection Methods

Feature selection is a key step in improving the predictive power of a model and avoiding overfitting. The purpose of feature selection is to select the most representative and informative subset from a large number of features and remove redundant or irrelevant features. The evolution of ice conditions is affected by a variety of factors, and there is often a complex nonlinear relationship between forecast objects and disaster-causing factors. Twelve candidate factors were identified through a comprehensive analysis of the geographic location and ice condition characteristics of the Heilongjiang Heihe section (see Table 1). Next, Min-Max normalization [44] was applied to eliminate the dimensional differences between different features. Four feature selection methods were used to identify the features that significantly impact the ice break-up date in the Heihe section.

Table 1. Candidate factors for predicting the river ice break-up date.

Candidate Factors	Code
Accumulated temperature during the break-up period (°C)	X_1
Date of positive temperature stabilization (d)	X_2
Average temperature before river ice break-up (°C)	X_3
Ice reserves (m^3)	X_4
Average wind speed during the break-up period (m/s)	X_5
Average snow depth during the break-up period (mm)	X_6
Precipitation prior to river freeze-up (mm)	X_7
Precipitation during the river freeze-up period (mm)	X_8
Precipitation before river ice break-up (mm)	X_9
Precipitation during the break-up period (mm)	X_{10}
Average cloud cover during the break-up period (0–1)	X_{11}
Maximum ice thickness (cm)	X_{12}

Note: 'Prior to river freeze-up' refers to October; 'during the river freeze-up period' refers to November through March of the following year; 'before river ice break-up' refers to March of the following year, and the 'break-up period' refers to April 1–20 of the following year.

2.5.1. Pearson Correlation Coefficient

The Pearson correlation coefficient (PCC) is used to measure the linear correlation between two sets of data. It is calculated as the covariance between the two variables, divided by the product of their standard deviations. Its value ranges from −1 to 1.

After calculating the Pearson correlation coefficient, a significance test can also be conducted to assess the statistical significance of the correlation coefficient. These significance tests are designed to evaluate whether the observed correlation coefficient is sufficiently large to reject the presence of correlation due to random sampling errors [45]. If the correlation is not significant, a high coefficient may be meaningless and due to chance. Generally, a p-value of less than 0.05 is required to conclude that a significant relationship exists between the two datasets.

2.5.2. Grey Relational Analysis

Grey Relational Analysis (GRA) is a multi-factor statistical method used to measure the degree of association between factors of two systems as they change over time or across different objects. This association, known as the Grey Relational Grade (GRG), is higher when the trends of the two factors are consistent, indicating synchronous changes. Conversely, if the trends are dissimilar, the GRG is lower. The threshold is set to 0.3.

This threshold is based on related studies [32,39], which suggest that a GRG value of 0.3 effectively excludes features with a low association with the river ice break-up date, thereby improving model accuracy, reducing computational complexity, and minimizing the risk of overfitting.

2.5.3. Mutual Information

Mutual Information (MI) is a concept in probability and information theory that is used to assess the degree of interdependence between two variables. Mutual information is calculated for each feature and the target variable, and the features are ranked according to the value of mutual information. A higher value of MI indicates a stronger dependency between the feature and the target variable.

For features with high MI values, permutation tests can be used to assess the statistical significance of the relationship. A small p-value (typically <0.05) indicates that the MI of the feature is significantly higher than expected under random conditions, confirming a strong relationship between the feature and the target variable.

2.5.4. Stepwise Regression

Stepwise Regression (SR) is a method for automatically screening variables that significantly affect the regression equation from a large set of available variables. SR begins with one independent variable and adds variables to the regression equation one by one, based on the significance of their effect on the dependent variable, y. After introducing a new variable, if any of the previously included variables are no longer significant, they will be removed. When previously introduced variables become insignificant due to the inclusion of subsequent variables, those variables are removed to ensure that the regression equation contains only significant variables before introducing new ones, ultimately leading to an optimal combination of variables.

In SR, the significance levels for adding (α) and removing (β) variables are typically set at 0.4. This threshold is based on related studies [32]. Studies indicate that the threshold of 0.4 can reflect the many-to-one relationships between various influencing factors and the river ice break-up dates.

2.6. *Machine Learning Algorithms*

ML refers to algorithms that automatically learn from data to make predictions or decisions. Datasets are typically divided into a training set and a test set. The training set is a dataset used to train an ML model so that it can learn the relationship between input features and target variables. The training process usually involves tuning the parameters of the model to minimize the model's prediction error on the training set. The test set, on the other hand, is the dataset used to evaluate the performance of the ML model. The test set verifies the model's ability to generalize to unseen data, i.e., its capacity to predict new, unknown data.

In our study, we utilized the 'random_state' parameter to randomly partition the dataset into a training set (80%) and a test set (20%). Models were then constructed using XGBoost, BPNN, RF, and SVR. The details of the training and test datasets are provided in Table 2.

Table 2. Dataset segmentation.

Dataset	Variable	Minimum	Maximum	Mean
Training Set	X_1 (°C)	−65.51	115.18	20.06
	X_2 (d)	−19	17	2
	X_3 (°C)	−14.51	−1.68	−8.69

Table 2. *Cont.*

Dataset	Variable	Minimum	Maximum	Mean
Training Set	X_4 (m^3)	2.55×10^8	3.21×10^8	2.88×10^8
	X_5 (m/s)	2.13	3.71	2.87
	X_6 (mm)	0	88.67	6.90
	X_7 (mm)	7.13	102.27	35.71
	X_8 (mm)	47.50	197.50	91.80
	X_9 (mm)	1.40	38.00	13.67
	X_{10} (mm)	1.5	151.30	34.53
	X_{11} (0–1)	30.54	76.11	57.87
	X_{12} (cm)	85	1.75	120
Test Set	X_1 (°C)	−38.64	84.12	29.39
	X_2 (d)	−11	18	3
	X_3 (°C)	−14.52	−3.84	−10.15
	X_4 (m^3)	2.79×10^8	3.18×10^8	2.98×10^8
	X_5 (m/s)	2.39	3.37	2.83
	X_6 (mm)	0	31.32	6.47
	X_7 (mm)	8.95	86.47	38.57
	X_8 (mm)	44.90	137	84.88
	X_9 (mm)	2.40	26.90	14.15
	X_{10} (mm)	4.10	72.6	33.24
	X_{11} (0–1)	42.43	65.54	54.83
	X_{12} (cm)	78	173	119

Note: The river ice break-up date was converted to a numeric value, which is shown as the number of days to 1 April.

2.6.1. Extreme Gradient Boosting

Extreme Gradient Boosting (XGBoost) is an efficient decision tree algorithm based on gradient boosting. It introduces several innovations to the traditional Gradient Boosting Decision Tree (GBDT) framework, markedly improving model performance and efficiency [46]. The core idea of XGBoost is to enhance the predictive capability of the model by incrementally optimizing the loss function. As a forward additive model, it makes predictions by combining multiple weak learners (i.e., decision trees) into a strong learner. During the training process, each new tree aims to fit the residuals (i.e., prediction errors) between the target values and the predictions of all existing trees. By continually reducing these residuals, XGBoost gradually enhances the prediction accuracy of the entire model. Figure 2 provides a schematic of the model.

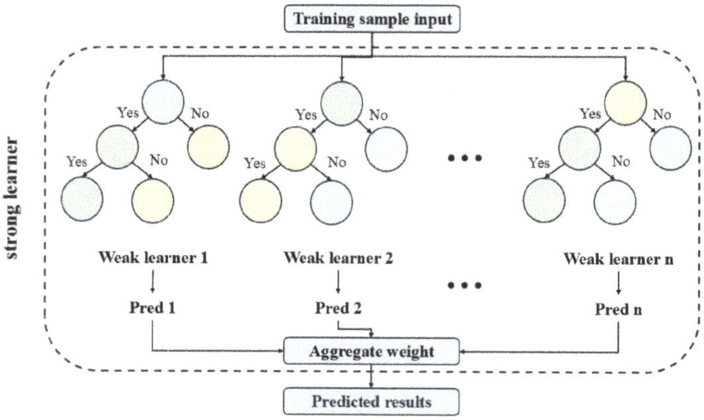

Figure 2. Extreme Gradient Boosting schematic.

2.6.2. Back Propagation Neural Network

The Back Propagation Neural Network (BPNN) is a type of artificial neural network that employs the backpropagation learning algorithm. It generally consists of an input layer, one or more hidden layers, and an output layer. During training, the network receives input data, calculates the output, and compares it to the expected result. The difference is used to compute an error signal, which is then backpropagated through the network to update the weights. Figure 3 provides a schematic of the model.

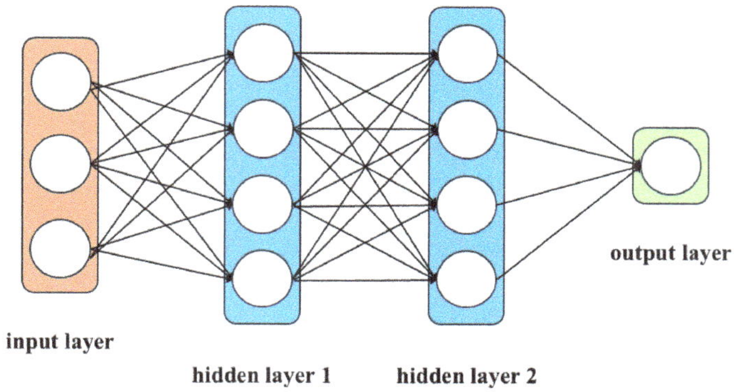

Figure 3. Neural network schematic.

2.6.3. Random Forest

The Random Forest (RF) model is an ML algorithm based on ensemble learning and random feature selection. It improves the accuracy and robustness of the decision trees by constructing multiple decision trees and combining their prediction results. In a Random Forest, each decision tree is trained on sub-samples randomly selected from the original dataset using bootstrap sampling. Moreover, whenever a decision tree splits a node, the algorithm randomly selects a subset of features and makes the split decision based only on those features. Through the double randomization of samples and features, RF effectively reduces the risk of overfitting that may occur with individual decision trees. For regression tasks, the final prediction is obtained by averaging the results from all decision trees, enhancing the model's stability and accuracy. Figure 4 provides a schematic representation of the model.

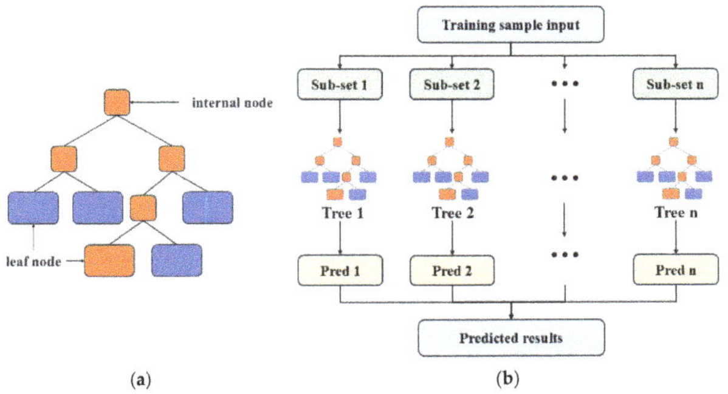

Figure 4. Random Forest schematic. (**a**) Decision tree diagram; (**b**) Random Forest model structure diagram.

2.6.4. Support Vector Regression

Support Vector Regression (SVR) is a regression analysis method based on the principles of Support Vector Machines (SVMs). The goal of SVR is to define a regression function that ensures that the prediction errors are within a predefined ε (tolerance) range without penalty while simultaneously minimizing the complexity of the model. This helps ensure that the model not only fits the training data well but also maintains good generalization ability. When dealing with nonlinear relationships, SVR uses a kernel function to map the data into a higher-dimensional space [47], where linear fitting can be applied to address the nonlinear regression problem.

2.7. Settings of Model Parameter

In ML, hyperparameters play a crucial role in the performance of models. To optimize the performance of each model, it is necessary to determine the optimal hyperparameter settings through iterative experiments that evaluate prediction errors between predicted and observed values. This study employed a cross-validated grid search to identify optimal parameters for enhancing model performance; a summary of the hyperparameter settings for the four models is presented in Table 3.

Table 3. ML training parameters.

Model	Parameters	PCC	GRA	MI	SR
XGBoost	learning_rate	0.1	0.01	0.01	0.01
	max_depth	11	5	9	9
	n_estimators	100	500	300	200
	subsample	0.6	0.6	1.0	0.6
	colsample_bytree	0.8	0.8	0.8	0.9
BPNN	hidden_layer_sizes	(50)	(50, 50)	(50)	(50, 50)
	activation	identity	relu	identity	tanh
	solver	lbfgs	lbfgs	lbfgs	adam
	alpha	0.0001	0.0001	0.0001	0.01
RF	n_estimators	100	200	100	100
	min_samples_split	4	5	10	5
	min_samples_leaf	2	4	2	4
	max_depth	10	20	10	20
SVR	C	10	10	10	0.1
	kernel	rbf	rbf	rbf	linear
	epsilon	0.2	0.2	0.2	0.1

2.8. Evaluation Metrics

After using ML models for predictions, it is essential to evaluate their performance to analyze their accuracy and practicality. This study employed commonly used evaluation metrics, including root mean square error (*RMSE*), Mean Absolute Error (*MAE*), Nash–Sutcliffe Efficiency (*NSE*), and Coefficient of Determination (R^2), to assess the performance of various ML models. The formulas for these evaluation metrics are as follows [19]:

$$RMSE = \sqrt{\frac{1}{n}\sum_{i=1}^{n}(O_i - S_i)^2}, \quad (4)$$

$$MAE = \frac{1}{n}\sum_{i=1}^{n}|O_i - S_i|, \quad (5)$$

$$NSE = 1 - \frac{\sum_{i=1}^{n} (O_i - S_i)^2}{\sum_{i=1}^{n} (O_i - \overline{O})^2}, \tag{6}$$

$$R^2 = \left[\frac{\sum_{i=1}^{n} (O_i - \overline{O})(S_i - \overline{S})}{\sqrt{\sum_{i=1}^{n} (O_i - \overline{O})^2} \sqrt{\sum_{i=1}^{n} (S_i - \overline{S})^2}} \right]^2, \tag{7}$$

where O_i and S_i represent the observed and predicted values of the river ice break-up date, respectively; $\overline{O}$ and $\overline{S}$ are the mean values of the observed and simulated data; n is the length of the sequence; and SD_O and SD_S are the standard deviations of the observed and predicted values, respectively. These statistics can be defined using the following formulas:

$$SD_O = \sqrt{\frac{\sum_{i=1}^{n} (O_i - \overline{O})^2}{n}} \tag{8}$$

$$SD_S = \sqrt{\frac{\sum_{i=1}^{n} (S_i - \overline{S})^2}{n}} \tag{9}$$

To visually represent the performance of different ML models, a Taylor diagram offers an effective visualization method. This diagram integrates the correlation coefficient (R), standard deviation (SD), and centered root mean square error ($CRMSE$) into a polar coordinate system using the cosine theorem. By plotting the R, SD, and $CRMSE$ of the simulated values against the observed values on the Taylor diagram, model performance can be represented as points. The relationships between these statistics are defined using the following formulas:

$$CRMSE^2 = {SD_O}^2 + {SD_S}^2 - 2SD_O SD_S R \tag{10}$$

Among them, the centered $CRMSE$ between observed and predicted values can be defined using the following equation:

$$CRMSE = \sqrt{\frac{1}{n}\sum_{i=1}^{n} [(O_i - \overline{O}) - (S_i - \overline{S})]^2} \tag{11}$$

The Taylor diagram can intuitively quantify the correlation between simulated and observed points. The Taylor Skill Score (TSS) is a numerical summary of the Taylor diagram and reflects a comprehensive measure of simulation skill. The calculation formula for the Taylor Skill Score is as follows:

$$TSS = \frac{4(1+R)^4}{\left(\frac{SD_S}{SD_O} + \frac{SD_O}{SD_S}\right)^2 (1 + R_0)^4} \tag{12}$$

where TSS represents the Taylor Skill Score, where a higher score indicates better model performance. R_0 is the maximum R between the predicted and observed river ice break-up date.

2.9. Model Simulations

The framework for predicting the river ice break-up date comprises four key steps: data collection and preprocessing, ML model construction, model simulation, and model evaluation. This framework efficiently processes input data, constructs appropriate prediction models, and systematically evaluates their performance to enhance the accuracy of river ice break-up date predictions.

Firstly, based on previous research [24,40], variables that could trigger IJFs were identified and collected. The collected data were subsequently normalized using Min-Max normalization to eliminate the impact of differing units and scales on model training. Different feature selection methods were used to filter out the feature combinations that have a significant impact on IJFs. These selected features were then fed into different models, which were trained using the data from the training set to learn the relationship between the features and the river ice break-up date. Grid search was employed to explore all possible hyperparameter combinations to find the best configuration. The performance of the models on the test set is evaluated using a variety of evaluation metrics (*MAE*, *RMSE*, R^2, *NSE*, and *TSS*) to determine the model's ability to generalize to unseen data. Figure 5 outlines the research framework.

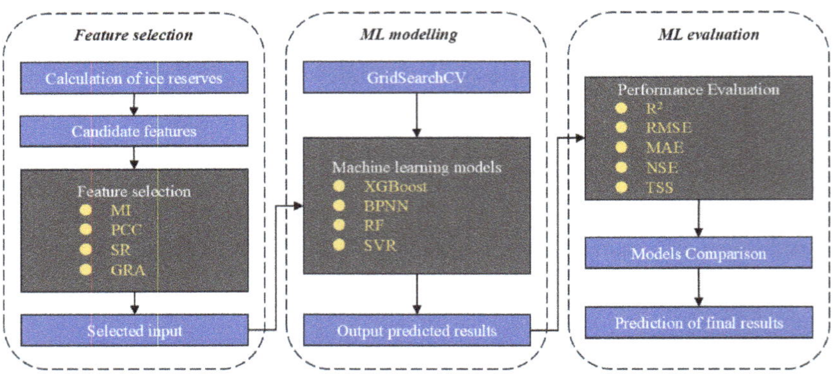

Figure 5. Flowchart of the prediction model.

3. Results and Analysis

3.1. Ice Conditions Change

Figure 6 shows that the average ice reserves in the upper reaches of the Heilongjiang River from 1956 to 2022 were 3.79×10^8 m^3, with maximum and minimum ice reserves recorded in 1966 and 2002, respectively, representing a difference of 1×10^8 m^3. The Mann–Kendall test produced a Z-value of -4.89, suggesting a significant downward trend with a rate of change of 5.2×10^5 m^3 per decade ($p < 0.05$).

To simplify the calculations and investigate the influence of ice reserves on the river ice break-up date, we analyzed the effects of river sections on the break-up date using MI, PCC, and GRA methods. The results indicate that, among the upper Heilongjiang River sections, the ice reserves in the segment from Oupu to Heihe have the most significant impact on the river ice break-up date at the Heihe section, with correlation coefficients, grey relational degrees, and MI values of 0.480, 0.479, and 0.176, respectively. Consequently, we chose the ice reserves in the section from Oupu to Heihe as a candidate feature for predicting the river ice break-up date at the Heihe section.

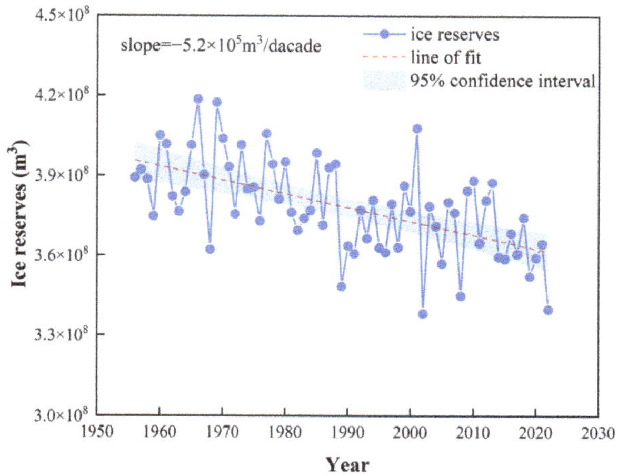

Figure 6. Interannual variability of ice reserves in the upper reaches of the Heilongjiang River.

3.2. Correlation Analysis

In this study, twelve candidate predictive factors were identified as potential input features. The significance of the relationship between the river opening date and these candidate input features was quantified using MI, PCC, SR, and GRA. Table 4 presents the final results after feature selection, while Figure 7 shows the detailed results.

Table 4. Feature selection results.

Study Area	PCC	GRA	MI	SR
Heihe section	X_1, X_2, X_3, X_4, X_6, X_8, X_{11}	X_1, X_2, X_3, X_4, X_6, X_{11}	X_1, X_2, X_3, X_4	X_1, X_3, X_5, X_7, X_9, X_{11}

In the feature selection based on PCC, the significance levels of X_5, X_7, X_9, X_{10}, and X_{12} were greater than 0.05, so these features were excluded from further analysis. X_1 and X_3 exhibited significant negative correlations with the river ice break-up date ($p < 0.05$), with correlation coefficients of −0.69 and −0.59, indicating that the river ice break-up date advances as accumulated temperature during the break-up period and average temperature before river ice break-up increase. On the other hand, X_2, X_4, X_6, X_8, and X_{11} exhibited significant positive correlations with the river ice break-up date ($p < 0.05$), with correlation coefficients of 0.65, 0.48, 0.35, 0.29, and 0.41, respectively. The increase in these positively correlated characteristic values will lead to a delay in the river's ice break-up date that year. This is consistent with the findings of Magnuson et al. on the relationship between climate change and river ice break-up dates [48–51], further confirming the impact of climate change on river ice conditions.

In the GRA-based feature selection process, a threshold of 0.3 was set for screening, which excluded candidate features X_5, X_7, X_8, X_9, X_{10}, and X_{12}. Additionally, in the MI feature selection, the significance levels of X_1 to X_4 were less than 0.05 based on permutation tests, so these features were selected for forecasting the river ice break-up date. In the Python 3.8 environment, SR was conducted using the ordinary least squares (OLS) method from the statsmodels library, with significance levels for introducing (α) and removing (β) variables set to 0.4. The analysis results indicated that factors X_1, X_{11}, X_3, X_7, X_9, and X_5 were sequentially included in the model. However, after introducing new variables, the influence of X_2 became non-significant and was subsequently excluded.

In the study at the Heihe section, four feature selection methods consistently identified "Accumulated temperature during the break-up period" (X_1) and "Average temperature before river ice break-up" (X_3) as key features, with strong predictive power for forecasting the river ice break-up date. Relatively, precipitation has a minor effect on the river ice break-up date, particularly "precipitation during the river opening period" (X_{10}), which was not selected in any of the four feature selection methods. The results suggest that different feature selection methods screen the data through different criteria, objectives, and algorithms and that these methods may focus on different characteristics during the selection process, and therefore, their final screening results may differ.

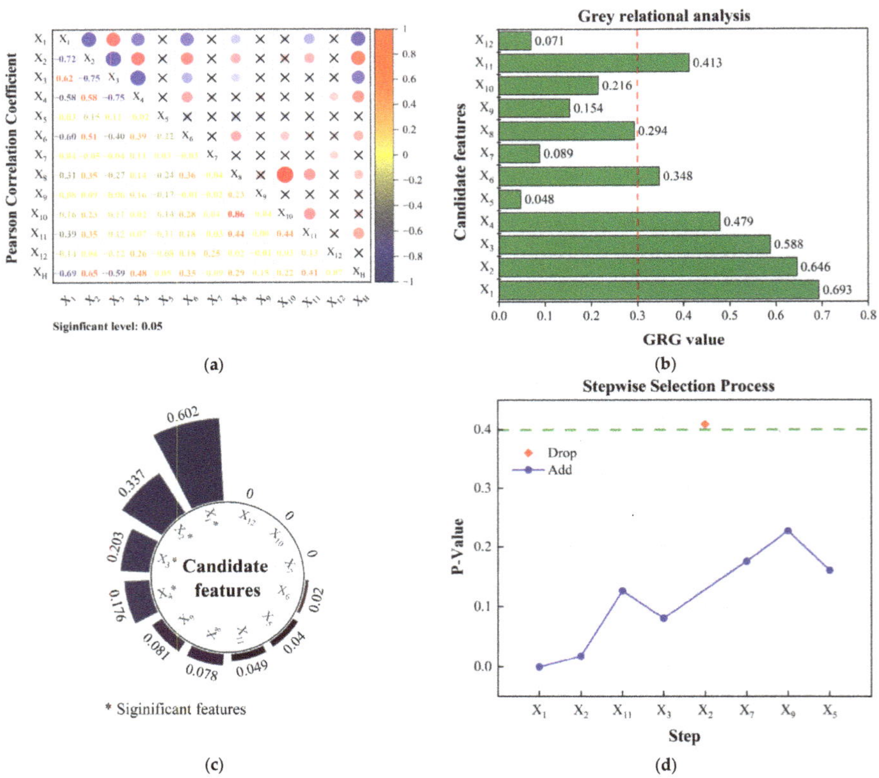

Figure 7. Feature importance analysis for features X_1 to X_{12}. (**a**) PCC; (**b**) GRA; (**c**) MI; (**d**) SR.

3.3. Results of Predicting River Ice Break-Up Dates

This study imports the features selected by different feature selection methods to different ML models for making predictions. Figures 8–11 show different ML model predictions based on different feature selection methods, respectively. In the figures, the blue dashed line represents the actual river ice break-up dates, while the solid line with hollow markers indicates the predictions for the training set, and the solid markers represent the predictions for the test set.

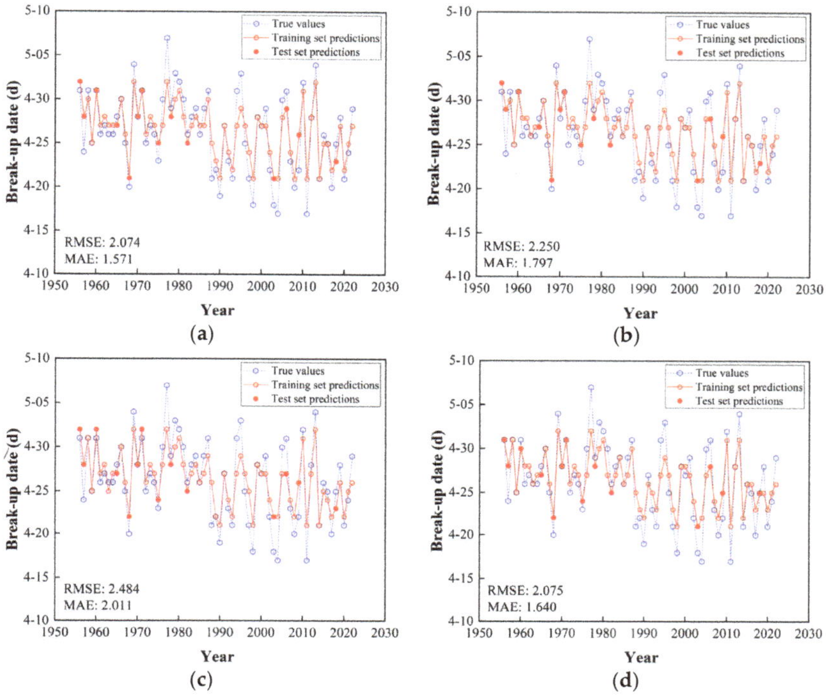

Figure 8. XGBoost model prediction results. (**a**) PCC; (**b**) GRA; (**c**) MI; (**d**) SR.

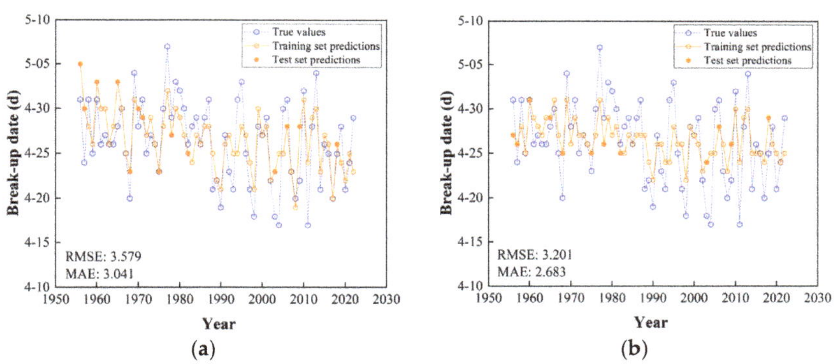

Figure 9. *Cont.*

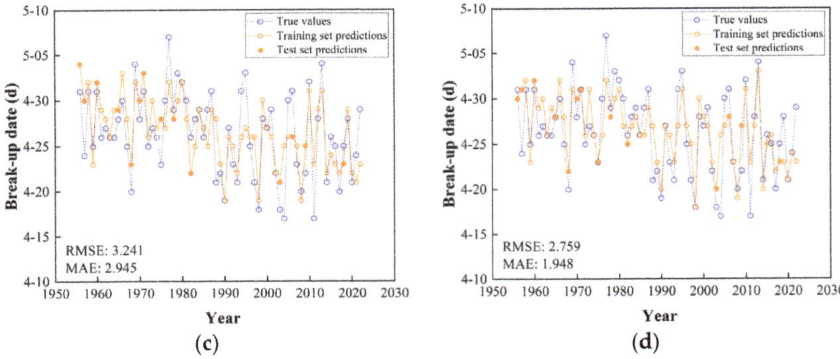

Figure 9. BPNN model prediction results. (**a**) PCC; (**b**) GRA; (**c**) MI; (**d**) SR.

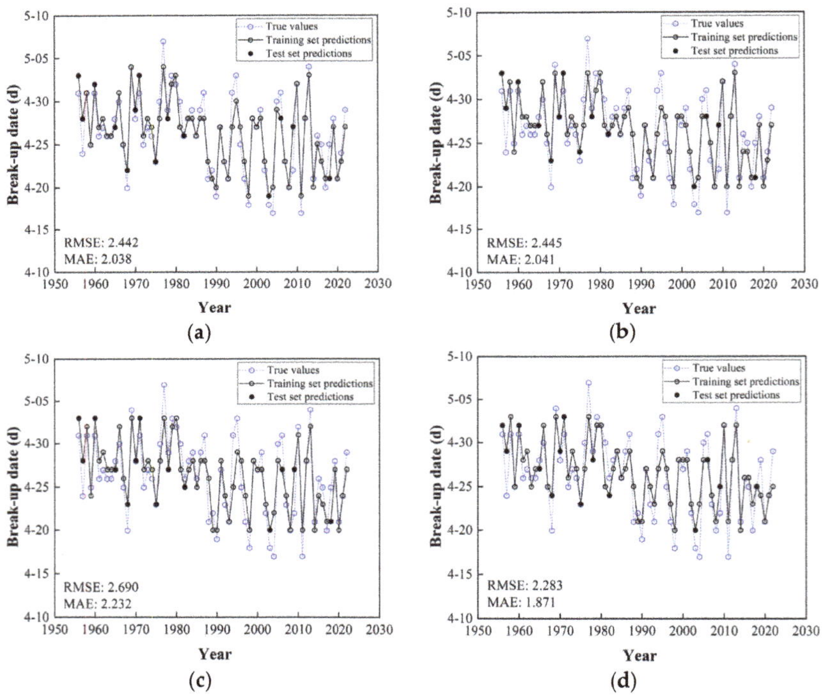

Figure 10. RF model prediction results. (**a**) PCC; (**b**) GRA; (**c**) MI; (**d**) SR.

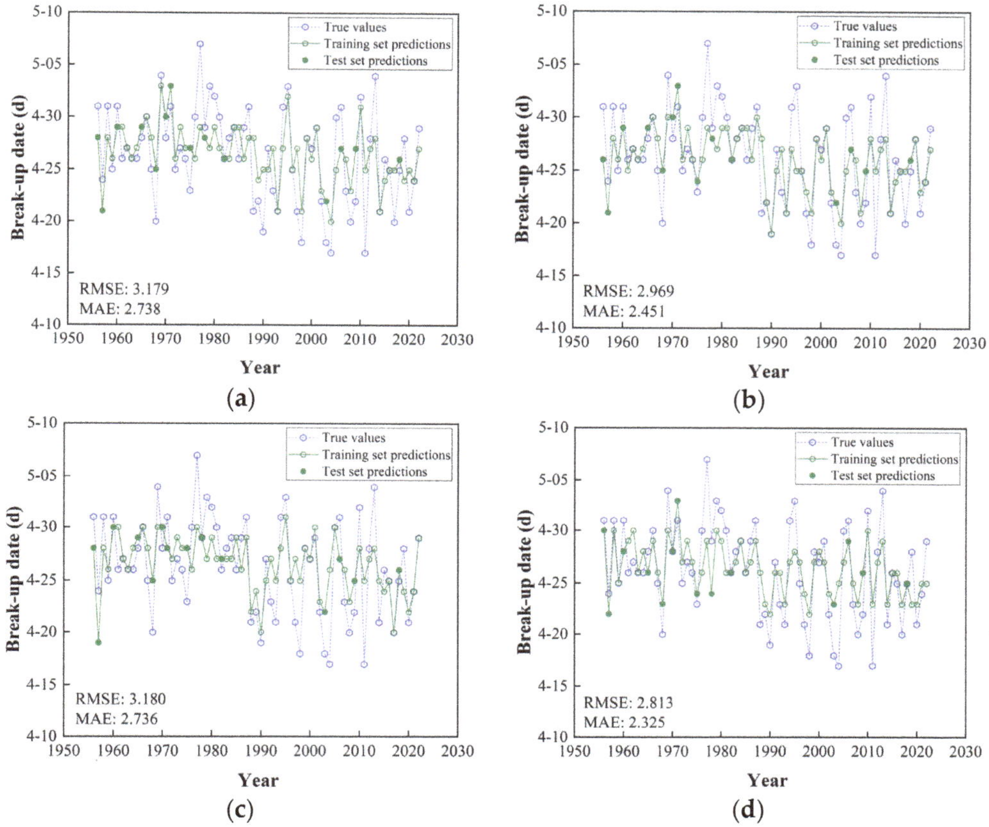

Figure 11. SVR model prediction results. (**a**) PCC; (**b**) GRA; (**c**) MI; (**d**) SR.

Between 1956 and 2022, the earliest and latest river ice break-up dates at the Heihe section occurred on 15 April 2004 and 5 May 1977, respectively, resulting in a 20-day interval. The Mann–Kendall test yielded a Z-value of −2.14, indicating a significant trend of advancement in the river ice break-up date at the Heihe section, with a rate of change of 0.682 days per decade ($p < 0.05$). Over the 67-year period, the river ice break-up date advanced by 4.57 days in total. The Mann–Kendall mutation test revealed that the river ice break-up date at the Heihe section experienced significant changes in 1969, 2005, 2011, and 2014.

From Table 5, we observe that different combinations of models and feature selection methods have a significant impact on the prediction performance. Overall, there are notable differences in the prediction performance of the models when different feature selection methods are used. Taking *RMSE* and *MAE* as examples, under different feature selection methods, the *RMSE* and *MAE* values of each model fluctuate, indicating that feature selection has an important impact on model prediction accuracy. Furthermore, specific to each model, we observe that the XGBoost model generally exhibits superior prediction performance. In particular, PCC-XGBoost achieves the best prediction accuracy and model fit. In addition, SR-BPNN exhibits relatively good performance. The BPNN's R^2 and *NSE* values improved when combined with SR, indicating that SR optimized the feature set of the RF to a certain extent, thus enhancing its predictive ability and model fit. The RF model, on the other hand, exhibits relatively stable predictive performance, with its *RMSE*, *MAE*, R^2, and *NSE* values remaining within a relatively stable range regardless of which feature selection method is used. Regarding the SVR model, its prediction performance tends to be relatively weaker across most feature selection methods; its R^2 and *NSE* values also increase under the SR method, indicating that SR has also optimized the feature set of the SVR to some extent.

Table 5. Model performance on training and test sets.

Model	Methods	Training Set					Test Set				
		RMSE	*MAE*	R^2	*NSE*	*TSS*	*RMSE*	*MAE*	R^2	*NSE*	*TSS*
XGBoost	PCC	1.950	1.537	0.907	0.823	0.848	2.074	1.571	0.784	0.756	0.950
	GRA	1.998	1.553	0.895	0.814	0.836	2.250	1.797	0.740	0.712	0.901
	MI	2.185	1.722	0.863	0.778	0.783	2.484	2.011	0.666	0.649	0.822
	SR	2.206	1.782	0.883	0.774	0.769	2.075	1.640	0.803	0.755	0.923
BPNN	PCC	3.140	2.645	0.549	0.539	0.521	3.579	3.041	0.514	0.272	0.682
	GRA	3.412	2.807	0.523	0.458	0.358	3.201	2.683	0.520	0.418	0.379
	MI	3.041	2.402	0.571	0.570	0.615	3.241	2.945	0.547	0.403	0.724
	SR	2.850	2.206	0.623	0.622	0.655	2.759	1.948	0.620	0.567	0.785
RF	PCC	1.475	1.116	0.914	0.898	0.974	2.442	2.038	0.708	0.661	0.907
	GRA	1.993	1.577	0.827	0.815	0.871	2.445	2.041	0.699	0.660	0.892
	MI	2.103	1.689	0.806	0.794	0.885	2.690	2.232	0.628	0.589	0.813
	SR	2.243	1.740	0.787	0.766	0.799	2.283	1.871	0.743	0.704	0.932
SVR	PCC	3.369	2.557	0.484	0.472	0.436	3.179	2.738	0.468	0.426	0.585
	GRA	3.242	2.309	0.537	0.511	0.476	2.969	2.451	0.511	0.499	0.647
	MI	3.413	2.602	0.476	0.458	0.412	3.180	2.736	0.441	0.425	0.543
	SR	3.121	2.560	0.631	0.547	0.461	2.813	2.325	0.551	0.550	0.680

In summary, the impact of feature selection methods on model performance is significant. Different feature selection methods can result in significant differences in model prediction accuracy and fit. Therefore, it is crucial to choose an appropriate feature selection method when performing model training. For a particular model and dataset, it may be necessary to experimentally compare the performance of different feature selection methods to select the most suitable feature set and model combination.

To comprehensively evaluate the prediction results of river ice break-up dates, this study employed five metrics: *RMSE*, *MAE*, R^2, *NSE*, and *TSS*. Given that different ML models exhibit varying performances on these metrics, we utilized Taylor diagrams as a visual representation of model performance. The Taylor diagram integrates the correlation coefficient, standard deviation, and *CRMSE* within a polar coordinate system, thereby illustrating the agreement between predicted and observed ice break-up dates. In the Taylor diagram, the radial distance from the origin to the predicted point represents the standard deviation. The cosine of the azimuthal angle signifies the correlation coefficient between the predicted and observed values; hence, an angle nearer to the origin direction indicates a correlation coefficient closer to +1, suggesting a stronger linear relationship between the two sets of values. The *CRMSE* is depicted by the distance between the predicted and observed points, with a shorter distance implying that the model's predictions are more consistent with the observed data, reflecting better model performance.

The Taylor diagram results are shown in Figure 12. Notably, when the same model is applied with different feature selection methods, there are discernible variations in the predictions of the river ice break-up dates. Similarly, when the same feature selection method is employed, significant performance disparities arise among different models. The distribution within the Taylor diagram is quite scattered, with notable differences. The correlation coefficients of the 16 model combinations range from 0.66 to 0.89. Furthermore, the worst-performing model combination is GRA-BPNN, situated farthest from the observation point, with a *TSS* of 0.379. In contrast, the best-performing combination is PCC-XGBoost, positioned closer to the observation point, boasting a *TSS* of 0.950. This indicates that PCC-XGBoost surpasses the other combinations in terms of prediction accuracy. Therefore, PCC-XGBoost was selected as the prediction model for the Heihe section.

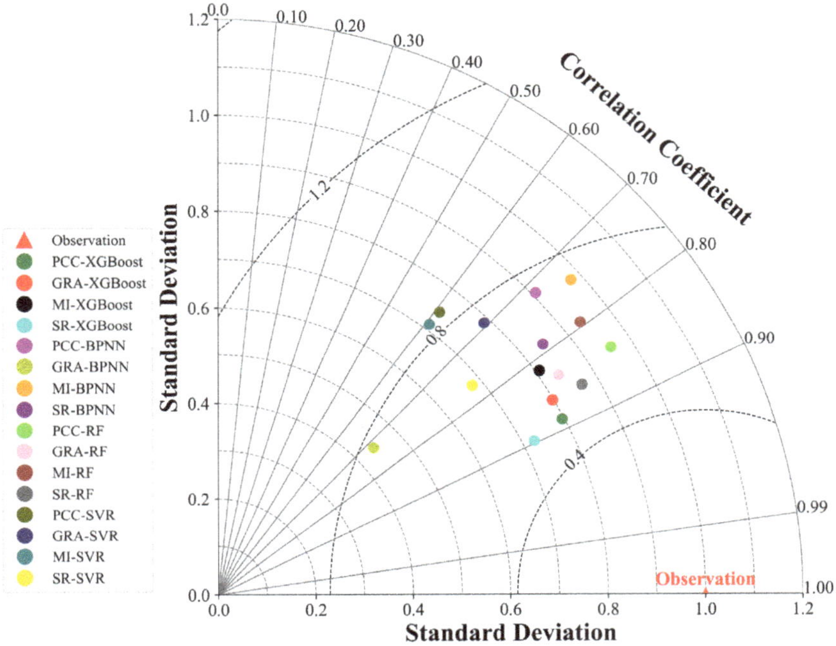

Figure 12. Taylor diagram of model predictions.

Based on the Standard for hydrological information and hydrological forecasting, for ice forecasts with a lead time of 6 to 10 days, a maximum permissible error of 3 days is established. Forecasts with an absolute error of less than 3 days are considered acceptable. The PCC-XGBoost model has a qualification rate of 85.71%, meeting the standard for first-class forecasting.

4. Discussion

The study results indicate a clear downward trend in the ice reserves volume in the upper reaches of the Heilongjiang River during the study period. This finding was confirmed using the Mann–Kendall test (Z = −4.89, $p < 0.05$), with a reduction of 5.2×10^5 cubic meters of ice reserves per decade. These findings are consistent with the global trend of decreasing ice masses due to climate warming [52]. Furthermore, methods such as PCC, GRA, and MI indicate that the ice reserve volume from Oupu to Heihe impacts the ice break-up date at the Heihe section. This supports previous studies suggesting that ice reserves are a key climatic factor that can provide valuable information for predicting the ice break-up date [53,54]. All four feature selection methods consistently identified the accumulated temperature during the ice break-up period and the average temperature before the ice break-up as key factors influencing the break-up date. The rise in both accumulated temperature during the break-up period and the average temperature before the break-up tends to lead to an earlier ice break-up. This further reinforces the importance of climatic factors in determining the river ice break-up date [55,56]. Ice break-up in the study area has significantly advanced, with an average advancement of 0.682 days per decade. The observed trend of earlier ice break-ups is likely a direct result of warming temperatures in the region, suggesting that climate change is accelerating the ice break-up date of the Heilongjiang River's Heihe section.

During the model development phase, input selection is often an overlooked key task. As shown in Table 5, a comparative analysis of different ML models reveals that feature selection has a significant impact on model performance. However, the best feature selection method varies depending on the structure of the ML model. This difference highlights the point that a "one-size-fits-all" feature engineering approach cannot optimize the performance of specific model structures. The four feature selection methods in this study are model-agnostic, aiming to identify predictive features independently of the underlying ML algorithm's structure. The best-performing model is PCC-XGBoost, with a prediction accuracy of 85.71%. Compared with methods proposed in the existing literature [16], the prediction accuracy improves by 13.71%. XGBoost is a powerful ensemble learning algorithm based on gradient boosting that is known for effectively handling nonlinear relationships. It outperforms other ML algorithms in prediction because it minimizes a regularized objective function by iteratively adding decision trees that correct residuals from previous iterations [57,58]. XGBoost's performance varies markedly across feature selection methods, with optimal results depending on scenario-specific input parameters. This aligns with the observations of Hussain and Khan [59]. XGBoost may overfit in imbalanced or extreme event data, and careful adjustment of regularization parameters is required [60]. The performance of BPNN is unstable. Although the best model can be obtained by iteratively adjusting parameters [61], BPNN has limited generalization ability in extreme climate event prediction, primarily due to the insufficient coverage of extreme scenarios in the training data rather than the network architecture itself [62,63]. This issue has also been confirmed by Dai et al. [64] through multiple experiments. As the model is trained, RF performs better than SVR and BPNN. Tyralis et al. [65] confirmed that RF has better stability and prediction performance compared with other ML models. However, RF is sensitive to noise or irrelevant features, which may reduce its generalization performance. Despite these limitations, ensemble and hybrid models generally improve prediction performance compared with independent ML methods, as clearly confirmed in the literature [35]. Compared with other machine learning models, SVR performs poorly in this study. Despite systematic hyperparameter tuning of SVR through grid search, its prediction accuracy remains significantly lower than that of other models. This phenomenon may be due to the highly non-stationary and asymmetric nonlinear relationships between the climate variables involved in the study, while SVR relies on kernel functions to map data into high-dimensional space for linear separation. Although theoretically, the RBF kernel can handle complex nonlinearity, its performance in practice is limited by challenges such as insufficient sample size and high feature dimensions [66]. When the training data fail to adequately cover the distribution range of extreme events, SVR struggles to learn a robust decision boundary from the limited samples.

In addition, this study has several limitations that need to be addressed and explored in future research. The results of the current ice reserve calculations are influenced by the accuracy of the freezing degree-day model and the precision of river freezing area calculations. In future research, we plan to utilize satellite remote sensing data to enhance the accuracy of ice reserve calculations. In ML, the quality of the data directly affects the model's performance and predictive ability. If the data contain noise, missing values, or outliers, the accuracy and robustness of the model will be compromised. In this study, although we made efforts to collect a broad dataset for training the ML model, the insufficient precision of the data may negatively impact the model's predictive performance. Additionally, we observed that significant errors in the meteorological information of the input data typically lead to decreased model prediction accuracy. This could be because the meteorological information serves as a key input feature for predicting the river ice break-up date, and its accuracy is directly related to the model's predictive performance. With the continuous expansion of future datasets and the emergence of high-precision data, we expect the model's accuracy to improve significantly, leading to more reliable predictions. Given the complexity involved in the formation of the IJF, further studies could incorporate additional feature engineering or consider integrating feature engineering with model architecture optimization during preprocessing. We will also focus on improving the interpretability of the model. This includes optimizing the selection of features and indicators, as well as applying the SHAP method to analyze the impact of different features on the river ice break-up date.

5. Conclusions

This study analysed and calculated river ice break-up dates and ice reserves in upper Heilongjiang from 1956 to 2022. In addition, key features of the break-up dates in the study area were screened based on multiple feature selection techniques (MI, PCC, SR, and GRA) to optimize the prediction accuracy of the ML models (XGBoost, BPNN, RF, and SVR). The results of the study show the follwoing:

1. The river ice break-up date of the Heihe section of the Heilongjiang River shows an early trend, with the river ice break-up date advancing by 0.682 days every 10 years. The ice reserves in the Oupu–Heihe section in the upper reaches of the Heilongjiang River have the most significant impact on the river ice break-up date in the Heihe section. The correlation coefficient, grey relational degree, and mutual information value were 0.480, 0.479, and 0.176, respectively.
2. The different feature sets obtained for feature selection using PCC, MI, GRA, and SR reflect the different data characteristics that these methods focus on. Accumulated temperature during the break-up period and average temperature before river ice break-up were considered to have a significant effect on the opening of the river for a wide range of criteria, with abrupt temperature changes being a key factor affecting the timing of the opening of the river.
3. The best feature selection method varies depending on the structure of the ML model. By comparing 16 different combinations, PCC-XGBoost resulted in the smallest bias, achieving a prediction accuracy of 85.71%. According to the Standard for Hydrological Information and Hydrological Forecasting, this is classified as a first-class solution and can be used for river opening date prediction.

Author Contributions: Conceptualization, H.H. and Y.L.; methodology, H.H., Z.L. and X.L.; formal analysis, Y.L. and X.L.; investigation, Z.L., Y.L., E.W. and X.L.; data curation, Z.L., H.H. and E.W.; writing—original draft preparation, Z.L. and H.H.; writing—review and editing, Y.L. and X.L.; funding acquisition, H.H. and E.W. All authors have read and agreed to the published version of the manuscript.

Funding: This research was supported by the Major Scientific and Technological Projects of the Ministry of Water Resources of China (No. SKS-2022017) and the Project to Support the Development of Young Talent by Northeast Agricultural University.

Data Availability Statement: The data are available upon request.

Conflicts of Interest: The authors declare no conflicts of interest.

References

1. Lindenschmidt, K.-E.; Rokaya, P. A stochastic hydraulic modelling approach to determining the probable maximum staging of ice-jam floods. *J. Environ. Inform.* **2019**, *34*, 45–54. [CrossRef]
2. Hicks, F.; Beltaos, S. River Ice. In *Cold Region Atmospheric and Hydrologic Studies. The Mackenzie GEWEX Experience*; Woo, M., Ed.; Springer: Berlin/Heidelberg, Germany, 2008; pp. 281–305. [CrossRef]
3. Rokaya, P.; Budhathoki, S.; Lindenschmidt, K.-E. Ice-jam flood research: A scoping review. *Nat. Hazards* **2018**, *94*, 1439–1457. [CrossRef]
4. Cunderlik, J.M.; Ouarda, T.B.M.J. Trends in the timing and magnitude of floods in Canada. *J. Hydrol.* **2009**, *375*, 471–480. [CrossRef]
5. Ghoreishi, M.; Lindenschmidt, K.-E. Unlocking effective ice-jam risk management: Insights from agent-based modeling and comparative analysis of social theories in Fort McMurray, Canada. *Environ. Sci. Policy* **2024**, *157*, 103731. [CrossRef]
6. Pagneux, E.; Gísladóttir, G.; Jónsdóttir, S. Public perception of flood hazard and flood risk in Iceland: A case study in a watershed prone to ice-jam floods. *Nat. Hazards* **2011**, *58*, 269–287. [CrossRef]
7. Das, A.; Lindenschmidt, K.-E. Current status and advancement suggestions of ice-jam flood hazard and risk assessment. *Environ. Rev.* **2020**, *28*, 373–379. [CrossRef]
8. Das, A.; Budhathoki, S.; Lindenschmidt, K.-E. Development of an ice-jam flood forecasting modelling framework for freeze-up/winter breakup. *Hydrol. Res.* **2023**, *54*, 648–662. [CrossRef]
9. Barrette, P.D.; Khan, A.A.; Lindenschmidt, K.-E. A glimpse at twenty-five hydraulic models for river ice. In Proceedings of the 27th IAHR International Symposium on Ice, Gdańsk, Poland, 9–13 June 2024. Available online: https://www.iahr.org/library/infor?pid=30349 (accessed on 20 May 2023).
10. Lindenschmidt, K.-E. RIVICE—A Non-Proprietary, Open-Source, One-Dimensional River-Ice Model. *Water* **2017**, *9*, 314. [CrossRef]
11. Fu, H.; Guo, X.; Yang, K.; Wang, T.; Guo, Y. Ice accumulation and thickness distribution before inverted siphon. *J. Hydrodyn.* **2017**, *29*, 61–67. [CrossRef]
12. Carson, R.; Beltaos, S.; Groeneveld, J.; Healy, D.; She, Y.; Malenchak, J.; Morris, M.; Saucet, J.-P.; Kolerski, T.; Shen, H.T. Comparative testing of numerical models of river ice jams. *Can. J. Civ. Eng.* **2011**, *38*, 669–678. [CrossRef]
13. Beltaos, S. Numerical modelling of ice-jam flooding on the Peace-Athabasca delta. *Hydrol. Process.* **2003**, *17*, 3685–3702. [CrossRef]
14. Ladouceur, J.-R.; Morse, B.; Lindenschmidt, K.-E. A comprehensive method to estimate flood levels of rivers subject to ice jams: A case study of the Chaudière River, Québec, Canada. *Hydrol. Res.* **2023**, *54*, 995–1016. [CrossRef]
15. Beltaos, S.; Burrell, B.C. Hydrotechnical advances in Canadian river ice science and engineering during the past 35 years. *Can. J. Civ. Eng.* **2015**, *42*, 583–591. [CrossRef]
16. Lei, Q.; Zhang, D.; Li, J. Discussion on the ice breakup forecasting and analysis method at Jiayin station, Heilongjiang. *Heilongjiang Hydr. Sci. Technol.* **2010**, *39*, 16–17. (In Chinese)
17. Lindenschmidt, K.-E.; Rokaya, P.; Das, A.; Li, Z.; Richard, D. A novel stochastic modelling approach for operational real-time ice-jam flood forecasting. *J. Hydrol.* **2019**, *575*, 381–394. [CrossRef]
18. Smith, J.D.; Lamontagne, J.R.; Jasek, M. Considering uncertainty of historical ice jam flood records in a Bayesian frequency analysis for the Peace-Athabasca Delta. *Water Resour. Res.* **2024**, *60*, e2022WR034377. [CrossRef]
19. Liu, M.; Wang, Y.; Xing, Z.; Wang, X.; Fu, Q. Study on forecasting break-up date of river ice in Heilongjiang Province based on LSTM and CEEMDAN. *Water* **2023**, *15*, 496. [CrossRef]
20. Shevnina, E.V.; Solov'eva, Z.S. Long-term variability and methods of forecasting dates of ice break-up in the mouth area of the Ob and Yenisei rivers. *Russ. Meteorol. Hydrol.* **2008**, *33*, 458–465. [CrossRef]
21. Madaeni, F.; Chokmani, K.; Lhissou, R.; Gauthier, Y.; Tolszczuk-Leclerc, S. Convolutional neural network and long short-term memory models for ice-jam predictions. *Cryosphere* **2022**, *16*, 1447–1468. [CrossRef]
22. Kalke, H.; Loewen, M. Support vector machine learning applied to digital images of river ice conditions. *Cold Reg. Sci. Technol.* **2018**, *155*, 225–236. [CrossRef]
23. Tom, M.; Prabha, R.; Wu, T.; Baltsavias, E.; Leal-Taixé, L.; Schindler, K. Ice monitoring in Swiss lakes from optical satellites and webcams using machine learning. *Remote Sens.* **2020**, *12*, 3555. [CrossRef]
24. Chen, S.; Ji, H. Fuzzy optimization neural network approach for ice forecast in the Inner Mongolia reach of the Yellow River. *Hydrol. Sci. J.* **2005**, *50*, 319–330. [CrossRef]
25. Ge, Q.; Wang, J.; Liu, C.; Wang, X.; Deng, Y.; Li, J. Integrating feature selection with machine learning for accurate reservoir landslide displacement prediction. *Water* **2024**, *16*, 2152. [CrossRef]

26. Paulson, N.H.; Kubal, J.; Ward, L.; Saxena, S.; Lu, W.; Babinec, S.J. Feature engineering for machine learning enabled early prediction of battery lifetime. *J. Power Sources* **2022**, *527*, 231127. [CrossRef]
27. Chicco, D.; Oneto, L.; Tavazzi, E. Eleven quick tips for data cleaning and feature engineering. *PLoS Comput. Biol.* **2022**, *18*, e1010718. [CrossRef]
28. Diebold, F.X.; Göbel, M.; Coulombe, P.G. Assessing and comparing fixed-target forecasts of Arctic sea ice: Glide charts for feature-engineered linear regression and machine learning models. *Energy Econ.* **2023**, *124*, 106833. [CrossRef]
29. Nichol, J.J.; Peterson, M.G.; Peterson, K.J.; Fricke, G.M.; Moses, M.E. Machine learning feature analysis illuminates disparity between E3SM climate models and observed climate change. *J. Comput. Appl. Math.* **2021**, *395*, 113451. [CrossRef]
30. Verdonck, T.; Baesens, B.; Óskarsdóttir, M.; vanden Broucke, S. Special issue on feature engineering editorial. *Mach. Learn.* **2024**, *113*, 3917–3928. [CrossRef]
31. Ji, H.; Zhang, A.; Gao, R.; Zhang, B.; Xu, J. Application of the break-up date prediction model in the Inner Mongolia Reach of the Yellow River. *Adv. Sci. Technol. Water Res.* **2012**, *32*, 42–45.
32. Li, C.; Guo, X.; Wang, Z. Determination on forecast factor of artificial intelligence ice-forecast. *Water Resour. Hydropower Eng.* **2012**, *43*, 9–13. (In Chinese) [CrossRef]
33. Rokaya, P.; Budhathoki, S.; Lindenschmidt, K.-E. Trends in the Timing and magnitude of ice-jam floods in Canada. *Sci. Rep.* **2018**, *8*, 5834. [CrossRef] [PubMed]
34. Das, A.; Lindenschmidt, K.-E. Modelling climatic impacts on ice-jam floods: A review of current models, modelling capabilities, challenges, and future prospects. *Environ. Rev.* **2021**, *29*, 378–390. [CrossRef]
35. Salimi, A.; Ghobrial, T.; Bonakdari, H. A comprehensive review of AI-based methods used for forecasting ice jam floods occurrence, severity, timing, and location. *Cold Reg. Sci. Technol.* **2024**, *227*, 104305. [CrossRef]
36. Das, A.; Rokaya, P.; Lindenschmidt, K.-E. The impact of a bias-correction approach (delta change) applied directly to hydrological model output when modelling the severity of ice jam flooding under future climate scenarios. *Clim. Change* **2022**, *172*, 19. [CrossRef]
37. Lubiniecki, T.; Laroque, C.P.; Lindenschmidt, K.-E. Identifying ice-jam flooding events through the application of dendrogeomorphological methods. *River Res. Appl.* **2024**, *40*, 191–202. [CrossRef]
38. Huang, F.; Shen, H.T. Dam removal effect on the lower St. Regis River ice-jam floods. *Can. J. Civ. Eng.* **2023**, *51*, 215–227. [CrossRef]
39. Li, Y.; Han, H.; Sun, Y.; Xiao, X.; Liao, H.; Liu, X.; Wang, E. Risk evaluation of ice flood disaster in the upper Heilongjiang River based on catastrophe theory. *Water* **2023**, *15*, 2724. [CrossRef]
40. Wang, T.; Liu, Z.; Guo, X.; Fu, H.; Liu, W. Prediction of breakup ice jam with Artificial Neural Networks. *J. Hydraul. Eng.* **2017**, *48*, 1355–1362. (In Chinese) [CrossRef]
41. Li, M.; Yang, D.; Hou, J.; Xiao, P.; Xing, X. Distributed hydrological model of Heilongjiang River basin. *J. Hydroelectr. Eng.* **2021**, *40*, 65–75. (In Chinese)
42. Cao, W.; Xiao, D.; Li, G. How to make ice dam forecasting for Heilongjiang River. *J. China Hydrol.* **2014**, *34*, 72–76. (In Chinese)
43. Chang, Y.; Liu, X.; Zhao, X.; Shen, Y. Multi-scale analysis of runoff variability and its influencing factors in the mountainous Hutuo River Basin. *J. Water Resour. Eng.* **2023**, *34*, 59–70. (In Chinese)
44. Massie, D.D.; White, K.D.; Daly, S.F. Application of neural networks to predict ice jam occurrence. *Cold Reg. Sci. Technol.* **2002**, *35*, 115–122. [CrossRef]
45. Rafi, M.N.; Imran, M.; Nadeem, H.A.; Abbas, A.; Pervaiz, M.; Khan, W.; Ullah, S.; Hussain, S.; Saeed, Z. Comparative influence of biochar and doped biochar with Si-NPs on the growth and anti-oxidant potential of *Brassica rapa* L. under Cd toxicity. *Silicon* **2022**, *14*, 11699–11714. [CrossRef]
46. Friedman, J.H. Greedy function approximation: A gradient boosting machine. *Ann. Stat.* **2001**, *29*, 1189–1232. [CrossRef]
47. Cortes, C.; Vapnik, V. Support-vector networks. *Mach. Learn.* **1995**, *20*, 273–297. [CrossRef]
48. Magnuson, J.J.; Robertson, D.M.; Benson, B.J.; Wynne, R.H.; Livingstone, D.M.; Arai, T.; Assel, R.A.; Barry, R.G.; Card, V.; Kuusisto, E.; et al. Historical trends in lake and river ice cover in the Northern Hemisphere. *Science* **2000**, *289*, 1743–1746. [CrossRef]
49. Prowse, T.D.; Beltaos, S. Climatic control of river-ice hydrology: A review. *Hydrol. Process.* **2002**, *16*, 805–822. [CrossRef]
50. Ji, X.; Zhao, J. Analysis of correlation between sea ice concentration and cloudiness in the central Arctic. *Acta Oceanol. Sin.* **2015**, *37*, 92–104. (In Chinese) [CrossRef]
51. Schweiger, A.J.; Lindsay, R.W.; Vavrus, S.; Francis, J.A. Relationships between Arctic sea ice and clouds during autumn. *J. Clim.* **2009**, *22*, 4799–4810. [CrossRef]
52. Jacob, T.; Wahr, J.; Pfeffer, W.T.; Swenson, S. Recent contributions of glaciers and ice caps to sea level rise. *Nature* **2012**, *482*, 514–518. [CrossRef]

53. Prowse, T.D.; Bonsal, B.R.; Duguay, C.R.; Lacroix, M.P. River-ice break-up/freeze-up: A review of climatic drivers, historical trends and future predictions. *Ann. Glaciol.* **2007**, *46*, 443–451. [CrossRef]
54. Rokaya, P.; Morales-Marín, L.; Bonsal, B.; Wheater, H.; Lindenschmidt, K.-E. Climatic effects on ice phenology and ice-jam flooding of the Athabasca River in western Canada. *Hydrol. Sci. J.* **2019**, *64*, 1265–1278. [CrossRef]
55. Burrell, B.C.; Beltaos, S.; Turcotte, B. Effects of climate change on river-ice processes and ice jams. *Int. J. River Basin Manag.* **2022**, *21*, 421–441. [CrossRef]
56. Yang, X.; Pavelsky, T.M.; Allen, G.H. The past and future of global river ice. *Nature* **2020**, *577*, 69–73. [CrossRef]
57. Chen, T.; Guestrin, C. XGBoost: A Scalable Tree Boosting System. In *Proceedings of the 22nd ACM SIGKDD International Conference on Knowledge Discovery and Data Mining (KDD '16), San Francisco, CA, USA, 13–17 August 2016*; Association for Computing Machinery: New York, NY, USA, 2016; pp. 785–794. [CrossRef]
58. Niazkar, M.; Menapace, A.; Brentan, B.; Piraei, R.; Jimenez, D.; Dhawan, P.; Righetti, M. Applications of XGBoost in water resources engineering: A systematic literature review (Dec 2018–May 2023). *Environ. Model. Softw.* **2024**, *174*, 105971. [CrossRef]
59. Hussain, D.; Khan, A.A. Machine learning techniques for monthly river flow forecasting of Hunza River, Pakistan. *Earth Sci. Inform.* **2020**, *13*, 939–949. [CrossRef]
60. Zhang, W.; Wang, H.; Lin, Y.; Jin, J.; Liu, W.; An, X. Reservoir inflow predicting model based on machine learning algorithm via multi-model fusion: A case study of Jinshuitan river basin. *IET Cyber-Syst. Robot.* **2021**, *3*, 265–277. [CrossRef]
61. Aksoy, H.; Mohammadi, M. Artificial neural network and regression models for flow velocity at sediment incipient deposition. *J. Hydrol.* **2016**, *541*, 1420–1429. [CrossRef]
62. Nielsen, M.A. *Neural Networks and Deep Learning*; Determination Press: San Francisco, CA, USA, 2015. Available online: http://neuralnetworksanddeeplearning.com/ (accessed on 29 December 2023).
63. Hu, H.; Zhang, J.; Li, T. A novel hybrid decompose-ensemble strategy with a VMD-BPNN approach for daily streamflow estimating. *Water Resour. Manag.* **2021**, *35*, 5119–5138. [CrossRef]
64. Dai, W.; Cai, Z. Predicting coastal urban floods using artificial neural network: The case study of Macau, China. *Appl. Water Sci.* **2021**, *11*, 161. [CrossRef]
65. Tyralis, H.; Papacharalampous, G.; Langousis, A. A Brief Review of Random Forests for Water Scientists and Practitioners and their Recent History in Water Resources. *Water* **2019**, *11*, 910. [CrossRef]
66. Deka, P.C. Support vector machine applications in the field of hydrology: A review. *Appl. Soft Comput.* **2014**, *19*, 372–386. [CrossRef]

water

MDPI

Article

Analysis of Open-Water Changes and Ice Microstructure Characteristics in Different River Channel Types of the Yellow River in Inner Mongolia Based on Satellite Images and Field Sampling

Yupeng Leng [1], Chunjiang Li [2,*], Peng Lu [1], Xiang Fu [1] and Shengbo Hu [1]

[1] State Key Laboratory of Coastal and Offshore Engineering, Dalian University of Technology, Dalian 116024, China; lengyupeng@mail.dlut.edu.cn (Y.L.)
[2] School of Energy and Environment, Inner Mongolia University of Science and Technology, Baotou 014010, China
* Correspondence: lichunjiang0405@imust.edu.cn

Abstract: The formation and evolution of ice in the Yellow River represent complex dynamic processes. To elucidate the structural characteristics of ice crystals and their governing mechanisms in the Inner Mongolia reach, this investigation utilized high-resolution Sentinel-2 satellite imagery to systematically monitor spatiotemporal variations in open-water formations across diverse channel morphologies throughout the ice regime period. Systematic ice sampling was conducted across diverse channel morphologies of the Yellow River to quantify critical parameters, including crystalline structure characteristics, equivalent diameter distributions, density variations, and sediment content profiles. The results indicate the transformation of open water resulting from various river configurations during the freezing season exhibits distinct characteristics, which are significantly influenced by temperature variations. Ice crystal characterization exhibits that the crystalline structure predominantly manifests as two primary forms: columnar and granular ice formations, with their distribution varying systematically across different channel configurations. Ice crystal morphology exhibits heterogeneity in both form and dimensional characteristics. Columnar ice consistently exhibits larger equivalent diameters compared to granular ice formations. A progressive enhancement in the equivalent diameter of crystals is observed along the vertical axis corresponding to the thickness of the ice during the growth process. The ranges of variation in ice crystal size, ice density, and mud content within ice exhibit differences contingent upon the specific crystal structures present. Observational studies and comparative analyses of ice samples from the Inner Mongolia reach of the Yellow River reveal that channel morphology, ambient thermal conditions, and hydrodynamic parameters are the primary determinants governing the variability in ice microstructure and its associated physical characteristics. This investigation provides fundamental scientific insights and quantitative data that advance our understanding of river ice microstructural characteristics.

Keywords: river ice; open water; crystal structure; Sentinel-2; equivalent diameter

1. Introduction

In high-latitude regions of northern China, rivers, lakes, and reservoirs are susceptible to severe ice-related hazards during winter periods due to extreme temperature conditions.

The Yellow River, as a major watercourse in northern China, presents particularly challenging ice-related phenomena due to its complex environmental conditions, extensive hazard zones, limited ice flood forecasting windows, and significant mitigation challenges. Consequently, ice-related disasters along the Yellow River represent one of the most critical and severe natural hazards in the region [1]. Ice floes encounter significant impediments at channel constrictions, shallow shoals, acute bends, meandering sections, and bridge pier structures, resulting in increased ice concentration densities across various river segments. These accumulations severely constrict the channel cross-section, leading to substantial upstream water level elevation. The subsequent formation of ice jams and ice dams presents significant flood hazard potential, posing substantial risks to riparian communities and infrastructure along both riverbanks [2–4]. River ice, as a natural composite material, constitutes a significant element within the field of Cryosphere Science [5]. The properties of river ice exhibit variability based on regional and temporal factors, primarily influenced by diverse environmental conditions. The processes governing the formation and transformation of ice on the Yellow River are notably intricate and dynamic. In recent years, this phenomenon has garnered extensive scholarly attention, with researchers employing various methodologies, including remote sensing, empirical measurement, and numerical simulation, to investigate it more thoroughly [6,7]. The ice microstructure, which constitutes the fundamental internal architecture, is characterized by three primary components: stratification patterns, crystalline structure, and incorporated impurities [8]. The microstructural configuration of ice serves as a fundamental determinant of its intrinsic characteristics, fundamentally governing its physical properties, including density parameters and sediment incorporation patterns [9]. The development and evolution of river ice are influenced by the presence of crystals. In addition to indirectly influencing the interaction and destructive power between ice and structures, the spatial distribution of ice crystals has a direct impact on the mechanical strength of the ice layer. As a result, studying the crystal structure of ice and the elements that influence it can theoretically support the microstructural properties of ice and serve as the foundation for research on ice engineering.

In relation to the related research on the microstructure of ice, Michel and Eicken [10, 11] identified the various forms of ice in the 1970s and 1980s, refined the various types of ice, and examined the size characteristics of various types of ice crystals. Langway proposed a technique for observing the microstructure of ice using a Fisher stage, which yielded parameters like the size distribution, c-axis orientation, bubble size and content in the ice, and crystal type [12]. Wilen [13] also suggested an automatic ice structure analyzer based on the principle of polarization to observe and differentiate ice crystals. Arnaud [14] and Azuma et al. [15] collected information on the crystal structure, particle size, and shape of the ice by using cameras to record its microscopic structure. Using an electron microscope and a camera, Kipfstuhl [16] created high-resolution photographs of ice crystals by applying the microstructure mapping method to their observation. Li analyzed the formation causes of various crystal types by observing crystals on reservoir ice and combining the ice growth process [17]. Currently, ice microstructure study and observation technology are becoming more sophisticated. Various new methods for observing microstructures, including CT [18,19] and nuclear magnetic resonance [20,21], have also been used to observe the microstructure of ice and have achieved good results. The crystal structure of ice determines its physical and mechanical properties to a large extent. Experimental studies on the mechanical properties of ice show that the brittle strength of granular ice is inversely proportional to the square root of its crystal diameter [22], and the ultimate compressive strength of polycrystalline ice is directly proportional to the inverse of the square root of its particle size [23]. Gao et al. [24] performed three-point bending tests on ice samples exhibiting various crystal structures obtained from the Yellow River. Subsequently, they utilized the relationship between crystal structure and mechanical properties to forecast the river's opening date. Deng used numerical simulation technology to find that there is a certain linear correlation between the uniaxial compression strength of ice and the grain size [25]. The microstructure of ice influences its acoustic, thermal, optical, and electrical properties, which in turn impact the ecological environment of subglacial waters. This microstructural influence forms the foundation for research in ice thermodynamics and remote sensing. Variations in ice types and crystal structures result in diverse thermal, optical, and electrical behaviors, thereby affecting the precision of remote sensing and radar observations [26,27]. Li et al. [28] employed ground-penetrating radar to measure the ice thickness of the Hongqipao Reservoir, revealing the growth patterns of reservoir ice and identifying that bubble content within the ice impacts radar measurement accuracy. Additionally, Li utilized Sentinel-2 remote sensing imagery and airborne radar to observe the ice thickness at the Shishifenzi Bend of the Yellow River across multiple scales, subsequently adjusting the radar dielectric constant to enhance monitoring accuracy for various ice types [29]. Cao et al. [30] utilized ground-penetrating radar to measure the ice thickness in two distinct river channel types: the Yellow River bend and areas around bridge piers. Their findings indicated that factors such as silt, entrapped bubbles, and ice temperature in the Yellow River ice contribute to variability in the propagation speed of radar waves through the ice layer. It is evident that the significance of microstructural features in observations across all scales is substantial. Investigating the structure and morphology of ice crystals, along with related parameters such as ice density, sediment content, and entrapped bubbles, enhances our understanding of the physical properties and damage mechanisms of ice. This research also provides essential theoretical support for the development of advanced ice physics models and the enhancement of remote sensing observation accuracy.

Advances in observational methodologies and technologies have enabled researchers to conduct extensive investigations into the microstructural characteristics of various ice formations, including lake ice [31,32], sea ice [33,34], reservoir ice [35,36], glacier ice [37,38], and polar ice in both Arctic and Antarctic regions [39,40], yielding significant findings. However, there are still few studies and quantitative analyses on river ice microstructures, such as ice crystals, ice density, mud content in ice, and bubbles. The composition, distribution, and morphological characteristics of impurities within ice matrices exhibit significant spatial heterogeneity across different geographical regions. The Yellow River, China's second-largest watercourse, is characterized by complex hydrodynamic conditions and intricate flow evolution patterns. As the river with the highest sediment concentration globally, its unique characteristics, coupled with complex flow dynamics, result in highly variable crystalline compositions within its ice formations.

Currently, there is little research on ice crystals in the Yellow River, and many issues still need further study, including how to effectively observe the microstructure of ice and how to understand and describe the distribution and evolution of the ice structure in the Yellow River. Based on this, this paper observes and analyzes the basic physical properties of the Yellow River ice, such as its crystal structure, ice density, and mud content. Ice samples were collected from different river channel locations in the Inner Mongolia section of the Yellow River, and parameters such as the shape and size of ice crystals, their distribution, ice density, and mud content in the ice were observed. This investigation integrates analysis of Yellow River ice formation processes with detailed examination of its microstructural characteristics and physical parameters across diverse channel locations, elucidating spatial variations and distinctive patterns. Through comprehensive analysis of ice microstructure and fundamental parameters across different channel morphologies, this research advances our theoretical understanding of ice formation and distribution patterns in the Yellow River system. Furthermore, these findings provide a robust scientific framework for enhancing ice detection methodologies and improving early warning systems for ice-related hazards, including ice jam formation and ice dam development.

2. Materials and Methods

2.1. Study Area

The Inner Mongolia section of the Yellow River, situated at the northernmost extremity of the Yellow River Basin, traverses one league, five cities, and seventeen banners within Inner Mongolia. This section spans a total length of 840 km, with an elevation drop of 162.5 m. It encompasses five hydrological stations: Shizuishan, BayanGaole, Sanhuhekou, Baotou, and Toudaoguai. Owing to its high-latitude location and intricate riverine conditions, this section represents the most complex and challenging area for ice-related phenomena along the Yellow River. In this study, during the winter freezing period of 2023–2024, five different river channel types including straight river channels, curved river channels, single-line bridge channels, double-line bridge channels, and Shaxintan river channels from upstream to downstream in the Inner Mongolia section of the Yellow River were selected for ice collection. The sampling points were the Sanhu estuary hydrological station, under the Dachengxi Yellow River Bridge, the front and middle part of the Shishifenzi Bend and the lower open-water area, before and after the Qianfangzi double-track bridge, and the right bank of the Shaxintan river channel next to Qianfangzi village. A total of 8 ice blocks were collected. The Yellow River Basin map and the locations of different types of river channels are shown in Figure 1. The locations of sampling points at different river channels and their basic information are shown in Figure 1 and Table 1.

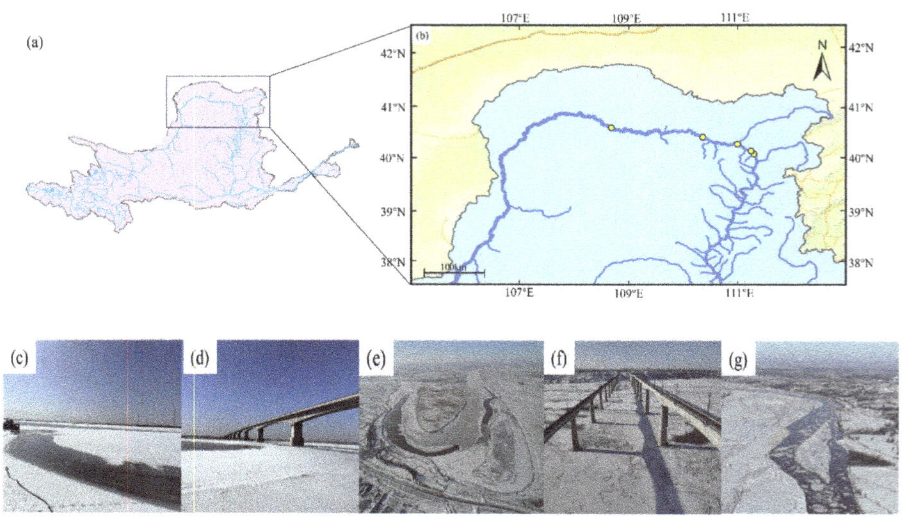

Figure 1. (**a**) Yellow River Basin; (**b**) the Inner Mongolia section of the Yellow River; (**c**) Sanhukou straight river channel; (**d**) Dachengxi Railway Bridge; (**e**) Shishifenzi Bend; (**f**) Qianfangzi double-track bridge; (**g**) Shaxintan river channel.

Table 1. Basic information of sampling points.

Sampling Point	Longitude (E)	Latitude (S)	River Channel Type	Ice Thickness (cm)
SHHK1#	108°46′24.70″	40°36′23.88″	Straight River	38
DXC1#	110°23′52.49″	40°25′33.00″	Single-track bridge	52
SSFZ1#	111°25′58.98″	40°17′42.67″	Bend River	68
SSFZ2#	111°26′44.16″	40°17′42.12″	Bend River	48
SSFZ3#	111°27′20.81″	40°17′40.48″	Bend River	32
QFZ1#	111°20′7.44″	40°7′12.48″	Double-track bridge	51
QFZ2#	111°20′8.97″	40°7′11.16″	Double-track bridge	49
SXT1#	111°19′19.63″	40°7′42.96″	Shaxintan river channel	64

2.2. Open-Water Change Monitoring

The open-water areas formed by the selected sampling points were observed using the Sentinel-2 satellite. Sentinel-2 consists of two satellites, 2A and 2B, which complement each other and have a revisit period of 5 days. The satellite is equipped with a multispectral imager, with an orbit altitude of 786 km, an image width of 290 km, and 13 spectral bands. High-resolution images that have been geometrically, radiometrically, and atmospherically corrected are available through the Copernicus Open Access Hub (https://dataspace.copernicus.eu). In order to better obtain the formation process of open waters and different types of ice at different locations, Sentinel-2 satellite images acquired between December 2023 and March 2024 were utilized. Level-2A data with less than 30 percent cloud cover and a spatial resolution of 10 m, in which the ice surface was clearly visible, were selected. The chosen images were temporally distributed to ensure the inclusion of 1–2 images per month, resulting in 14 images per observation point over the four-month period. Remote sensing image processing was conducted using ArcMap 10.8 and ENVI 5.6 software. The post-classification comparison method was employed to continuously analyze changes in the ice surface and open-water areas based on the remote sensing imagery.

2.3. Ice Sample and Crystal Acquisition Method

Ice samples were extracted from stable ice surfaces utilizing chain saws, manual panel saws, and hand ice drill. To extract the ice sample, first use a chain saw to create grooves and cuts on the ice surface. If the ice sample does not separate, employ a manual panel saw to cut the edge of the ice surface. Finally, extract the ice sample vertically and steadily using a hand ice drill. The samples were subsequently preserved in cold storage facilities. To examine the microstructure, samples were vertically sectioned into uniform-thickness segments using a precision bone saw. Each ice segment was processed by planing one surface to achieve flatness, then adhered to a glass sheet to ensure optimal contact. To facilitate microscopic examination, the segments were further reduced to approxim no problemately 1 mm thick sections using a precision planer. The glass sheets were heated slightly, which let the ice sample adhere to the glasses easily. The thinner the ice crystal, the brighter its appearance under an orthogonal polarizing microscope, which conveniently distinguishes crystal boundaries and clarifies the shape of the individual crystal within the ice. These thin sections were analyzed using a polarizing microscope mounted on the Ferris Observatory to observe crystallographic structures. During the acquisition of the crystal image, a calibration scale was established. Utilizing various edge detection techniques based on the Canny operator, the dimensions of the crystals were extracted through the application of Adobe Photoshop, MATLAB, and Image J. Additionally, pertinent parameters, including calculations of equivalent diameter and other associated microstructural metrics, were determined.

3. Results

3.1. Open-Water Observations from Different Sampling Points Based on Sentinel-2

The unfrozen open-water areas on the Yellow River's winter ice surface exhibit complex spatiotemporal dynamics, reflecting the synergistic effects of hydrodynamic conditions, meteorological parameters, and anthropogenic influences. During initial ice formation, localized freezing patterns emerge on the river surface. However, heterogeneous flow conditions and thermal gradients maintain certain areas in an unfrozen state, resulting in distinct open-water zones. Utilizing Sentinel-2 imagery, this study monitored the formation and evolution of unfrozen water areas proximate to sampling locations during the winter of 2023–2024. Among the five sampling sites, open-water formations were observed at the Dachengxi Railway Bridge and the Shishifenzi Bend. According to the location of the open-water, three different position sections were selected to measure the change in the open water relative to the shore, so as to analyze the lateral change in the open water. Figure 2 shows the changes in the open-water area of the Dachengxi Railway Bridge. The remote sensing imagery captured on specific dates, as illustrated in Figure 2, provides a comprehensive overview of the alterations occurring in the open-water area created by the single-track bridge channel. While the open-water area predominantly exhibits a rectangular shape, variations in the front and rear sections, as well as the overall positioning, are evident. The formation of open-water beneath the bridge is particularly pronounced, likely attributable to alterations in water velocity induced by the presence of the pier. Consequently, the freezing processes at various locations exhibit distinct characteristics, with sections situated further from the pier demonstrating greater variability in their changes.

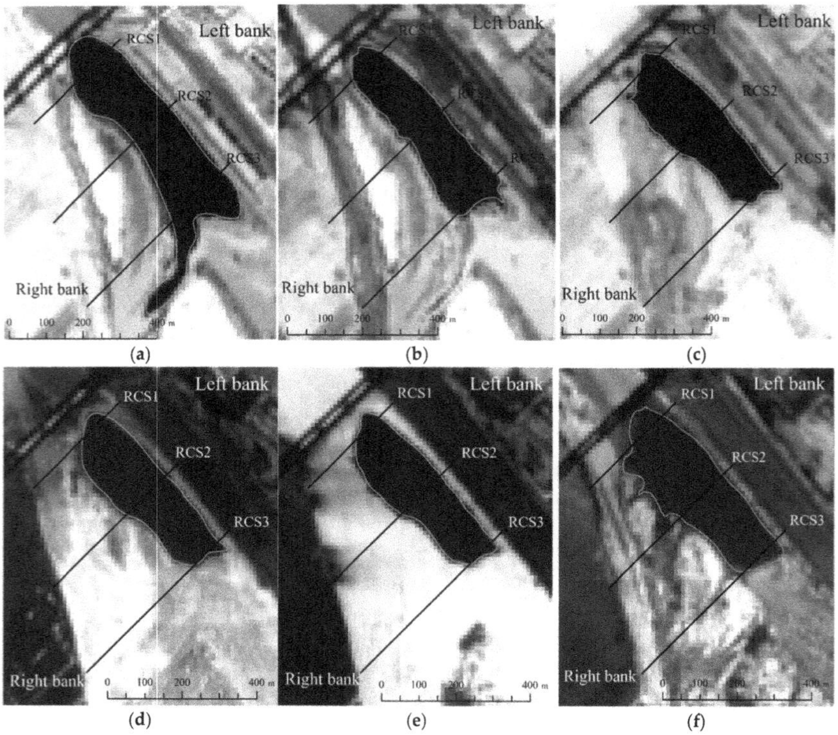

Figure 2. Changes in the open−water area of the Dachengxi Railway Bridge. (**a**) 19 December 2023; (**b**) 3 January 2024; (**c**) 28 January 2024; (**d**) 17 February 2024; (**e**) 3 March 2024; (**f**) 18 March 2024.

To quantitatively analyze the temporal evolution of open-water areas, measurements were conducted and normalized to establish standardized values. The ratio of open-water length to total riverbank section length was calculated to characterize relative temporal variations. Additionally, distances between open-water boundaries and both riverbanks were measured systematically. Comparative analysis of these spatial relationships relative to both banks enabled comprehensive understanding of open-water positioning and ice surface dynamics during the freeze period. The detailed quantitative results of these analyses are presented in Figure 3. Figure 3a illustrates the correlation between open-water areas and average temperature. During winter, the temperature at the Dachengxi Railway Bridge sampling site initially decreases, then stabilizes, and eventually increases. Correspondingly, the open-water area first contracts, then stabilizes, and finally expands. This pattern is primarily attributed to decreasing temperatures, which lead to increased ice coverage, gradually encroaching upon open-water areas. When temperatures remain below freezing, changes in open-water areas diminish and stabilize. By March, as temperatures rise above 0 °C, ice begins to melt, causing a sharp increase in open-water areas. The melting of ice due to temperature variations impacts the delineation of open water. This phenomenon highlights temperature as the principal determinant affecting the dynamics of open water and ice. The variation in temperature serves as a reliable indicator of alterations in open-water conditions. Figure 3b shows the change in the length of the open-water area. Due to the irregular shape of the open-water area, the change in the cross section will also be different. The expanse of open-water surrounding the Dachengxi Railway Bridge exhibits a shape that approximates a rectangle. It can be seen that the changes in the first two sections selected for the Dachengxi Railway Bridge are basically the same, both increasing first and then stabilizing. This may be due to the water flow conditions that caused the section to not freeze and the section length to gradually increase. However, the change in the third section is relatively small, mainly because the river channel width of the third section is large, the water flow conditions have changed, and the section is at the bottom of the ditch and far away from the bridge, which has little impact on the formation of ice. Figure 3c reveals that open-water areas are generally situated closer to the left bank and more distant from the right bank. Additionally, during ice formation and melting, the right bank experiences relative changes, whereas the left bank remains relatively stable. The open-water region is not situated at the center of the river, suggesting that the flow dynamics are a contributing factor to the alterations observed in the open-water area.

Figure 4 illustrates the evolution of the open-water area at the Shishifenzi Bend. Unlike the open-water areas formed by single-line bridge channels, the Shishifenzi Bend features a long, strip-shaped open-water area that follows the curvature of the bend. The open-water area is narrow at the head and bottom of the bend, widening in the middle. As temperatures fluctuate and ice accumulates within the bend, portions of the upstream open water gradually freeze, leading to a reduction in the total open-water area. As illustrated in Figure 4, the transformation process involved the freezing of certain areas of open water as temperatures progressively decreased. The unique curvature of the bend contributed to an uneven velocity distribution within the open-water, resulting in a greater extent of freezing at the bend's apex. Concurrently, as temperatures fluctuated and ice accumulation occurred within the bend, portions of the upstream open water began to freeze, leading to a gradual reduction in the overall area of open-water. In March, with a gradual increase in temperature, melting commenced in the central regions of the open-water, progressing downstream. Consequently, the ice began to melt, resulting in an increase in the area of open-water until the river was fully opened.

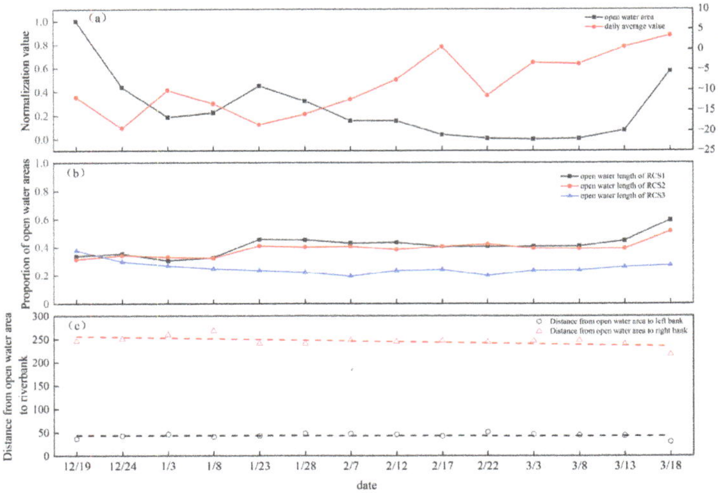

Figure 3. (**a**) Open−water area and temperature change process of Dachengxi Railway Bridge; (**b**) the proportion of open−water length relative to different riverbank sections; (**c**) the distance of open water from the left and right banks of the riverbank section.

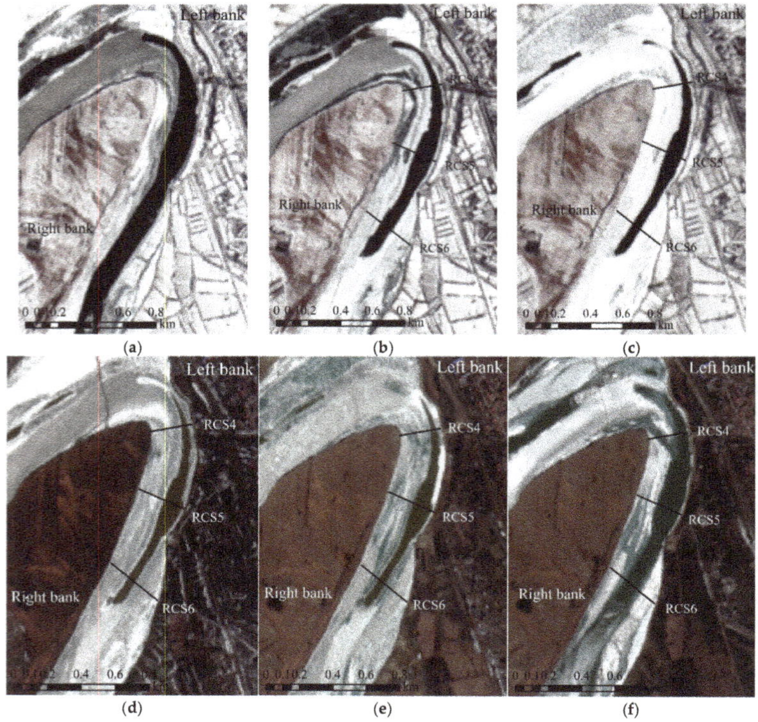

Figure 4. Changes in the open−water area in Shishifenzi Bend. (**a**) 19 December 2023; (**b**) 3 January 2024; (**c**) 28 January 2024; (**d**) 17 February 2024; (**e**) 3 March 2024; (**f**) 18 March 2024.

Figure 5 shows the quantitative changes in the open-water area of the Shisifenzi Bend. As illustrated in Figure 5a,b, the characteristics of the open-water in the bend differ from those of the open-water located beneath the bridge, exhibiting distinct patterns of change. Throughout the freezing period, fluctuations in the open-water area correspond with variations in temperature. Specifically, the open water in the Shisifenzi Bend initially decreases, subsequently stabilizes, and ultimately increases, with the open-water area comprising approximately one-third of the total river course. The unique configuration of the bend channel results in varying lengths of different sections within the open water, with the middle section demonstrating a more pronounced trend of change compared to the sections at both the head and tail of the bend. The analysis presented in Figure 5c facilitates an understanding of the relationship between the open-water boundary and the riverbank. Furthermore, the observed alterations in the boundary indicate a preferential modification of the ice–water contact surface in contrast to the stable ice surface. The newly added ice cover in the open-water is mainly distributed on the right bank, and the change on the left bank is very small because the left bank is the mainstream area with a larger flow velocity. Due to the current conditions in the bend and the absence of lower and longer negative air temperatures to promote ice sheet growth, the lateral growth rate of the right bank ice sheet slowed down over time, resulting in less fluctuation in the area change in open-water. During the melting stage, it can be seen that the right bank still changes more significantly, indicating that open-water areas generally begin to melt from the right bank, and water flow conditions are also the main reason affecting the changes.

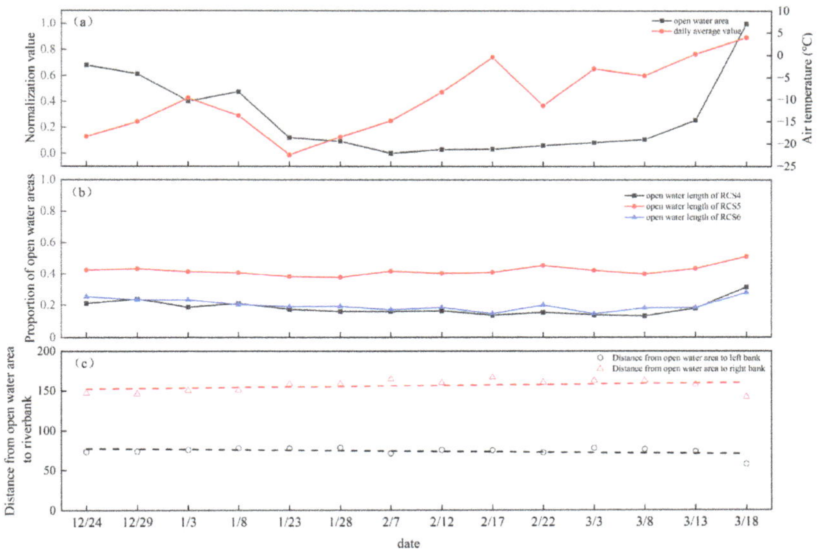

Figure 5. (**a**) The open−water area and temperature changes in the Shishifenzi Bend; (**b**) the length of the open−water area relative to different riverbank sections; (**c**) the distance of the open−water area from the left and right banks of the riverbank section.

3.2. *Ice Crystal Structure Observation Results*

Crystal observations were conducted on ice samples taken from different locations in the Inner Mongolia section of the Yellow River. The observation results were compared, analyzed, and sorted out with reference to the classification methods for different types of ice in references [9–11]. The ice surface and vertical sections were photographed under an orthogonal polarizer. In the resulting images, adjacent grains appear in different colors, with each color region representing a single grain. For boundaries where the grain structure was not clearly defined, visual identification was used to supplement the interpretation, and the crystal boundaries were manually delineated. This approach allowed us to obtain the crystal structures corresponding to different positions. Based on the obtained ice crystal images and grain size and shape, the Yellow River ice of different river types was classified. Through observation, it was found that the main types of ice crystals in the Inner Mongolia section of the Yellow River are columnar ice and granular ice. Snow-ice will exist on the surface of some ice blocks, and special frazil ice crystal structures will appear in some river channels. In order to better distinguish ice samples at different locations, we also collected ice samples from open waters that froze after the river was closed. DXC1# and SSFZ2# were both post-freezing ice samples of the open waters formed. Figure 6 shows the ice samples and crystal structures formed by the freezing of the Dachengxi Railway Bridge and the Shishifenzi Bend after a period of time in the open-water area. Analysis reveals that ice formations in open-water regions predominantly manifest as columnar ice structures, characterized by rectangular crystalline configurations. These crystals exhibit distinct boundary demarcations, with their longitudinal axes extending continuously along the vertical growth direction of ice thickness. Its freezing process is mainly affected by water flow conditions and heat conduction. In the single-track bridge channel at Dachengxi, the surface stratum of the ice formation displays distinctive snow-ice crystalline characteristics. Snow-ice formations exhibit high granular density with diminutive particle dimensions. The crystalline morphology predominantly consists of circular and granular microstructures. These formations typically originate from accumulated snow deposits that undergo multiple cycles of compression, followed by freeze–thaw processes in conjunction with the underlying ice. There are gaps inside, and the freezing process is greatly affected by hydrodynamics and environmental factors. Both the middle and lower layers are obvious columnar ice, the crystal shape is close to a rectangle, the boundaries are obvious, and the crystal growth direction is consistent with the direction of ice thickness growth. The crystal structure of the ice sample at SSFZ2# is mostly columnar ice, and in the middle and lower layers, some crystals are mixed with some ice crystals with larger particle sizes, and its crystal structure is frazil ice. The crystal shapes of frazil ice are extremely irregular, mostly dendritic and serrated. Most ice crystals are small in size, there are wrapping crystals on the edges of the grains, and there are many gaps between the ice crystals. The freezing process is mainly controlled by environmental interference and hydrodynamic conditions.

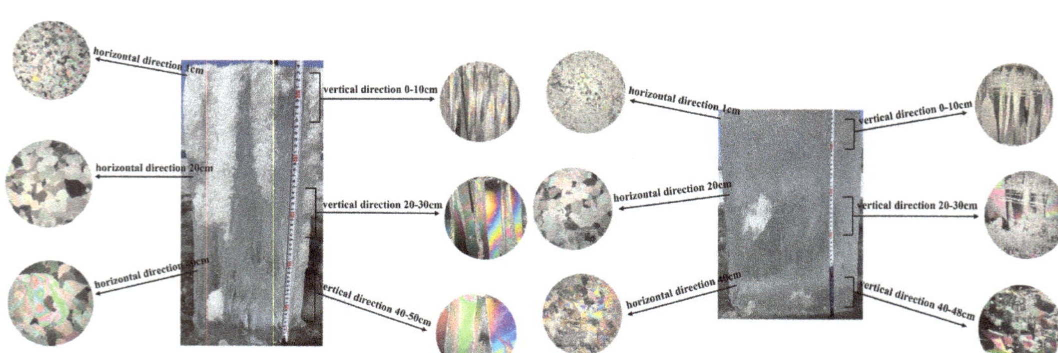

(**a**) (**b**)

Figure 6. Types of ice crystal structures after freezing in open water. (**a**) DXC1#; (**b**) SSFZ2#.

Figure 7 shows the crystal structures of different river channel types along the longitudinal section of the Yellow River in Inner Mongolia. It can be seen from the vertical direction of the longitudinal section that the ice crystal structure of the vertical section of different types of rivers is very different. The ice in the straight river channel of Sanhuhekou and the Dachengxi Railway Bridge is columnar ice grown purely by thermodynamics, with only a small amount of granular ice and snow-ice on the surface. Crystal dimensional analysis reveals a progressive increase in particle size along the vertical profile. This pattern indicates that ice formation in straight channel segments and bridge-adjacent sections experiences minimal perturbation, resulting in simplified crystalline architectures predominantly characterized by distinct columnar formations. In contrast, the Shishifenzi Bend exhibits markedly different crystallographic characteristics. The complex hydrodynamic conditions characteristic of curved channels, coupled with ice accumulation phenomena at bend locations, result in more intricate crystal distribution patterns. Furthermore, significant variations in crystalline structure are observed across different positions within the bend configuration. The ice crystal structure at the head of the bend is mostly obvious granular ice, and the particle size of the crystal is very small. The closer to the open-water area, the closer the ice is to pure thermodynamic natural growth. The bottom of the bend close to the open water is mostly columnar ice. In the crystal structure of SSFZ1#, there is frazil ice frozen in the granular ice in the upper and middle ice layers. This may be due to the complex boundary conditions of the curved river channel. Under the action of the water flow, the ice flowers dived and accumulated, and finally froze in the ice layer, thus changing the original crystal structure. For the ice crystal structure of SXT1#, 0–15 cm is granular ice, 15–23 cm is columnar ice, and below 23 cm is mainly granular ice. There is mud and sand in the ice of Shaxintan and some frazil ice is frozen at the bottom of the columnar ice along with mud and sand. The attachment of frazil ice and mud affects the natural growth of the original ice layer and will have different effects on the composition of the crystal structure. This phenomenon occurs predominantly in meandering channels and alluvial reaches characterized by complex hydrodynamic regimes. The crystal structure of the QFZ2# double-track bridge is columnar ice, and there is frazil ice embedded between the columnar ice layers at a depth of 30–40 cm. The double-track bridge piers significantly influence local hydrodynamic conditions, facilitating rapid consolidation of frazil and drift ice to form interstitial frazil layers, which subsequently affect columnar ice development. The predominance of granular ice formations upstream of the double-track bridge indicates that the pier structures impede ice transport and modify flow dynamics. Ice accumulation phenomena near bridge piers result in ice formations that deviate from pure thermodynamically grown columnar structures, instead exhibiting heterogeneous crystalline compositions comprising both granular and columnar ice. This evidence substantiates channel morphology as a primary determinant of crystallographic structure.

3.3. Ice Crystal Distribution

All of the ice crystals at different river channel locations and different horizontal depths were classified and counted. A total of 45 ice crystal images were obtained, and the size data of the ice crystals were obtained based on the scales arranged during observation. Due to the irregular shape of ice particles, the equivalent particle size is used to describe the size of ice particles. Figure 8 shows the equivalent diameter distribution of columnar ice and granular ice. It can be seen that there is a clear difference in the equivalent diameter distribution of the two crystal types. The equivalent diameters of columnar ice and granular ice both increase first and then decrease. The equivalent diameter of columnar ice is mainly distributed in the range of 6–12 mm, accounting for 40.09% of the total calculated amount.

Among them, 8–10 mm accounts for the largest proportion, accounting for 13.98% of the total amount of columnar ice. The columnar ice is relatively large in size, with more large-sized grains, and the proportion of equivalent grains greater than 20 mm accounts for 10.62%. The number of grains with equivalent diameter of 10–20 mm decreases with the increase of equivalent diameter of grains, and the decrease in grain number is relatively gentle. There are more large-sized ice crystals in columnar ice, and the size distribution is more uniform. The equivalent diameter of granular ice is mainly distributed in the range of 2–8 mm, and the proportion of crystals from 0 to 6 mm gradually increases. The amount of 4–6 mm accounts for the largest proportion, accounting for 19.87% of the total calculated granular ice. The equivalent diameter of columnar ice crystals is unevenly distributed, with small sizes being the main ones. The minimum equivalent particle size of columnar ice in the ice sample is 1.12 mm, and the maximum is 42.3 mm. The equivalent diameter range of granular ice is 0.39–19.86 mm, and the equivalent diameter of granular ice is significantly smaller than that of columnar ice.

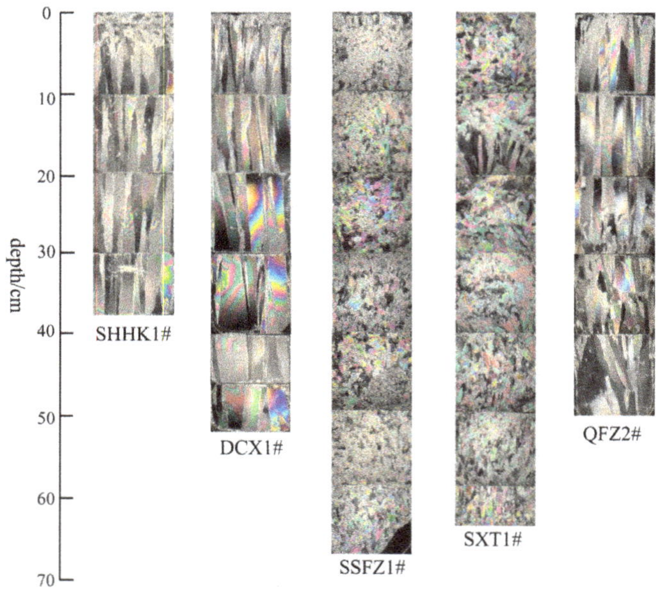

Figure 7. Ice crystal structures at different river locations in the Inner Mongolia section.

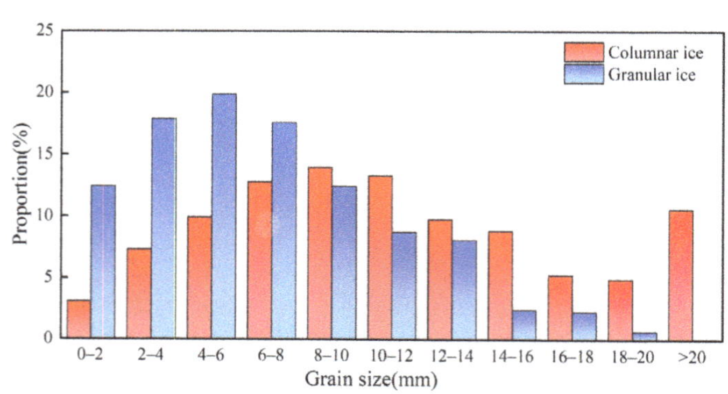

Figure 8. Distribution of ice crystal equivalent diameter.

Two different types of river channels and different types of crystals were selected for ice blocks. The equivalent diameter distribution ratio of each horizontal slice is shown in Figure 9. As can be seen from Figure 9, the equivalent diameters of columnar ice at depths of 10 cm and 20 cm are concentrated in the range of 5–10 mm, and the proportion of 5–10 mm columnar ice at a depth of 10 cm is as high as 50.4%. With increasing depth, the equivalent diameter distribution shifts significantly, predominantly ranging from 10 to 15 mm at the 30 cm depth stratum. At greater depths of 40 cm and 50 cm, crystals with equivalent diameters exceeding 25 mm constitute 19.5% and 13.8% of their respective populations. This shows that the equivalent diameter of columnar ice grown purely by thermodynamics increases with depth, and the proportion of columnar ice grains increases with depth. The proportion of columnar ice grains in the 50 cm layer is relatively uniform, and the difference is not significant. The change trend of the crystal proportion of granular ice at different depths is the same, which is to increase first and then decrease. Unlike columnar ice, the crystal diameter proportion will drop sharply after it is larger than 12 mm, and the proportion of crystals larger than 15 mm is very small, less than 10%. The main grain size of each layer is concentrated between 6 and 12 mm. In the 10 cm layer, the grain size of 6–9 mm accounts for 39.8%; in the 30 cm layer, the grain size of 9–12 mm accounts for 34.1%.

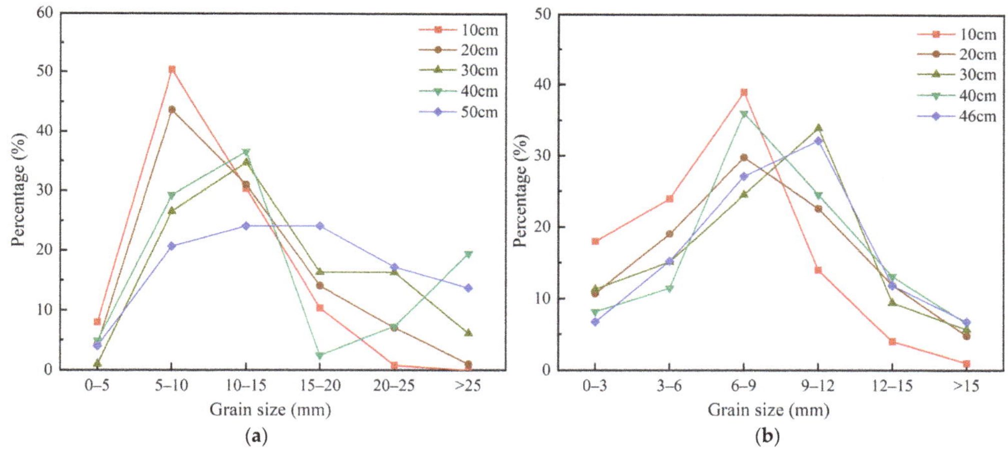

Figure 9. The proportion of different types of ice crystals at different depths. (**a**) Columnar ice; (**b**) granular ice.

The crystal equivalent diameters of ice samples at different river channel locations in the Inner Mongolia section of the Yellow River were statistically analyzed, and the variation in grain size under different crystal structures along the direction of ice thickness growth was obtained, as shown in Figure 10. The equivalent diameters of various ice crystal types exhibit similar trends, progressively increasing with depth. However, variations in channel morphology and local environmental conditions at different sites further influence crystal growth dynamics. The straight river channel of Sanhu River estuary is columnar ice with pure thermodynamic growth, and the crystal size varies from 7.23 mm to 23.12 mm along the direction of ice thickness growth. The Dachengxi Railway Bridge is a single-track bridge river channel, and its crystal structure is columnar ice with a size variation range of 5.62–17.01 mm, which gradually increases along the depth direction. Notably, for columnar ice, the crystal diameters at the Sanhu River estuary are significantly larger than those observed at the Dachengxi Railway Bridge. This indicates that the bridge piers have an impact on the flow rate of the river. Therefore, the formation process of ice crystals under the bridge is affected to a certain extent compared with the unobstructed straight river channel, resulting in relatively smaller crystal size. SSFZ1# is situated at the ice accumulation zone at the head of the bend, where the crystal structure is predominantly granular ice, with crystal sizes ranging from 5.20 mm to 10.22 mm. SSFZ2#, located in the middle of the bend and formed from post-freeze open water, is adjacent to an existing open-water area. This sample exhibits a columnar ice structure, resulting in larger crystal sizes compared to SSFZ1#, with a size range of 6.77 mm to 17.17 mm and considerable variability in crystal dimensions. The bends in the river are complex and have a great influence on the flow and accumulation of ice, so the types of crystals formed at different locations in the bends are also different. The crystal size range of the Shaxintan channel is 7.08–15.06 mm. In the Shaxintan River channel, the ice contains a lot of sediment, and the growth of crystals is also affected by the sediment, causing the crystal area to fluctuate to a certain extent as the ice thickness increases. Owing to the prevalence of bifurcated channel morphologies in the Shaxintan region, the formation processes of ice are also affected by the boundary conditions of the river channels, and the crystal type is mostly granular ice.

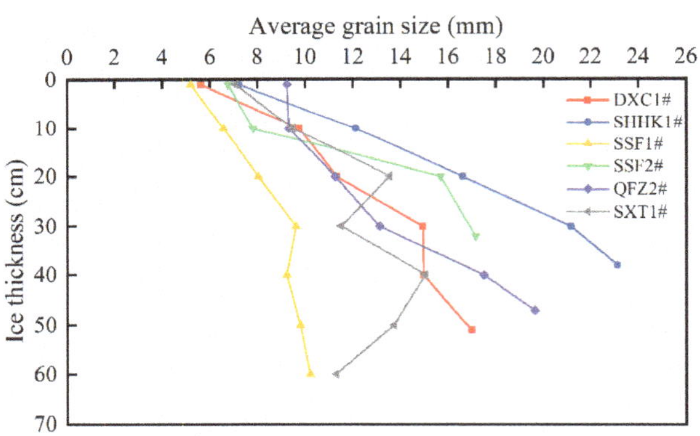

Figure 10. Variation in equivalent diameter of crystal structure along depth in different channel positions of the Yellow River in Inner Mongolia.

3.4. Ice Density and Sediment Concentration

The ice density and sediment concentration of the ice samples were measured and the results were statistically analyzed, as shown in Figure 11. The density of the ice samples varied from 837 to 918 kg/m^3. The ice density across various river channel types in the Inner Mongolia reach of the Yellow River exhibits substantial variability, reflecting the complexity of channel boundary conditions and hydrodynamics. The presence of numerous bubbles, pores, and impurities within the ice significantly influences its density. Analysis of ice density at various locations indicates that columnar ice consistently exhibits higher density than granular ice; however, no clear stratified distribution pattern is observed along the vertical profile of ice thickness. In the Shishifenzi Bend and Shaxintan channel, the density of granular ice displays pronounced fluctuations, likely attributable to channel morphology, which promotes increased sediment incorporation as well as a higher prevalence of pores and bubbles within the granular ice. These factors contribute to marked density stratification and heterogeneous variation. The density variation in columnar ice at the Dachengxi Railway Bridge and Sanhu River estuary is minimal, and its density consistently exceeds that of granular ice. This suggests that columnar ice formed under purely thermodynamic conditions contains fewer pores and impurities, resulting in a more homogeneous density profile. Given the high sediment load of the Yellow River, sediment is frequently incorporated into the ice matrix during freezing, making sediment concentration a key physical parameter in characterizing Yellow River ice. As can be seen from Figure 11b, the distribution of mud content in the ice along the depth direction is uneven, with large variations and lack of certain regularity, indicating that the sediment content brought by water flow at different river channel locations is different. Columnar ice generally grows under the action of thermodynamics. The ice is denser and has fewer pores than granular ice. The amount of sediment that can be directly frozen in the ice is limited. Granular ice, on the other hand, is greatly affected by the scouring of water flow. Sediment generally freezes in granular ice, which has a great impact on the ice density. Most of the sediment is frozen in the middle and lower parts of the ice layer, and the mud content of granular ice is greater than that of columnar ice. While sediment content in river ice varies across different sampling points, comparative analysis reveals that the sediment concentration in the main channel ice is significantly higher than in nearshore ice. Additionally, sediment levels exhibit considerable variability at different depths within the ice layer, accompanied by complex changes in the crystalline structure.

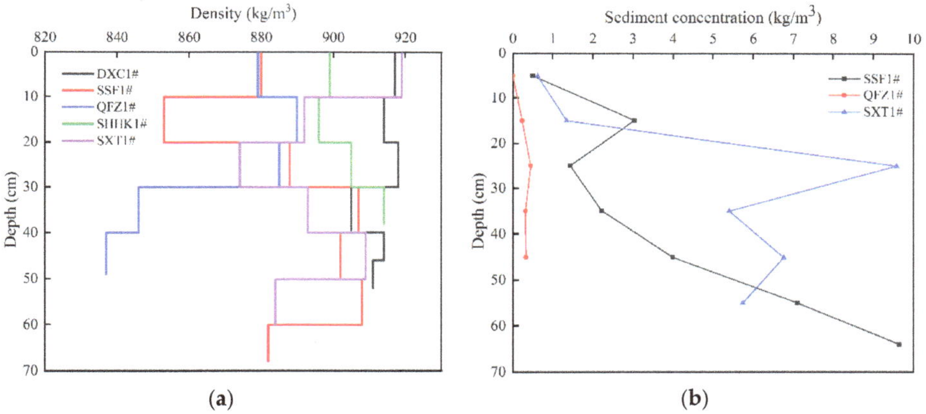

Figure 11. Ice density and sediment concentration in the Inner Mongolia section of the Yellow River (**a**) Density; (**b**) sediment concentration.

4. Discussion

Using the acquired ice sample images, distinct grayscale values were assigned to various regions of the samples. These grayscale values were then correlated with the ice sample's density and sediment content, allowing for comprehensive statistical analysis and integration. The results are shown in Figure 12. Figure 12a shows the relationship between grayscale value and mud content. The grayscale value range is 0–255. After grayscale conversion of different ice sample images, the values are mostly between 100 and 200. As the sediment content increases, the color of the ice sample becomes darker, so the grayscale value of this area will be significantly lower than that of the area with low sediment content, indicating that the mud content has a great influence on the grayscale value. As the sediment content increases, the grayscale value shows a downward trend. Figure 12b shows the relationship between ice density and sediment content. It can be found that the distribution of sediment content and density is relatively discrete, the correlation is not significant, and the fluctuation range of density and mud content is large. However, if the mud content in the ice increases, the density of the ice will still increase accordingly, which shows that the mud content is only one of the factors affecting the change in ice density. The density will also be affected by multiple factors such as bubbles, pores, unfrozen water, and frazil ice attachment in the ice. Based on this, detailed analysis of gray value and sediment concentration can also be used as a future research direction. Due to the numerous and intricate factors that influence ice density across various river sections, conducting quantitative research on these determinants represents a significant avenue for future investigation.

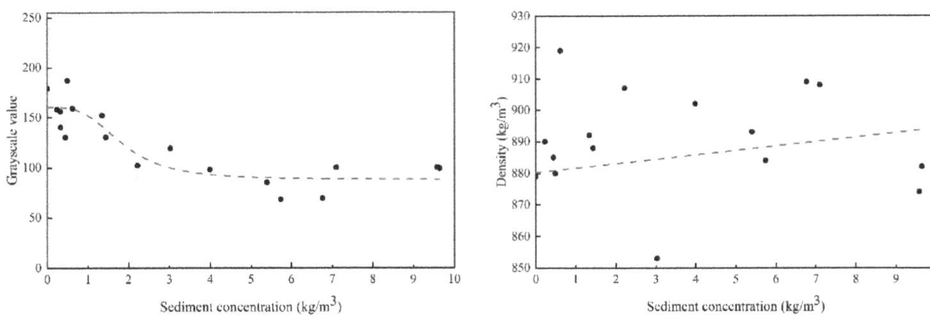

(a) (b)

Figure 12. (**a**) Variation in grayscale value of ice samples at different river positions with sediment concentration; (**b**) relationship between ice sample density and sediment concentration.

Comparative analysis of crystals from various river channel types reveals that the formation of different crystal types is influenced by multiple factors, including hydraulic dynamics, thermodynamic conditions, channel morphology, and anthropogenic activities. The variable channel morphology and intricate hydrodynamic conditions of the Yellow River contribute to the complexity of its ice crystal structures. The types of ice crystal structures vary across different sampling points, and even within the same river section, distinct variations are observed. Additionally, the proportion of granular and columnar ice within the total ice thickness differs by location. The structural types of ice crystals exhibit significant variability across different temporal and spatial scales. Within the same river section, ice crystal structures at various sampling locations also demonstrate distinct characteristics, including irregular shapes and considerable variations in size. Furthermore, the proportions of granular ice and columnar ice relative to the total ice thickness differ among the various sampling sites [41,42]. The development of columnar ice is primarily influenced by thermodynamic factors, resulting in a relatively stable growth process. In contrast to river ice, which exhibits greater complexity, lake ice predominantly consists of columnar ice [31]. Consequently, the intricate nature of river ice crystals constitutes a significant focus of this study and will remain a pertinent area of investigation in future research. In contrast, granular ice growth is affected by a multitude of factors, including crushed ice, drifting ice, impurities, and temperature fluctuations. Generally, the crystal area of granular ice is smaller than that of columnar ice, with most ice samples containing numerous air bubbles, pores, and impurities. Due to the complex river channel shape, during the supercooling process of water, many ice crystals will form in the river channel. The ice crystals can be suspended in the water flow and constantly adsorb and aggregate with other ice crystals to form frazil ice. Under the action of the water flow, the frazil ice and drift ice from the upstream can dive under the ice sheet due to the drag force of the water flow. The diving frazil ice and drift ice will be inserted or adsorbed on the bottom of the downstream ice layer [43,44]. In the observation of ice crystal structure, it was found that many crystals were attached with frazil ice, and there were also many frazil ice particles accumulated. When frazil ice and drift ice freeze in the original ice layer of the river under the action of thermodynamics, it not only increases the thickness of the original ice layer, but also changes the structural characteristics of the original ice layer, resulting in the discontinuity of the crystal structure of the Yellow River ice. There are fewer ice crystals with a single structure, and it is mainly manifested as the alternation of ice crystals with multiple structures, and the emergence of a special frazil ice crystal structure [9]. Different sampling points and different river channel types have different freezing depths and ice thicknesses, which are mainly related to the original thickness of the ice and the effect of the water flow under the ice. Higher flow rates exert a more significant impact on frazil ice formation and enhance the water flow's ice-carrying capacity. When frazil ice particles are small, their friction with the underside of the ice layer is minimal, allowing them to be transported downstream by the current. Conversely, larger frazil ice particles are more likely to adhere to the ice bottom and, as temperatures decrease, become incorporated into the basal ice layer through further freezing [45]. Due to the complex hydrodynamic conditions in the Yellow River, the subglacial flow velocities at different locations in the river are different, resulting in different thicknesses of frazil ice accumulated under the ice at different locations in the river. As a result, it was found that the thickness of frazil ice frozen at the bottom of the ice layer at different locations in the river was different, which ultimately made the crystal structure of the Yellow River ice layer complex. Frazil ice found in the river can be collected utilizing specialized equipment designed for frazil ice harvesting [46]. Furthermore, the particle size of the frazil ice can be assessed through laboratory measurements and comparative analyses of relevant influencing factors [47,48]. Therefore, further study on the special frazil ice composition and crystal structure of different river sections is also a future research direction.

5. Conclusions

Through the analysis and comparisons of the changing process of open water at different river channel locations in the Inner Mongolia section of the Yellow River as well as the formation and microstructure of ice samples, the results indicate that the open waters created by the Dachengxi Railway Bridge and the Shishifenzi Bend exhibit distinct shapes and dynamic changes. The ice in the mainstream areas freezes rapidly, whereas the ice in non-mainstream areas develops more slowly under the same temperature conditions. During the melting phase, the changes observed on the right bank are considerably more pronounced than those on the left bank. The ice formation in the Yellow River is characterized predominantly by two morphologically distinct types: granular and columnar ice structures. The crystalline composition exhibits remarkable heterogeneity across diverse channel configurations. In straight river segments and sections incorporating single-line bridges, columnar ice formations demonstrate clear predominance. Notably, in curved river segments and alternative channel configurations, where complex boundary conditions exert significant influence, the ice manifests primarily as granular formations or presents a sophisticated alternating pattern of columnar and granular ice architectures. In regions where freezing occurs subsequent to open-water formation, the ice crystals predominantly manifest as columnar structures. Observational and statistical analyses of ice crystals at various locations along the Yellow River reveal that granular ice constitutes the majority of the ice formations, whereas columnar ice represents a comparatively smaller proportion. The growth of columnar ice is relatively stable, primarily influenced by thermodynamic factors. Compared to granular ice, columnar ice features fewer grain boundaries, reduced porosity, and less sediment accumulation. Additionally, certain river channels exhibit the presence of frazil ice and snow-derived ice formations. The equivalent diameter of granular ice predominantly ranges from 2 to 8 mm, while columnar ice exhibits equivalent diameters primarily between 6 and 12 mm. With increasing depth, the ice layer demonstrates a progressive enhancement in the proportion of larger crystal formations, characterized by alternating sequences of granular and columnar ice strata, and the crystal dimensions display distinct maxima at various depths. The mean equivalent diameter of granular ice is demonstrably smaller than that of columnar ice, with a notably lower frequency of large-scale crystal formations. In contrast, columnar ice exhibits a higher proportion of large-scale crystals and demonstrates a more uniform size distribution pattern. The vertical distribution of ice density and sediment content in ice samples collected from various sampling sites along the Inner Mongolia reach of the Yellow River exhibits no discernible patterns with depth. The crystal equivalent diameter, ice density, and sediment content demonstrate variability across different crystalline structures, primarily influenced by channel morphology and ice formation environmental conditions. Analysis of ice sample images reveals that distinct regions are characterized by varying grayscale values, demonstrating a significant inverse correlation between grayscale intensity and sediment concentration. Specifically, lower grayscale values correspond to higher sediment content within the ice. Research indicates that columnar ice consistently exhibits higher density values compared to granular ice formations. While sediment content represents one contributing factor to ice density variations, the observed density fluctuations are attributed to a complex interplay of multiple variables.

Author Contributions: Conceptualization, Y.L. and C.L.; methodology, Y.L.; software, Y.L. and C.L.; validation, Y.L., X.F., and P.L.; formal analysis, Y.L.; investigation, Y.L., X.F., and S.H.; resources, C.L. and P.L.; data curation, C.L.; writing—original draft preparation, Y.L.; writing—review and editing, Y.L., C.L., and P.L.; visualization, Y.L. All authors have read and agreed to the published version of the manuscript.

Funding: This research was supported by Joint Funds of the National Natural Science Foundation of China (U23A2012), Natural Science Foundation of Inner Mongolia Autonomous Region of China (2025MS05122), Keju Plan of Inner Mongolia University of Science and Technology (KJJH2024910).

Data Availability Statement: The data presented in this study are available upon request from the first author.

Acknowledgments: The authors are grateful to the members of the research group who helped collect ice samples. The authors thank the editor and anonymous reviewers for their valuable comments and suggestions to this paper.

Conflicts of Interest: The authors declare no conflicts of interest.

References

1. Gao, G.; Deng, Y.; Tian, Z.; Liu, S.; Zhang, B. Brief Introduction and Prospect of Recent Ice Research in the Yellow River. *Yellow River* **2019**, *41*, 77–81+108. (In Chinese)
2. Bhambri, R.; Hewitt, K.; Kawishwar, P.; Kumar, A.; Verma, A.; Tiwari, S.; Misra, A. Ice-dams, outburst floods, and movement heterogeneity of glaciers, Karakoram. *Glob. Planet. Change* **2019**, *180*, 100–116. [CrossRef]
3. Tuo, Y.; Deng, Y.; Li, J.; Li, N.; Li, K.; Wei, L.; Zhao, Z. Effects of dam reconstruction on thermal-ice regime of Fengman Reservoir. *Cold Reg. Sci. Technol.* **2018**, *146*, 223–235. [CrossRef]
4. Zhao, S.; Li, C.; Li, C.; Shi, X.; Zhao, S. Processes of river ice and ice-jam formation in Shensifenzi Bend of the Yellow River. *J. Hydraul. Eng.* **2017**, *48*, 351–358. (In Chinese) [CrossRef]
5. Yang, X.; Pavelsky, T.M.; Allen, G.H. The past and future of global river ice. *Nature* **2020**, *577*, 69–73. [CrossRef]

6. Liu, B.; Ji, H.; Zhai, Y.; Luo, H. Estimation of river ice thickness in the Shisifenzi reach of the Yellow River with remote sensing and air temperature data. *IEEE J. Sel. Top. Appl. Earth Obs. Remote Sens.* **2023**, *16*, 5645–5659. [CrossRef]
7. Wang, T.; Guo, X.; Liu, J.; Chen, Y.; She, Y.; Pan, J. Ice Process Simulation on Hydraulic Characteristics in the Yellow River. *J. Hydraul. Eng.* **2024**, *150*, 05024001. [CrossRef]
8. Huang, W.; Li, Z.; Han, H.; Niu, F.; Lin, Z.; Leppäranta, M. Structural analysis of thermokarst lake ice in Beiluhe Basin, Qinghai–Tibet Plateau. *Cold Reg. Sci. Technol.* **2012**, *72*, 33–42. [CrossRef]
9. Zhang, Y.; Li, Z.; Li, C.; Zhang, B.; Deng, Y. Microstructure characteristics of river ice in Inner Mongolia section of the Yellow River and its influencing factors. *J. Hydraul. Eng.* **2021**, *52*, 1418–1429. [CrossRef]
10. Michel, B.; Ramseier, R.O. Classification of river and lake ice. *Can. Geotech. J.* **1971**, *8*, 36–45. [CrossRef]
11. Eicken, H.; Lange, M.A. Development and properties of sea ice in the coastal regime of the southeastern Weddell Sea. *J. Geophys. Res. Ocean.* **1989**, *94*, 8193–8206. [CrossRef]
12. Langway, C.C. Ice fabrics and the universal stage. In *Technical Report 62*; Department of Defense, Department of the Army, Corps of Engineers, Snow Ice and Permafrost Research Establishment: Wilmette, IL, USA, 1959.
13. Wilen, L.A.; Diprinzio, C.L.; Alley, R.B.; Azuma, N. Development, principles, and applications of automated ice fabric analyzers. *Microsc. Res. Tech.* **2023**, *62*, 2–18. [CrossRef] [PubMed]
14. Arnaud, L.; Gay, M.; Barnola, J.M.; Duval, P. Imaging of firn and bubbly ice in coaxial reflected light: A new technique for the characterization of these porous media. *J. Glaciol.* **1998**, *44*, 326–332. [CrossRef]
15. Azuma, N.; Wang, Y.; Yoshida, Y.; Narita, H.; Hondoh, T.; Shoji, H.; Watanabe, O. Crystallographic analysis of the Dome Fuji ice core. In *Physics of Ice Core Records*; Hokkaido University: Sapporo, Japan, 2000; pp. 45–61. Available online: https://eprints.lib.hokudai.ac.jp/dspace/bitstream/2115/32461/1/P45-61.pdf (accessed on 23 June 2025).
16. Kipfstuhl, S.; Hamann, I.; Lambrecht, A.; Freitag, J.; Faria, S.H.; Grigoriev, D.; Azuma, N. Microstructure mapping: A new method for imaging deformation-induced microstructural features of ice on the grain scale. *J. Glaciol.* **2006**, *52*, 398–406. [CrossRef]
17. Li, Z.; Huang, W.; Jia, Q.; Leppäranta, M. Distributions of crystals and gas bubbles in reservoir ice during growth period. *Water Sci. Eng.* **2011**, *4*, 204–211.
18. Kaempfer, T.U.; Schneebeli, M.; Sokratov, S.A. A microstructural approach to model heat transfer in snow. *Geophys. Res. Lett.* **2005**, *32*, 1. [CrossRef]
19. Golden, K.M.; Eicken, H.; Heaton, A.L.; Miner, J.; Pringle, D.J.; Zhu, J. Thermal evolution of permeability and microstructure in sea ice. *Geophys. Res. Lett.* **2007**, *34*, L16501. [CrossRef]
20. Mercier, O.R.; Hunter, M.W.; Callaghan, P.T. Brine diffusion in first-year sea ice measured by Earth's field PGSE-NMR. *Cold Reg. Sci. Technol.* **2005**, *42*, 96–105. [CrossRef]
21. Cho, H.; Shepson, P.B.; Barrie, L.A.; Cowin, J.P.; Zaveri, R. NMR investigation of the quasi-brine layer in ice/brine mixtures. *J. Phys. Chem. B* **2002**, *106*, 11226–11232. [CrossRef]
22. Schulson, E.M. The brittle compressive fracture of ice. *Acta Metall. Mater.* **1990**, *38*, 1963–1976. [CrossRef]
23. Peng, W.; Zhu, Y.; Zhang, J. Effects of grain size on the unixal compressive strength of polycrystalline ice. *Mech. Eng.* **1998**, *4*, 37–39. (In Chinese)
24. Gao, Z.; Deng, Y.; Zhang, P.; Wang, J. Evaluation of fracture behavior of Yellow River ice based on three-point bending test and PSO-BP model. *Theor. Appl. Fract. Mech.* **2022**, *122*, 103644. [CrossRef]
25. Deng, Y.; Wang, J.; Li, Z. Numerical simulation of uniaxial compression failure process of river ice. *J. Hydraul. Eng.* **2018**, *49*, 1339–1345. (In Chinese) [CrossRef]
26. Liu, H.; Ji, H.; Mou, X.; Gao, G.; Zhang, B. Experimental study on Yellow River ice thickness monitoring testbased on UAV Radar technology. *South-to-North Water Transf. Water Sci. Technol.* **2020**, *18*, 217–224. (In Chinese)
27. Li, Z.; Li, C.; Yang, Y.; Zhang, B.; Deng, Y.; Li, G. Physical scheme and parametrization by using air temperature and ice thickness to improve accuracy of GPR propagation velocity in Yellow River ice. *J. Hydraul. Eng.* **2022**, *53*, 902–913. (In Chinese) [CrossRef]
28. Li, Z.; Jia, Q.; Zhang, B.; Leppäranta, M.; Lu, P.; Huang, W. Influences of gas bubble and ice density on ice thickness measurement by GPR. *Appl. Geophys.* **2010**, *7*, 105–113. [CrossRef]
29. Li, C.; Li, Z.; Huang, W.; Zhang, B.; Deng, Y.; Li, G. Morphology dynamics of ice cover in a river bend revealed by the UAV-GPR and Sentinel-2. *Remote Sens.* **2023**, *15*, 3180. [CrossRef]
30. Cao, X.; Li, C.; Yan, X.; Wu, Y.; Li, Z. Measuring ice thickness around the curve and piers in the Yellow River with ground penetrating radar. *South-to-North Water Transf. Water Sci. Technol.* **2016**, *14*, 91–95. (In Chinese)
31. Huang, W.; Li, Z.; Han, H.; Jia, Q. Seasonal evolution of static freshwater lake ice microstructures and the effects of growth processes. *J. Glaciol. Geocryol.* **2016**, *38*, 699–707. (In Chinese)
32. Huang, W.; Li, Z.; Niu, F.; Lin, Z. Structural features of the rmokarst lake ice in Beiluhe Basin of Qinghai-Tibet Plateau. *J. Hydraul. Eng.* **2011**, *42*, 1277–1282. (In Chinese) [CrossRef]

33. Cole, D.M. The microstructure of ice and its influence on mechanical properties. *Eng. Fract. Mech.* **2001**, *68*, 1797–1822. [CrossRef]
34. Lieb-Lappen, R.M.; Golden, E.J.; Obbard, R.W. Metrics for interpreting the microstructure of sea ice using X-ray micro-computed tomography. *Cold Reg. Sci. Technol.* **2017**, *138*, 24–35. [CrossRef]
35. Li, Z.; Jia, Q.; Huang, W.; Li, C. Characteristics of ice crystals air bubbles and densities of fresh ice in a reservoir. *J. Hydraul. Eng.* **2009**, *40*, 1333–1338.
36. Huang, W.; Lei, R.; Ilkka, M.; Li, Q.; Wang, Y.; Li, Z. The physical structures of snow and sea ice in the Arctic section of 150°-180° W during the summer of 2010. *Acta Oceanol. Sin.* **2013**, *32*, 57–67. [CrossRef]
37. Li, Y.; Kipfstuhl, S.; Huang, M. Ice microstructure and fabric of Guliya ice cap in Tibetan plateau, and comparisons with vostok3G-1, EPICA DML, and north GRIP. *Crystals* **2017**, *7*, 97. [CrossRef]
38. Shi, L.; Li, Z.; Niu, F.; Huang, W.; Lu, P.; Feng, E.; Han, H. Thermal diffusivity of the rmokarst lake ice in the Beiluhe basin of the Qinghai–Tibetan Plateau. *Ann. Glaciol.* **2014**, *55*, 153–158. [CrossRef]
39. Faria, S.H.; Weikusat, I.; Azuma, N. The microstructure of polar ice. Part II: State of the art. *J. Struct. Geol.* **2014**, *61*, 21–49. [CrossRef]
40. Li, Z.; Kang, J. Crystals and Fabrics of an Arctic Multi-Year Ice Sample. *J. Glaciol. Geocryol.* **2001**, *23*, 383–388.
41. Zhang, Y.; Gao, G.; Deng, Y.; Li, Z.; Li, G.; Guo, W. Investigation on Ice Crystal, Density and Sediment Content in Ice at Different Positions in Bayannaoer Section of the Yellow River. *Yellow River* **2018**, *40*, 44–48. (In Chinese)
42. Deng, Y.; Li, Z.; Wang, J.; Xu, L. The microstructure of Yellow River ice in the freezing period. *Crystals* **2019**, *9*, 484. [CrossRef]
43. Beltaos, S. Progress in the study and management of river ice jams. *Cold Reg. Sci. Technol.* **2008**, *51*, 2–19. [CrossRef]
44. Shen, H.T. Mathematical modeling of river ice processes. *Cold Reg. Sci. Technol.* **2010**, *62*, 3–13. [CrossRef]
45. Luo, H.; Ji, H.; Gao, G.; Zhang, B.; Mou, X. Study on the characteristics of flow and ice jam in Shisifenzi bend in the Yellow River during the freeze-up period. *J. Hydraul. Eng.* **2020**, *51*, 1089–1100. [CrossRef]
46. Zhang, Y.; Li, Z.; Xiu, Y.; Li, C.; Zhang, B.; Deng, Y. Microstructural Characteristics of Frazil Particles and the Physical Properties of Frazil Ice in the Yellow River, China. *Crystals* **2021**, *11*, 617. [CrossRef]
47. McFarlane, V.; Loewen, M.; Hicks, F. Measurements of the evolution of frazil ice particle size distributions. *Cold Reg. Sci. Technol.* **2015**, *120*, 45–55. [CrossRef]
48. Morse, B.; Richard, M. A field study of suspended frazil ice particles. *Cold Reg. Sci. Technol.* **2009**, *55*, 86–102. [CrossRef]

water

MDPI

Article

Designing the Engineering Parameters of the Sea Ice Based on a Refined Grid in the Southern Bohai Sea

Ge Li [1,2], Song Gao [1,2], Xue Chen [1,2], Yan Jiao [1,2,*], Linfeng Wang [3], Qiaokun Hou [1,2], Donglin Guo [1,2], Yiding Zhao [1,2], Chengqing Ruan [1,2] and Qingkai Wang [4,*]

1 North China Sea Marine Forecasting Center of State Oceanic Administration, Qingdao 266000, China; lige@ncs.mnr.gov.cn (G.L.); gaosong@ncs.mnr.gov.cn (S.G.); chenxue@ncs.mnr.gov.cn (X.C.); qd_hdk@163.com (Q.H.); gdlcici@163.com (D.G.); dingdingdingdin@126.com (Y.Z.); ruanchengqing@126.com (C.R.)

2 Shandong Key Laboratory of Marine Ecological Environment and Disaster Prevention and Mitigation, Qingdao 266000, China

3 Shengli Oilfield Company, China Petroleum & Chemical Corporation, Dongying 257001, China; wanglinfeng.slyt@sinopec.com

4 State Key Laboratory of Coastal and Offshore Engineering, Dalian University of Technology, Dalian 116024, China

* Correspondence: jiaoyan87@163.com (Y.J.); wangqingkai@dlut.edu.cn (Q.W.)

Abstract: The current standard for sea ice engineering in the Bohai Sea implements a 1/4° grid method, which cannot satisfy the safety of oil and gas activities in the southern Bohai Sea, and therefore more detailed information on ice conditions and a more refined ice zone division are necessary. In the present study, up to 1/12° resolution sea ice characteristic data (period, thickness, concentration, and strength) were obtained based on the NEMO-LIM2 ice–ocean coupling model. On this basis, the design sea ice strength parameters were derived with different return periods from 1 to 100 years. Among the total of 53 grids, the mean ice periods in the southern Bohai Sea from 1951 to 2022 were 2–35 days, the mean ice concentration values were 8.3–64.6%, and the mean ice thicknesses were 2–15 cm. The design uniaxial compressive strengths and shear strengths at almost all grids exceeded 2.00 MPa and 1.00 MPa for return periods over 20 years, respectively. The design flexural strengths for the 100-year return period ranged from 463 to 594 kPa. For the 100-year return period scenario, all grids exhibited design tensile strengths exceeding 200 kPa. Across the southern Bohai Sea, the most severe ice conditions occur in nearshore zones, and the ice conditions display a distinct spatial gradient with Bohai Bay > offshore deep-water areas > Laizhou Bay. The mean ice thickness, concentration, design flexural and tensile strengths derived in this study were lower compared to the ice parameters suggested in the current standard, and design uniaxial compressive and shear strengths derived here were comparable to those suggested in the current standard. The refined grid used here captures more detailed spatial variations in the design strength values of sea ice engineering parameters in the southern Bohai Sea, providing more accurate data support for the anti-ice design of marine structures.

Keywords: sea ice; refined grid division; ice engineering parameters; return period design; southern Bohai Sea

1. Introduction

The Bohai Sea serves as a critical industrial zone in China, supporting dense concentrations of infrastructure for maritime transportation, petroleum operations, and power generation. Located at relatively high latitudes, the region experiences frequent winter incursions of cold air masses, triggering annual ice formation. Recent intensification of extreme climate events, combined with rapid coastal development of ports and oilfields, has significantly altered the sea ice spatiotemporal distribution.

For offshore structures operating in ice-covered waters, understanding the mechanical properties of sea ice is essential for evaluating ice loads and ensuring structural safety [1]. Ice–structure interactions depend strongly on the ice strength and failure behavior, which vary with temperature, salinity, and internal structure of ice [2]. Early research by Li et al. provided a foundational overview of Bohai Sea ice mechanical properties, summarizing long-term observational data to define key design parameters such as strength values and recurrence intervals [3]. Since then, a range of mechanical properties, including uniaxial compressive, flexural, tensile, and shear strength, have been systematically investigated through both laboratory and in situ experiments. Ji et al. conducted flexural and shear experiments across Bohai coastal sites, revealing rate-dependent and brine volume-related strength degradation [4,5]. Uniaxial compressive strength was comprehensively investigated by Wang et al., who conducted over 400 tests across 12 Bohai coastal sites, establishing empirical relationships with temperature, brine volume, and stress rate [6]. In response to climate-induced changes in sea ice formation, particularly increased air content within sea ice, porosity has recently emerged as a critical internal factor in strength modeling [7]. Li et al. proposed a three-dimensional framework linking uniaxial compressive strength with sea ice porosity and strain rate [8]. Chen et al. used Brazilian splitting tests with digital image correlation to assess tensile strength in Bohai Sea ice, finding a clear negative correlation of sea ice tensile strength with porosity [9]. However, these experimental studies were conducted at relatively isolated sites. Although they provided valuable insights into the intrinsic mechanical behavior of sea ice, they have not been systematically linked to large-scale spatial patterns or regional ocean–atmosphere conditions in the Bohai Sea.

With continuous advancements in remote sensing monitoring and numerical modeling, research on the distribution and seasonal evolution of sea ice in the Bohai Sea has become increasingly refined. Sun et al. integrated multi-source datasets, including AVHRR, MODIS, and GOCI, to systematically analyze sea ice area and thickness across major sub-regions in the Bohai Sea over the past 30 years [10]. Their results highlighted the dominant role of Liaodong Bay in regional ice formation and demonstrated strong correlations between ice conditions and cumulative freezing/melting degree days. Yan et al. reported a slight upward trend (~1.38% yr^{-1}) in Bohai Sea ice area from 1988 to 2015, contrary to the general decline observed in polar regions [11]. Ouyang et al. found strong negative correlations between mean ice thickness and mean air temperature, and between median ice extent and negative accumulated temperature in the Bohai Sea [12]. Yan et al. used a multivariate regression approach to extend sea ice area reconstruction to 1958–2015 [13]. Based on long-term sea ice severity data, Zheng et al. revealed a regime shift from heavy to light ice conditions of the Bohai Sea [14]. Most existing studies concentrate on the variations in ice thickness and area, and there remains a lack of distribution of sea ice strength data for engineering applications.

Marine structures in ice-covered waters require information on key ice parameters (e.g., thickness, period, strength) in planning, design, and construction. Current Bohai Sea ice engineering applications derive design ice thickness and strength primarily from the guideline of "Sea Ice Conditions and Application Regulations for China Seas" issued by the

China National Offshore Oil Corporation (CNOOC) in 2002 [15]. Over the past two decades, significant alterations in the spatiotemporal distribution of sea ice in the Bohai Sea have probably emerged, driven by extensive coastal development with consequent shoreline modifications and amplified climate change effects. With numerous platforms approaching service-life limits, new constructions and replacements demand urgent attention. This necessitates precise regional assessment of engineering ice parameters across oil and gas development zones, establishing partition-specific design criteria for ice loads. Such refinement ensures winter operational safety and optimizing structural costs in Bohai ice management strategies.

To support the design of offshore infrastructure in ice-covered waters of the Bohai Sea with high spatial resolution, previous work partitioned the region into a refined $1/12^{\circ}$ grid and derived key sea ice parameters, including ice period, thickness, concentration, temperature, and salinity [16]. However, the grid-level sea ice strength data remains unavailable. Indeed, the lack of gridded sea ice strength data extends beyond the Bohai Sea to the Arctic Ocean, obscuring the spatiotemporal evolution of Arctic sea ice strength and complicating engineering design for Arctic offshore structures. Arctic sea ice strength exhibits significant seasonal variability. Sea ice strength is typically obtained by field and experiment measurements. Based on numerous mechanical tests, scholars have proposed influencing factors and empirical relationships for sea ice strength [17–19]. In recent years, researchers have attempted to investigate the spatiotemporal distribution of sea ice flexural strength across the Arctic Ocean. Chai et al. utilized freezing degree-days to derive ice thickness and thermohaline properties, subsequently applying empirical formulae to assess the statistical characteristics of level-ice flexural strength at four critical locations along the Northeast Passage during winter [20]. Tarovik et al. integrated meteorological, hydrological, and sea ice thermodynamic data to estimate brine volume fraction and flexural strength, evaluating spatiotemporal distributions across nine Arctic sea areas along the Northeast Passage route [21]. These studies of Arctic ice gridded strength provide valuable scientific references for developing analogous gridded sea ice strength models in the Bohai Sea.

Building on previous work [16], the present study addresses the impacts of climate change on engineering in the Bohai Sea ice zone and provides gridded sea ice mechanical parameters for the region. Focusing specifically on the southern Bohai Sea, where critical infrastructure like the Shengli Oilfield is located, the key engineering parameters of sea ice are derived including flexural strength, uniaxial compressive strength, shear strength, and tensile strength. Crucially, actionable design benchmarks for ice mechanical properties are also established here, advancing the research toward direct engineering implementation.

2. Materials and Methods

2.1. Study Area

The Bohai Sea ($37^{\circ}70'$–$41^{\circ}00'$ N, $117^{\circ}35'$–$121^{\circ}10'$ E), situated in the mid-latitude monsoon zone of the Northern Hemisphere, represents a semi-enclosed inland sea extending northwestward from the Yellow Sea. Ice formation in the Bohai Sea initiates in late November to early December, progressing southward from shallow coastal zones to deeper offshore areas. The ablation phase commences in mid-to-late February with northward retreat from deep southern basins to shallow northern coasts, culminating in complete ice clearance by mid-March, yielding a 3–4-month ice season.

The southern Bohai Sea (Bohai Bay and Laizhou Bay) hosts numerous offshore oil and gas platforms, notably within the Shengli Oilfield operational waters. To ensure the safety of the platforms during ice season, CNOOC guidelines partition the southern Bohai Sea into sub-grids with assigned design ice parameters [15]. However, the standardized grid system, originally developed for pan-Bohai Sea ice conditions, proves inadequate for the Shengli Oilfield waters due to its insufficient resolution for local operational demands. Our study implements a multi-criteria grid refinement scheme (Figure 1) tailored to the critical area. Deep-water zones (grids 1–6, 47, 51) maintain quarter-degree ($1/4° \times 1/4°$) grids aligned with the current guideline. Nearshore shallow waters (grids 7–14, 43–50, 52–53) adopt $1/12°$ resolution, with adjacent grids merged based on spatial homogeneity in hydrological and ice regime characteristics. High-priority operation areas implement non-merged $1/12°$ grids to preserve critical spatial details (grids 15–42). The relevant results of ice condition characteristics in the southern Bohai Sea will be presented based on this new ice zone division hereafter.

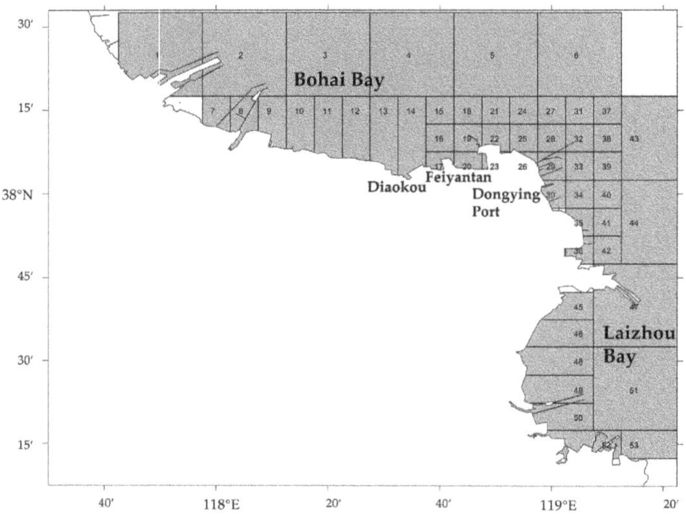

Figure 1. Grid division in the southern Bohai Sea. The pink part represents land. Grids 1–6, 47, 51 are deep-water zones, grids 7–14, 43–50, 52–53 are nearshore shallow waters, grids 15–42 are high-priority operation areas.

2.2. Sea Ice Concentration, Thickness, and Duration

Sea ice concentration, level ice thickness, and sea ice duration were derived based on the NEMO-LIM2 ice–ocean coupling model. The model was generated here to provide detailed information on the ice characteristics in the Bohai Sea from 1951 to 2022 with a refined ice zone division [16]. The model domain is located in a spatial range of 35–41° N, 117–127° E, covering the entire Bohai Sea with a spatial resolution of $1/12°$. The topographic data in the NEMO model uses the ETOPO1 public data set, with a spatial resolution of 1 km, which is interpolated into the NEMO model grid through bilinear interpolation. The ice–ocean coupling model can finely simulate the ocean and sea ice environments near complex islands and shorelines.

It should be stated that compared to LIM2 model, LIM3 model employs a five-category ice thickness scheme, which aims to improve performance in high-latitude ice-covered regions by better representing ice thickness heterogeneity [22]. However, the Bohai Sea represents the lowest-latitude ice-covered region in the Northern Hemisphere. All sea ice here is first-year ice, and ice thickness rarely exceeds 0.5 m. Consequently, the complex ice thickness categorization scheme of LIM3 cannot be fully utilized in the Bohai Sea.

2.3. Sea Ice Strength

Currently, there are no theoretical formulae for calculating sea ice strength; existing formulae are empirical. These empirical equations are in simple forms but exhibit direct physical meaning, clearly indicating that sea ice strength is directly related to specific ice parameters. To guide oil and gas production operations in ice-covered regions of the Bohai Sea, the CNOOC guideline provides recommended methodologies for calculating sea ice load parameters in the Bohai Sea [15]. The sea ice uniaxial compressive strength, flexural strength, shear strength, and tensile strength were calculated here based on the regulation.

The sea ice uniaxial compressive strength was determined using Equation (1).

$$\sigma_c = 1.77 - 0.13T_i \tag{1}$$

where T_i is sea ice temperature (°C), and σ_c is maximum uniaxial compressive strength of level ice loaded horizontally to ice surface (MPa).

The sea ice shear strength was determined using Equation (2).

$$\tau = \frac{\sigma_c}{2} \tag{2}$$

where τ is sea ice shear strength (MPa).

The sea ice flexural strength was determined using Equation (3).

$$\sigma_f = 340 - 64T_i \tag{3}$$

where σ_f is sea ice flexural strength (kPa).

The sea ice tensile strength was determined using Equation (4).

$$\sigma_t = 95 + 50|T_i| \tag{4}$$

where σ_t is sea ice tensile strength (kPa).

As shown in Equations (1)–(4), sea ice temperature is a key parameter to determine sea ice strength. In the CNOOC regulation [15], according to the characteristics of sea ice in the Bohai Sea, the ice temperature was determined using Equation (5).

$$T_i = (T_{ia} + T_{iw})/2 - 273.15 \tag{5}$$

where T_{ia} and T_{iw} are sea ice temperatures (K) at the top and bottom surfaces of the ice cover determined by Equations (6) and (7).

$$T_{ia} = \frac{\lambda(-1.4 - T_a)}{4.7} + T_a \tag{6}$$

$$T_{iw} = -1.4 - \frac{\lambda(-1.4 - T_a)}{125} \tag{7}$$

where λ is thermal conductivity coefficient of sea ice ($kJm^{-2} \cdot h^{-1} \cdot K^{-1}$) determined using Equation (8), T_a is air temperature (K). T_a was adopted from Climate Forecast System Reanalysis data issued from the National Centers for Environmental Prediction (NCEP).

$$\lambda = \left(0.221 + \frac{H}{2}\right)^{-1} \tag{8}$$

where H is ice thickness (m).

2.4. Sensitivity Analysis

The primary inputs of NEMO-LIM2 model include shortwave/longwave radiation, 2 m air temperature, 2 m humidity, 10 m wind fields, and precipitation. Comprehensive sensitivity tests have been performed in the previous work of [23]. Through observational comparisons, sea ice thermodynamic and dynamic parameters were calibrated to ensure model stability and applicability in the Bohai Sea. Sea ice strength values were calculated using standardized formulae recommended by established specifications. As these represent fundamental principles rather than adjustable parameters, sensitivity analysis would not yield meaningful insights.

2.5. Model Verification

The details on the configuration of the model were reported in the previous work [16]. Comparisons between observed and simulated sea ice thickness and duration showed high accuracy of the model. The correlation coefficient between the hindcasting sea ice thickness and observations was above 0.60, with an absolute error of less than 4.0 cm. The hindcasting data could better reflect the distribution of ice periods in mild and severe ice years. Sea ice strength values were not directly validated, as they were derived from ice thickness and air temperature. Therefore, we focused our verification efforts on ice thickness, which plays a key role in the estimation of sea ice strength. Sea ice concentration validation was also omitted due to methodological differences between conventional coastal station observations and model-derived estimates. It was found that there is a good consistency between ice thickness and concentration, so the validation effect of ice thickness can also represent the accuracy of other sea ice parameters to a certain extent [16].

2.6. Return Period Design

The sea ice strength was designed using the Pearson-III (P-III) distribution with different return periods (100, 50, 25, 20, 10, 5, 2, 1 years). When determining the design ice engineering parameters, an excessively small estimate may lead to damage to marine structures, while an excessively large estimate can result in increased costs. The biggest advantage of the P-III distribution is its large elasticity. In most cases, it can be fitted well with the theoretical curve and empirical frequency points by repeatedly fitting the line or adjusting the coefficient of variation and mean appropriately. The probability density function of P-III distribution is Equation (9).

$$f(x) = \frac{\beta^{\alpha}}{\Gamma(\alpha)}(x - x_0)^{\alpha-1}\exp[-\beta(x - x_0)] \tag{9}$$

where $\Gamma(\cdot)$ is the Gamma function, and α, β as well as x_0 are the shape, scale, and position coefficients, respectively, related to the statistical parameters of the random variables:

$$\alpha = \frac{4}{C_s^2} \tag{10}$$

$$\beta = \frac{2}{\overline{x}C_v C_s} \tag{11}$$

$$x_0 = \overline{x}(1 - \frac{2C_v}{C_s}) \tag{12}$$

where C_s is skewness coefficient, C_v is variation coefficient, and $\overline{x}$ is the average.

Therefore, the determination of the probability density function of P-III distribution is transformed into the determination of the statistical parameters. In general, the observations of ice parameters are not long enough, the empirical distribution based on observations must be extended to determine the design ice parameters. The preliminary values of the statistical parameters were obtained first using the observation data, and an empirical distribution curve was then depicted. Afterwards, the statistical parameters were adjusted until the corresponding empirical distribution curve fitted well with the observations. The statistical parameters after adjustment were selected to determine the final P-III distribution, and the design ice parameters were determined from corresponding cumulative frequency density curve according to the return period.

3. Results

3.1. Ice Duration

Accurate assessment of ice duration is important for engineering design and operational safety in ice-covered waters. The ice duration is defined as the temporal span between initial ice formation and final ice clearance (measured in days). Figure 2 presents the spatial distributions of average ice duration across study regions. The ice duration exhibited a decreasing gradient across the study area, with the longest persistence at the coastal Bohai Bay, intermediate duration at the coastal Laizhou Bay, and the shortest in central deep basins. Table 1 provides the maximum, minimum, and average ice duration of each grid division. The Diaokou nearshore area demonstrated prolonged severe ice periods compared to adjacent waters, attributable to shallow bathymetry facilitating ice grounding and obstruction effects from the eastern No.168 artificial island limiting ice drift. The western sectors showed longer ice durations than eastern counterparts. The maximum ice period days and the average ice period days near Dongying Port were short.

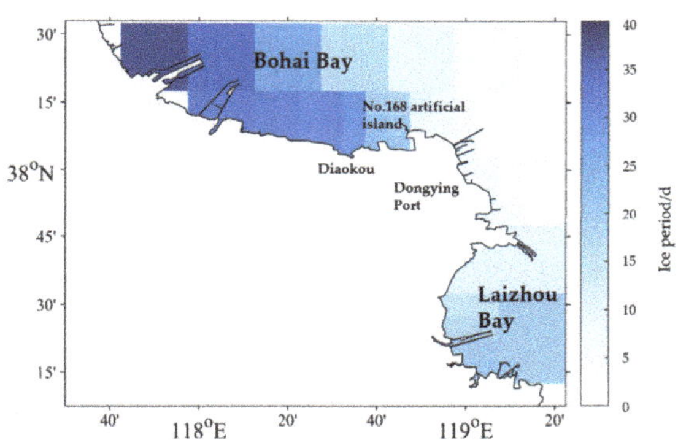

Figure 2. Average ice period in the southern Bohai Sea from 1951 to 2022.

Table 1. The maximum, minimum, and average ice periods from 1951 to 2022 in each grid division.

Grid No.	Maximum/d	Minimum/d	Average/d	Grid No.	Maximum/d	Minimum/d	Average/d
1	80	0	35	28	64	0	5
2	80	0	33	29	64	0	5
3	74	0	25	30	64	0	5
4	71	0	15	31	61	0	3
5	69	0	9	32	61	0	3
6	63	0	4	33	63	0	4
7	81	0	33	34	59	0	6
8	82	0	32	35	59	0	6

Table 1. *Cont.*

Grid No.	Maximum/d	Minimum/d	Average/d	Grid No.	Maximum/d	Minimum/d	Average/d
9	82	0	31	36	59	0	7
10	81	0	31	37	52	0	2
11	81	0	30	38	52	0	3
12	79	0	30	39	62	0	3
13	80	0	31	40	63	0	4
14	78	0	29	41	59	0	4
15	73	0	22	42	59	0	6
16	73	0	22	43	51	0	2
17	73	0	22	44	47	0	3
18	73	0	19	45	59	0	10
19	73	0	19	46	58	0	10
20	73	0	19	47	60	0	10
21	73	0	22	48	62	0	14
22	73	0	22	49	67	0	16
23	73	0	22	50	67	0	18
24	66	0	7	51	67	0	18
25	65	0	6	52	66	0	19
26	60	0	5	53	68	0	18
27	64	0	5				

3.2. Level Ice Thickness

Precise quantification of ice thickness is critical for structural integrity assessments, as it directly governs ice load on offshore infrastructure. Figure 3 presents the spatial distributions of maximum ice thickness and mean ice thickness in the southern Bohai Sea from 1951 to 2022. The maximum ice thickness across the study area was 13–36 cm, and the mean ice thickness was 2–15 cm. The southwestern Bohai Bay and western Laizhou Bay exhibited significantly greater maximum ice thickness. Diaokou coastal waters also showed a high maximum ice thickness. The southern Bohai Bay consistently maintained a mean ice thickness > 10 cm, exceeding the mean ice thickness in the western Laizhou Bay.

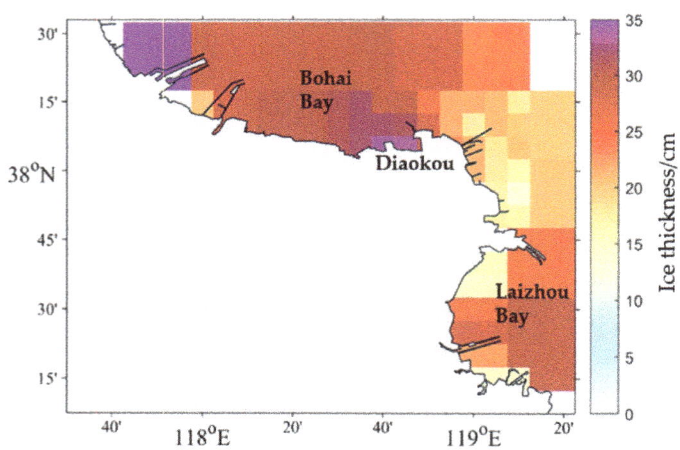

(**a**)

Figure 3. *Cont.*

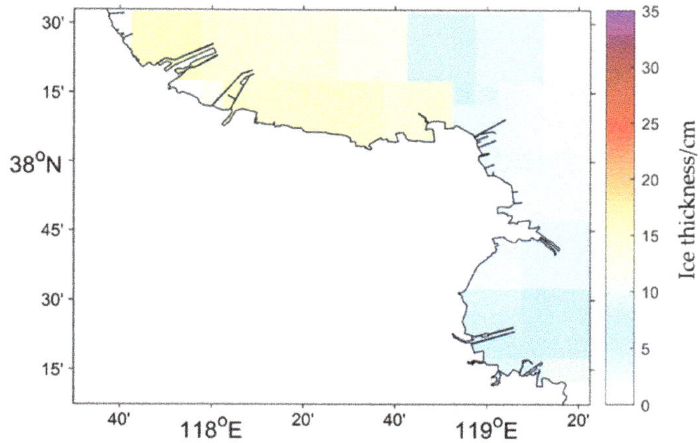

(b)

Figure 3. (**a**) Maximum and (**b**) average ice thicknesses from 1951 to 2022 in each grid division.

The refined grid division in this study corresponds to grids No. 9 and No. 10 in the original regulation. Under the original zoning, both grids No. 9 and No. 10 had a maximum ice thickness of 35 cm, with average thicknesses of 18 cm and 15 cm, respectively. Figure 4 shows the changes in ice thickness between the new zoning and original regulations, revealing that both maximum and average ice thicknesses decreased under the refined zoning, with relatively more significant reductions along the western coastal areas of Laizhou Bay.

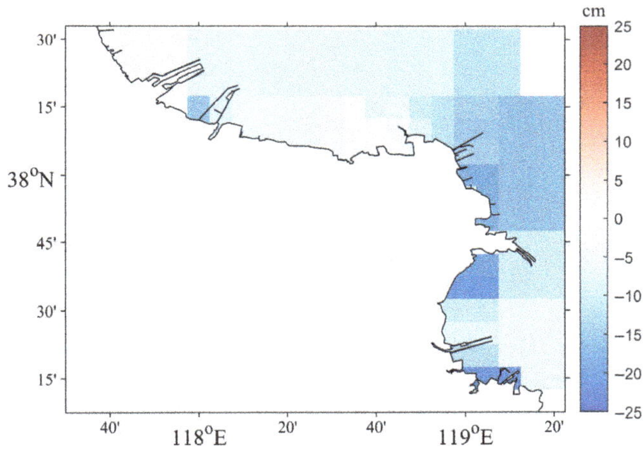

(a)

Figure 4. *Cont.*

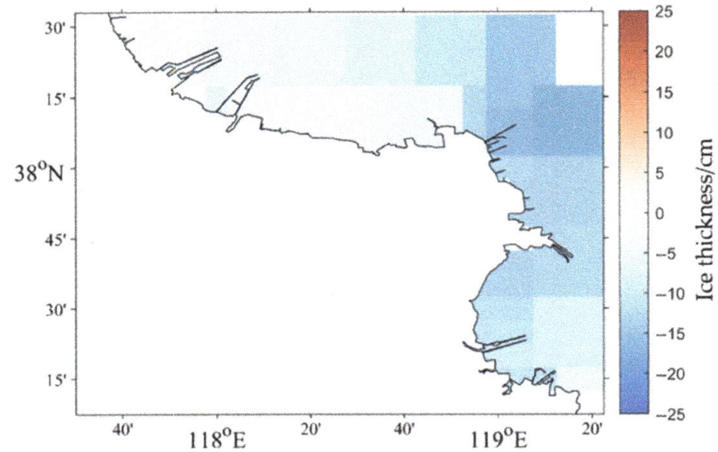

(b)

Figure 4. The differences of (**a**) maximum and (**b**) average ice thicknesses at the refined grids compared to those at the grids in the CNOOC standard.

3.3. Sea Ice Concentration

As a primary driver of ice dynamics, accurate spatial mapping of sea ice concentration is critical for quantifying ice–structure interaction on platforms, wind turbines, and subsea pipelines. Figure 5 presents the spatial distributions of maximum and mean sea ice concentration in the southern Bohai Sea from 1951 to 2022. Bohai Bay exhibited higher sea ice concentrations, followed by Laizhou Bay. In the southern Bohai Bay and southwestern Laizhou Bay, the maximum sea ice concentration exceeded 90% across most areas, with sea ice concentrations reaching up to 99.8% in the Feiyantan offshore area. Regarding average sea ice concentration, the Bohai Bay region generally showed higher values than the Laizhou Bay region.

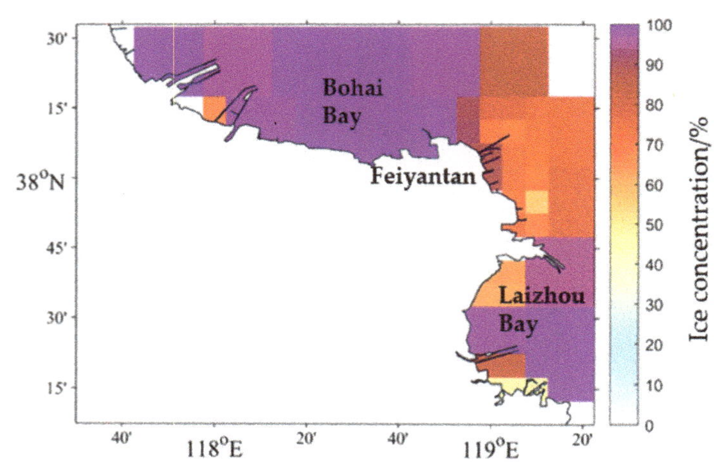

(a)

Figure 5. *Cont.*

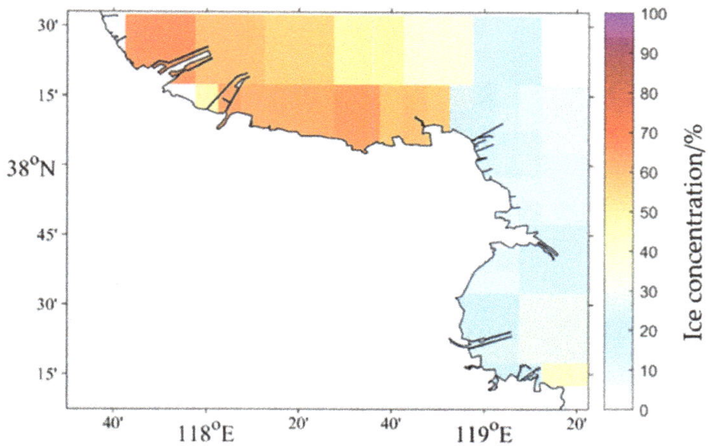

(b)

Figure 5. (a) Maximum and (b) average sea ice concentrations from 1951 to 2022 in each grid division.

According to the original regulation, the maximum sea ice concentrations for grids No. 9 and No. 10 are 94.4% and 85.7%, respectively, with average concentrations of 63.4% and 43.8%. As shown in Figure 6, in this study, the maximum sea ice concentration in the southern Bohai Bay showed values essentially consistent with the original standard; the maximum sea ice concentration in the eastern sections of the study area were slightly below the original standards; and the maximum sea ice concentration in the southwestern Laizhou Bay exceeded the standard values. For average sea ice concentration, all zones in the southern Bohai Sea demonstrated lower values than the original standard, and the most significant reduction occurred near the Dongying Port area.

The decrease in sea ice concentration near the Dongying Port area compared to the values in the original standard is primarily caused by two reasons. First, the CNOOC standard was implemented and published in 2002. Over the past 15 years, sea ice severity in the China Seas has exhibited an overall declining trend. This regional trend is a key factor contributing to the generally lower average sea ice concentrations found in our results for the southern Bohai Sea compared to the baseline values established in the original standard. Second, the CNOOC standard utilized shore-based observation data from the Diaokou location. Diaokou is situated within a shallow river channel. Sea ice concentration within such confined, shallow riverine environments is typically higher than in the adjacent open sea. Consequently, the sea ice concentration values adopted in the original standard for the Dongying Port area, based on Diaokou data, were inherently biased upwards due to this localized effect. In contrast, the data used in our analysis was derived entirely from NEMO numerical model results. Therefore, our results for Dongying Port reflected the open-water sea ice conditions more accurately, thus avoiding the overestimation presented in the original standard data source for this specific area.

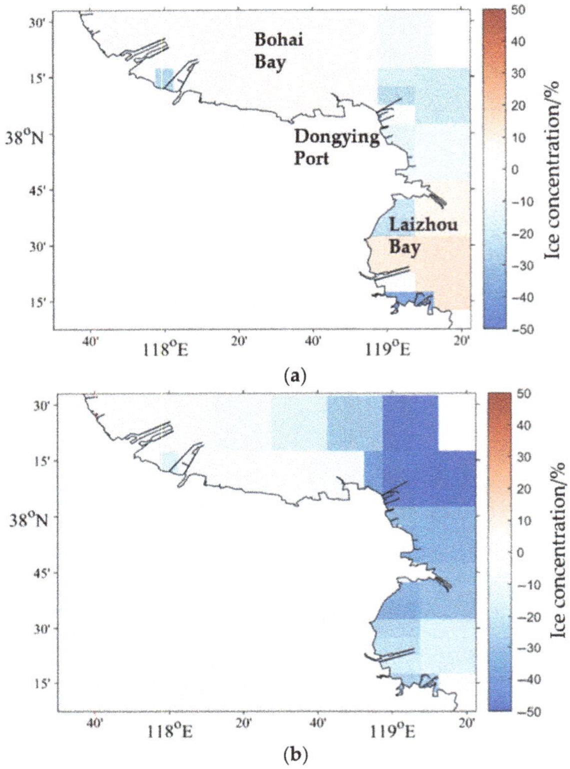

Figure 6. The differences of (**a**) maximum and (**b**) average sea ice concentrations at the refined grids compared to those at the grids in the CNOOC standard.

3.4. Sea Ice Uniaxial Compressive Strength

Sea ice uniaxial compressive strength governs the ice loads against vertical structures in ice-covered waters. Figure 7 presents the spatial distributions of design sea ice maximum uniaxial compressive strength in the southern Bohai Sea. The design values of maximum sea ice uniaxial compressive strength exhibited a distinct spatial gradient, decreasing sequentially from the Bohai Bay (highest values) through deep-water area (medium values) to the Laizhou Bay (lowest values), with the Bohai Bay bottom and Dongying Port areas showing the most severe strength values. Table A1 in Appendix A specifies the design strength values for different return periods. The design strength values at almost all grids exceeded 2.00 MPa for return periods over 20 years. The design strength values in the Bohai Bay and Dongying Port areas maintained >2.00 MPa for 10-year and 5-year return periods while the design strength values at other regions fell below 2.00 MPa. For 2-year and 1-year return periods, the design strength values at almost all grids were <2.00 MPa.

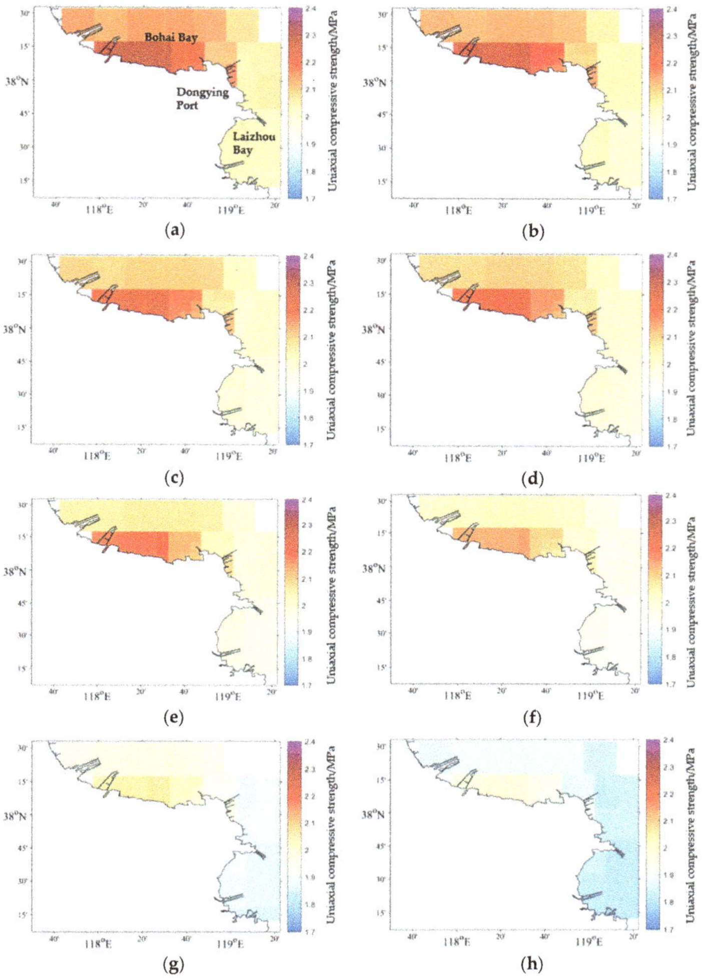

Figure 7. The design sea ice uniaxial compressive strength with different return periods of (**a**) 100, (**b**) 50, (**c**) 25, (**d**) 20, (**e**) 10, (**f**) 5, (**g**) 2, and (**h**) 1 years.

3.5. Sea Ice Shear Strength

Sea ice shear strength is another important parameter determining the ice load during ice–structure interaction. Figure 8 presents the spatial distributions of design sea ice shear strength in the southern Bohai Sea. Since the shear strength was determined based on the uniaxial compressive strength, the spatial variation pattern of design sea ice shear strength in the southern Bohai Sea was consistent with the distribution characteristics of the design uniaxial compressive strength. Table A2 in the Appendix A specifies the design shear strength values for different return periods. The design shear strength values at almost all grids exceeded 1.00 MPa for return periods over 20 years. The design strength values in the Bohai Bay and Dongying Port areas maintained >1.00 MPa for 10-year and 5-year return periods while the design strength values at other regions fell below 1.00 MPa. For design values with 2-year and 1-year return periods, the strength values at almost all grids were <1.00 MPa.

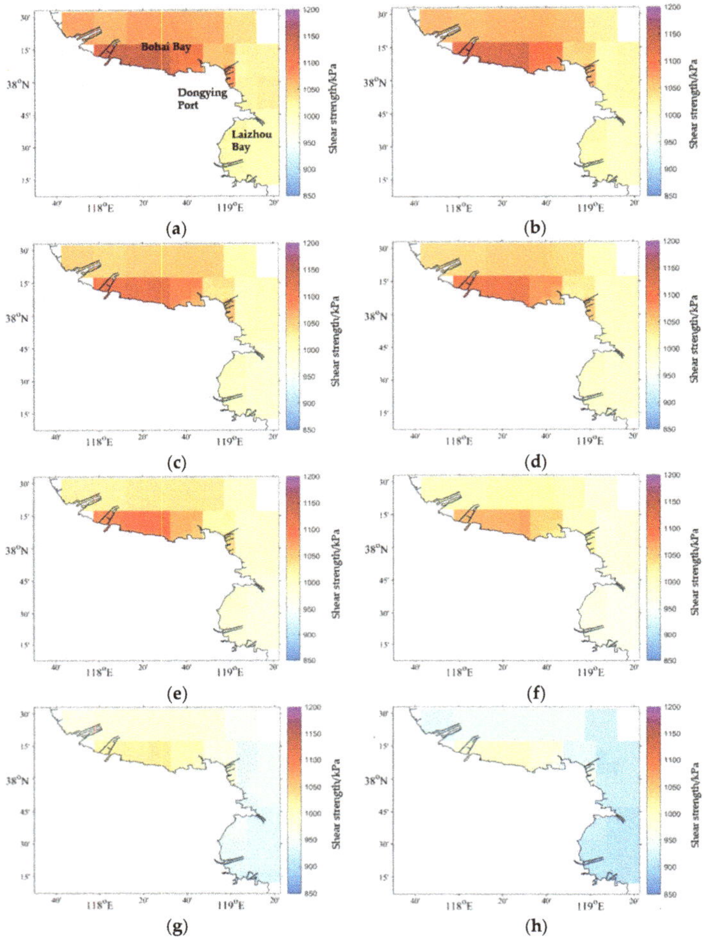

Figure 8. The design sea ice shear strength with different return periods of (**a**) 100, (**b**) 50, (**c**) 25, (**d**) 20, (**e**) 10, (**f**) 5, (**g**) 2, and (**h**) 1 years.

3.6. Sea Ice Flexural Strength

As the critical parameter controlling ice failure on sloped structures, quantification of sea ice flexural strength is essential for the design of ice-breaking cones. Figure 9 presents the spatial distributions of design sea ice flexural strength in the southern Bohai Sea. The design values of sea ice flexural strength exhibited a distinct spatial gradient, decreasing sequentially from the Bohai Bay (highest values) through deep-water area (medium values) to the Laizhou Bay (lowest values), with the Bohai Bay bottom and Dongying Port areas showing the most severe strength values. Table A3 in the Appendix A specifies the design flexural strength values for different return periods. The design flexural strength for the 100-year return period ranged from 463 to 594 kPa. The design values decreased with shorter return periods. For return periods of 5 years and above, the minimum and maximum design flexural strengths remained greater than 400 kPa and 500 kPa, respectively. At the 2-year return period, the design flexural strength decreased to 386 kPa and 480 kPa.

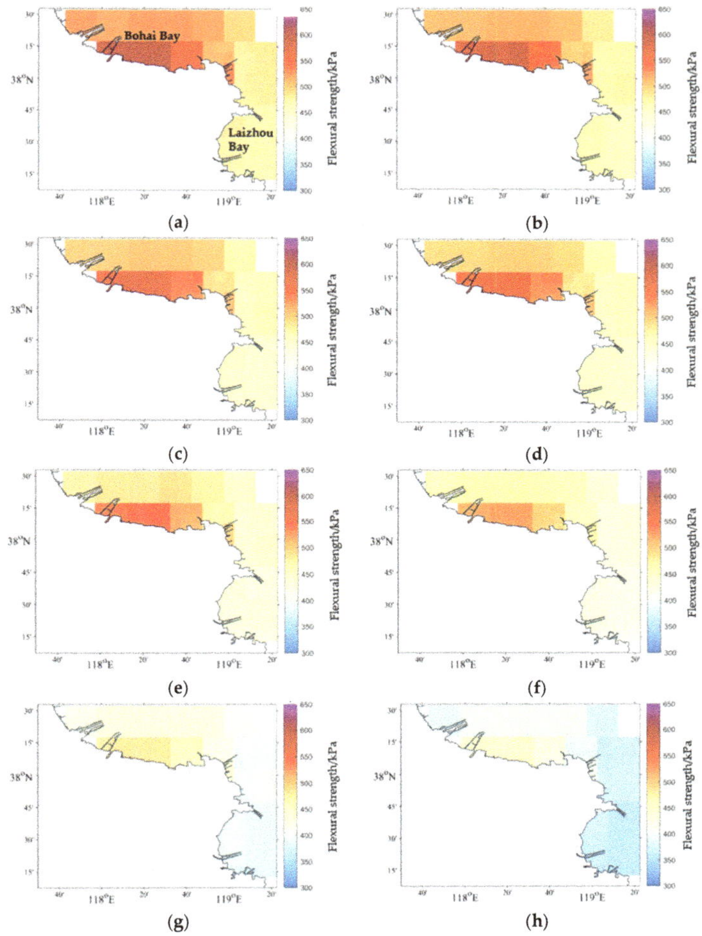

Figure 9. The design sea ice flexural strength with different return periods of (**a**) 100, (**b**) 50, (**c**) 25, (**d**) 20, (**e**) 10, (**f**) 5, (**g**) 2, and (**h**) 1 years.

3.7. Sea Ice Tensile Strength

Despite its low magnitude compared to uniaxial compressive strength, sea ice tensile strength influences the crack formation of large ice cover. Figure 10 presents the spatial distributions of sea ice tensile strength in the southern Bohai Sea. The design values of sea ice tensile strength exhibited a distinct spatial gradient, decreasing sequentially from the Bohai Bay (highest values) through deep-water area (medium values) to the Laizhou Bay (lowest values), with the Bohai Bay bottom and Dongying Port areas showing the most severe strength values. Table A4 in the Appendix A specifies the design tensile strength values for different return periods. For the 100-year return period scenario, all grids exhibited design tensile strengths exceeding 200 kPa. For return periods between 50 and 20 years, the Laizhou Bay area showed design tensile strengths below 200 kPa, and other locations maintained values above 200 kPa. Below the 20-year return period, only several severe grids demonstrated tensile strengths surpassing 200 kPa. Under 1-year return period conditions, all grids displayed design tensile strengths less than 200 kPa.

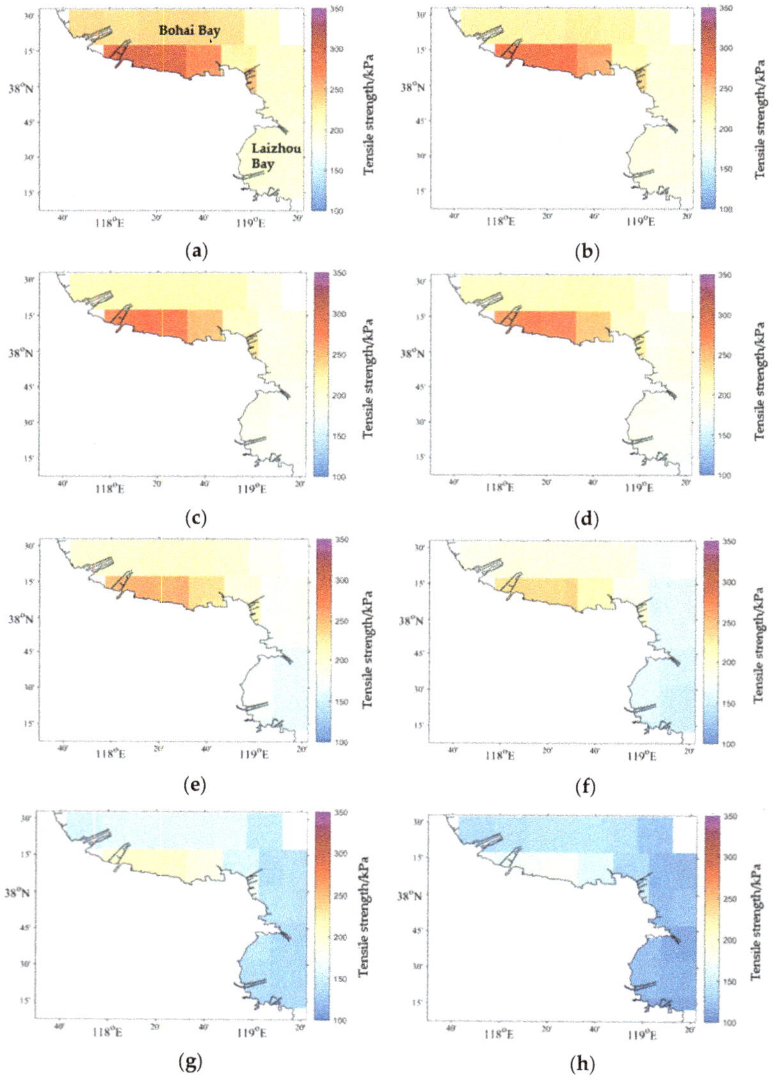

Figure 10. The design sea ice tensile strength with different return periods of (**a**) 100, (**b**) 50, (**c**) 25, (**d**) 20, (**e**) 10, (**f**) 5, (**g**) 2, and (**h**) 1 years.

4. Discussion

4.1. The Selection of Sea Ice Strength Formulae

Numerous uniaxial compressive strength tests have been conducted on Bohai Sea ice. However, these tests have predominantly focused on the Liaodong Bay, where ice conditions are relatively severe, whereas data availability for the Laizhou Bay has been limited. Both of the Chinese ice-region engineering design codes [15,24] recommend that, in the absence of measured data, the peak value of sea ice uniaxial compressive strength be calculated using Equation (1).

The National Marine Environmental Monitoring Center (NMEMC) historically conducted uniaxial compressive strength tests on sea ice in the Yellow River Estuary area near the Laizhou Bay [25]. These tests served as design references for offshore oil production structures of the Shengli Oilfield, providing the relationship between the design value of uniaxial compressive strength of level ice in the Yellow River Estuary and ice temperature, as expressed by Equation (13).

$$\sigma_c = 1.84 - 0.161T_i \tag{13}$$

Recently, during the 2020/2021 winter, ice blocks were collected at different sites in the southern Bohai Sea (Feiyantan, No. 168 artificial island, and Zhuangxi 106). Ice samples were then machined to carry out uniaxial compressive strength tests of sea ice at different ice temperatures in the laboratory. Test results showed that as the temperature of the ice samples decreased, the uniaxial compressive strength of the sea ice increased. Previous studies have typically employed a linear relationship to fit the variation in the peak uniaxial compressive strength of ice with temperature. Although the linear fitting yielded good results, it fails to account for the fact that the ice strength approaches zero as the ice temperature nears the freezing point. Therefore, a nonlinear relationship was adopted for fitting, as shown in Equations (14)–(16).

For sea ice in the Zhuangxi 106 site,

$$\sigma_c = 1.11 \ln(|T_i|) + 0.45 \tag{14}$$

For sea ice in the No. 168 artificial island,

$$\sigma_c = 0.7 \ln(|T_i|) + 0.52 \tag{15}$$

For sea ice in the Feiyantan site,

$$\sigma_c = 0.61 \ln(|T_i|) + 1.21 \tag{16}$$

The relationship curves between the peak compressive strength of the sea ice and ice temperature, as derived from the aforementioned formulae, were compared in Figure 11. It can be observed that the formula for the sea ice in the Yellow River Estuary area (Equation (13)) and the code-specified formula (Equation (1)) all yield higher values than the recent laboratory-derived relationships between peak strength and ice temperature (Equations (14)–(16)). The discrepancy can be attributed to the impact of climate warming on ice conditions in China. Climate warming has made extremely low air temperatures less frequent, resulting in higher air temperatures during sea ice growth periods. The warmer environment increases the brine and gas content within the ice, leading to reduced strength. Compared to Equation (13), Equation (1) shows closer agreement with the field measurement data obtained in the 2020/2021 winter. The ice sampling sites for deriving Equation (13) were closer to the Yellow River Estuary, where sea ice has lower salinity and consequently higher strength. Therefore, Equation (1) was selected as the calculation formula for uniaxial compressive strength in this paper.

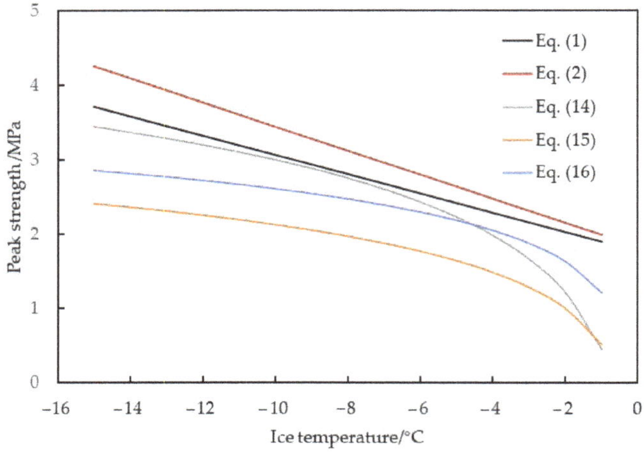

Figure 11. Comparisons between the varying trends of peak compressive strength of sea ice with ice temperature derived from different formulae.

4.2. Comparions with Original CNOOC Guideline

The CNOOC Guideline was issued in 2002, reflecting sea ice conditions under prevailing climatic-hydrological regimes of the previous era. Comparing the sea ice parameters derived here with the values proposed in the CNOOC guideline provides climate-driven modifications in Bohai Sea ice conditions to some degree. As shown in Figures 4b and 5b, the southern Bohai Sea exhibits marked declines in both average ice thickness and concentration.

Compared with the sea ice uniaxial compressive strength in the original standards, our refined grid division shows generally comparable design sea ice strength values across the southern Bohai Sea. The most notable differences occurred for 100-year, 50-year and 1-year return periods, where the original design values are slightly higher, but only by 0.05 MPa. For sea ice shear strength, there is no obvious difference either. The maximum difference was 43 kPa occurring under the 10-year return period design condition, while the minimum difference was 11 kPa which was observed for the 1-year return period scenario.

The flexural strength and tensile strength of sea ice show significant differences. Compared with the design sea ice flexural strength in the corresponding grids of the original standards, our refined grid yielded lower values. The original guideline provided design flexural strengths of 642.5–23.5 kPa for return periods varying from 100 to 1 years. The maximum differences reached 132 kPa for the 100-year return period, and the minimum difference was 85 kPa for the 5-year return period.

Compared with the design sea ice tensile strength in the corresponding grids of the original standards, our refined grid yielded lower values. The original guideline provided design flexural strengths of 331.5–237.5 kPa for return periods varying from 100 to 1 years. The maximum differences reached 103 kPa for the 100-year return period, and the minimum difference was 66 kPa for the 5-year return period.

Moreover, the original guideline characterizes southern Bohai Sea ice conditions using only two grid cells, which is not enough to resolve critical differences between nearshore and offshore ice regimes. The refined grid here successfully captures more detailed spatial variations in the design strength values of sea ice engineering parameters throughout the study waters. The most severe ice conditions occur in nearshore zones, exhibiting maximum values in thickness, concentration, and design ice strength. Across the southern

Bohai Sea, the ice conditions display a distinct spatial gradient with Bohai Bay > offshore deep-water areas > Laizhou Bay.

4.3. Potential Impacts of Climate Change on Bohai Sea Ice Engineering

The Bohai Sea ice management authority in China classifies ice severity into five levels (Level 1–5) based on floating ice extent, where Level 1 represents the mildest conditions and Level 5 the most severe. With global warming, according to the China Marine Disaster Bulletin, from 1951 to 2023, the ice severity in the Bohai Bay has decreased from Level 4 towards Level 2 in general, and Laizhou Bay has declined from Level 3 towards Level 2 typically.

This changing ice regime is further evidenced by comparing our results with the original CNOOC guideline. Compared to the previous standards, our analysis yields lower values for key sea ice parameters, including ice thickness, concentration, and flexural and tensile strength. These changed ice parameters may carry significant practical implications for engineering in the Bohai Sea. The reduced ice thickness and ice strength will reduce ice loads on structure. The reduced ice loads translate directly into lighter and more economical structures, lowering material requirements and construction costs. It is also necessary to update design sea ice parameters to reflect the climate-driven reduction in ice severity, avoiding unnecessary over-engineering while maintaining safety margins appropriate for the actual, diminished ice hazard.

4.4. Limitations of the Study

There are several limitations in this study. First, the high-resolution sea ice strength model was specifically developed for the southern Bohai Sea and may be less applicable to other ice-covered regions due to unique local characteristics. Bohai Sea is the lowest-latitude freezing sea in the Northern Hemisphere, exhibiting significant interannual ice variability with maximum annual ice thickness not exceeding 0.5 m. Its shallow bathymetry, high water temperatures, substantial river discharge, and high brine/gas porosity in sea ice have led to the development of region-specific parameterization methods for ice strength (the approach adopted in this study). While regionally constrained, the core framework of nesting mechanical parameterization within climatic/oceanographic models provides transferable methodology for gridded ice strength assessment in other regions, including the Arctic.

Second, gridded sea ice strength values were calculated using standardized formulae recommended by established specifications, and were not validated by field measurements. Validating gridded strength data presents critical technical challenges. Sea ice strength measurements rely on in situ tests and exhibit diurnal variations due to changing ice temperatures. Achieving spatiotemporal resolution matching between field measurements and model outputs in the future may be an effective method.

5. Conclusions

Based on our previous high-resolution study of sea ice parameters in the Bohai Sea [16], this study extends the focus to the southern Bohai Sea where intensive oil and gas operations are located and sea ice strength properties are critical for engineering design.

As the most intensive offshore oil-gas development area in China, the southern Bohai Sea undergoes annual sea ice encroachment, creating unique operational challenges. Current design standards characterize this region with only two coarse-resolution grids for ice thickness, concentration, and strength parameters. Our study employed high-resolution meshing (minimum 1/12° grids) to partition the southern Bohai Sea into 51 zones. Utilizing the NEMO-LIM2 sea ice–ocean coupled model, the grid-specific outputs for ice thickness, concentration, and period were obtained, and through return-period analysis, design values of sea ice strength parameters were further derived for each grid. The analysis reveals pronounced spatial variations in ice conditions across the southern Bohai Sea. Nearshore zones exhibit the most severe ice conditions, demonstrating maximum values in thickness, concentration, and design ice strength. A clear spatial gradient of ice conditions is displayed in the southern Bohai Sea with Bohai Bay > offshore deep-water areas > Laizhou Bay.

Compared to the current standards of CNOOC, the mean ice thickness and concentration derived here were reduced. For design sea ice strength values, the flexural and tensile strength derived here were lower than those in the current standards. The reductions in sea ice thickness, concentration, and flexural strength compared to current standards yield significant engineering benefits. First, the lower ice loads enable substantial cost savings through optimized structural designs. Second, the mitigated ice–structure interactions directly reduce operational risks. Third, these findings support design enhancements such as revising ice load zoning maps for Bohai Sea offshore blocks.

Author Contributions: Conceptualization, G.L.; methodology, S.G.; software, X.C.; validation, Q.H. and D.G.; formal analysis, Q.H., D.G. and Y.Z.; investigation, G.L.; resources, L.W.; data curation, G.L.; writing—original draft preparation, G.L., Y.J. and Q.W.; writing—review and editing, Y.J. and C.R. All authors have read and agreed to the published version of the manuscript.

Funding: This research was funded by the National Natural Science Foundation of China (42206222 and 42276242) and the Open Fund of State Key Laboratory of Coastal and Offshore Engineering (SL2507).

Data Availability Statement: The original contributions presented in this study are included in the article. Further inquiries can be directed to the corresponding authors.

Acknowledgments: The authors thank the editor and anonymous reviewers for their valuable comments and suggestions to this paper.

Conflicts of Interest: Author Linfeng Wang was employed by the company China Petroleum & Chemical Corporation. The remaining authors declare that the research was conducted in the absence of any commercial or financial relationships that could be construed as a potential con-flict of interest.

Appendix A

The design strength values in the southern Bohai Sea for different return periods are shown in Tables A1–A4.

Table A1. The design uniaxial compressive strength (in MPa) of sea ice for different return periods.

Grid No.	Return Period/Year							
	100	50	25	20	10	5	2	1
1	2.14	2.12	2.09	2.08	2.05	2.01	1.93	1.88
2	2.13	2.11	2.08	2.07	2.05	2.01	1.95	1.89
3	2.14	2.12	2.09	2.08	2.05	2.02	1.95	1.89
4	2.14	2.12	2.09	2.08	2.06	2.02	1.95	1.89
5	2.14	2.11	2.09	2.08	2.05	2.02	1.95	1.89
6	2.06	2.04	2.02	2.02	1.99	1.96	1.90	1.85
7	2.27	2.24	2.21	2.20	2.16	2.12	2.04	1.96
8	2.28	2.25	2.22	2.21	2.17	2.13	2.05	1.97

Table A1. *Cont.*

Grid No.	Return Period/Year							
	100	50	25	20	10	5	2	1
9	2.28	2.25	2.22	2.21	2.18	2.13	2.05	1.97
10	2.28	2.26	2.23	2.21	2.18	2.13	2.05	1.98
11	2.28	2.25	2.22	2.21	2.18	2.13	2.05	1.98
12	2.28	2.26	2.23	2.22	2.18	2.13	2.05	1.98
13	2.29	2.26	2.23	2.22	2.18	2.14	2.06	1.98
14	2.22	2.19	2.16	2.15	2.12	2.08	2.01	1.94
15	2.22	2.19	2.17	2.16	2.12	2.09	2.01	1.95
16	2.21	2.19	2.16	2.15	2.12	2.08	2.01	1.94
17	2.21	2.19	2.16	2.15	2.12	2.08	2.01	1.94
18	2.22	2.19	2.17	2.16	2.13	2.09	2.01	1.95
19	2.22	2.19	2.17	2.16	2.13	2.09	2.01	1.95
20	2.10	2.08	2.05	2.05	2.02	1.99	1.92	1.87
21	2.10	2.08	2.06	2.05	2.02	1.99	1.93	1.87
22	2.10	2.08	2.05	2.05	2.02	1.99	1.92	1.87
23	2.10	2.08	2.06	2.05	2.02	1.99	1.93	1.87
24	2.10	2.08	2.06	2.05	2.03	1.99	1.93	1.88
25	2.10	2.08	2.06	2.05	2.02	1.99	1.93	1.87
26	2.10	2.08	2.06	2.05	2.02	1.99	1.93	1.87
27	2.11	2.09	2.06	2.06	2.03	2.00	1.93	1.88
28	2.10	2.08	2.06	2.05	2.02	1.99	1.93	1.87
29	2.17	2.15	2.12	2.11	2.08	2.04	1.97	1.91
30	2.17	2.15	2.12	2.11	2.08	2.04	1.97	1.91
31	2.04	2.02	2.00	2.00	1.97	1.94	1.88	1.83
32	2.04	2.02	2.00	1.99	1.97	1.94	1.88	1.83
33	2.04	2.02	2.00	1.99	1.96	1.93	1.88	1.83
34	2.05	2.03	2.01	2.00	1.97	1.94	1.88	1.83
35	2.05	2.03	2.01	2.00	1.97	1.94	1.88	1.83
36	2.04	2.03	2.01	2.00	1.97	1.94	1.88	1.83
37	2.04	2.03	2.01	2.00	1.97	1.94	1.89	1.83
38	2.04	2.02	2.00	1.99	1.97	1.94	1.88	1.83
39	2.04	2.02	2.00	1.99	1.97	1.94	1.88	1.83
40	2.05	2.03	2.01	2.00	1.98	1.95	1.89	1.84
41	2.05	2.03	2.01	2.00	1.98	1.95	1.89	1.84
42	2.05	2.03	2.01	2.00	1.97	1.94	1.89	1.83
43	2.05	2.03	2.01	2.00	1.98	1.95	1.89	1.84
44	2.05	2.03	2.01	2.01	1.98	1.95	1.89	1.84
45	2.05	2.03	2.01	2.00	1.97	1.94	1.89	1.84
46	2.05	2.03	2.01	2.00	1.98	1.94	1.89	1.84
47	2.02	2.00	1.98	1.97	1.95	1.92	1.87	1.82
48	2.04	2.02	2.00	1.99	1.97	1.94	1.88	1.83
49	2.04	2.02	2.00	1.99	1.97	1.94	1.88	1.83
50	2.04	2.02	2.00	1.99	1.97	1.94	1.88	1.83
51	2.02	2.00	1.98	1.98	1.95	1.92	1.87	1.82
52	2.02	2.00	1.98	1.97	1.95	1.92	1.86	1.81
53	2.02	2.00	1.98	1.97	1.95	1.92	1.86	1.82

Table A2. The design shear strength (in kPa) of sea ice for different return periods.

Grid No.	Return Period/Year							
	100	**50**	**25**	**20**	**10**	**5**	**2**	**1**
1	1071	1059	1045	1040	1024	1004	965	938
2	1063	1053	1041	1037	1023	1006	973	945
3	1069	1058	1046	1042	1027	1009	976	947
4	1070	1059	1047	1042	1028	1010	977	947
5	1068	1057	1045	1040	1026	1008	974	944
6	1032	1022	1012	1008	996	980	951	926
7	1133	1119	1104	1099	1081	1059	1018	981
8	1140	1126	1111	1105	1087	1065	1023	986
9	1141	1127	1112	1106	1088	1066	1024	987
10	1142	1128	1113	1107	1089	1067	1025	988
11	1141	1127	1112	1107	1088	1067	1025	988
12	1142	1128	1113	1108	1089	1067	1026	989
13	1143	1130	1115	1109	1091	1069	1028	991
14	1108	1096	1082	1077	1061	1042	1004	972
15	1109	1096	1083	1078	1062	1043	1005	973
16	1107	1094	1081	1076	1060	1041	1003	971
17	1107	1094	1081	1076	1060	1041	1003	971
18	1109	1097	1084	1079	1063	1043	1006	973
19	1109	1097	1084	1079	1063	1043	1006	973
20	1049	1039	1027	1023	1010	993	962	934
21	1051	1041	1029	1025	1012	995	964	936
22	1049	1039	1027	1023	1010	993	962	934
23	1051	1041	1029	1025	1012	995	964	936
24	1052	1042	1031	1027	1013	997	965	938
25	1050	1040	1028	1025	1011	994	963	935
26	1050	1039	1028	1024	1010	994	963	935
27	1053	1043	1032	1028	1014	998	966	939
28	1051	1041	1030	1026	1012	996	964	937
29	1085	1073	1060	1056	1040	1022	986	955
30	1085	1073	1060	1056	1040	1022	986	955
31	1021	1012	1001	998	985	970	941	916
32	1019	1010	999	996	983	968	940	914
33	1018	1009	999	995	982	967	939	913
34	1023	1014	1003	999	987	972	942	917
35	1023	1013	1003	999	986	971	942	917
36	1022	1013	1003	999	986	971	942	917
37	1022	1013	1003	999	986	971	943	917
38	1021	1011	1001	997	985	970	941	916
39	1020	1011	1000	997	984	969	940	915
40	1025	1015	1005	1001	989	973	944	919
41	1024	1015	1004	1000	988	973	943	918
42	1023	1014	1003	1000	987	972	943	917
43	1024	1014	1004	1000	988	973	944	919
44	1027	1017	1007	1003	990	975	946	921
45	1024	1014	1004	1000	987	972	943	918
46	1024	1014	1004	1000	988	972	943	918
47	1010	1001	991	987	975	961	933	909
48	1019	1010	999	996	983	968	939	914
49	1019	1010	999	996	983	968	939	914
50	1019	1010	1000	996	984	969	940	914
51	1010	1001	991	988	976	961	934	910
52	1008	999	989	986	974	959	932	907
53	1009	1000	990	986	974	960	932	908

Table A3. The design flexural strength (in kPa) of sea ice for different return periods.

Grid No.	Return Period/Year							
	100	50	25	20	10	5	2	1
1	520	508	495	490	474	455	419	389
2	517	506	494	490	476	459	427	399
3	522	511	499	495	480	463	430	400
4	523	512	500	496	481	464	430	401
5	515	504	492	488	474	458	428	401
6	487	477	467	463	450	435	405	380
7	584	571	556	551	533	512	471	435
8	591	577	562	557	539	517	475	439
9	592	578	563	558	540	518	476	440
10	593	579	564	559	541	519	477	441
11	592	578	563	558	540	519	478	442
12	593	579	564	559	541	520	478	442
13	594	581	566	561	543	522	480	444
14	559	547	534	529	513	494	458	425
15	560	548	535	530	514	495	458	426
16	558	546	533	528	512	493	456	424
17	558	546	533	528	512	493	456	424
18	561	549	536	531	515	496	459	427
19	561	549	536	531	515	496	459	427
20	504	494	482	478	464	448	416	388
21	507	497	485	481	467	450	418	389
22	504	494	482	478	464	448	416	388
23	507	497	485	481	467	450	418	389
24	499	490	479	476	463	448	419	394
25	506	495	484	480	466	449	417	389
26	505	495	483	479	465	448	416	388
27	501	491	481	477	465	449	420	395
28	508	497	486	482	468	451	418	390
29	537	525	513	508	493	475	440	409
30	537	525	513	508	493	475	440	409
31	474	464	454	451	438	424	395	371
32	471	462	452	448	436	422	394	369
33	470	461	451	447	435	421	393	368
34	474	465	455	452	439	425	397	372
35	474	465	455	451	439	424	396	372
36	474	465	454	451	439	424	396	372
37	475	466	456	452	440	425	397	372
38	473	464	454	450	438	423	395	371
39	472	463	453	449	437	422	394	370
40	477	468	457	454	442	427	398	373
41	476	467	456	453	441	426	398	373
42	475	466	456	452	440	425	397	372
43	477	468	457	454	442	427	398	373
44	479	470	460	456	444	429	400	375
45	475	466	456	452	440	425	397	373
46	475	466	456	452	440	426	397	373
47	466	457	446	443	431	416	388	363
48	471	462	452	448	436	421	393	369
49	471	462	452	448	436	421	393	369
50	471	462	452	449	436	422	394	369
51	467	458	447	444	431	416	388	363
52	463	454	444	441	429	414	386	361
53	464	455	445	441	429	414	386	362

Table A4. The design tensile strength (in kPa) of sea ice for different return periods.

Grid No.	Return Period/Year							
	100	50	25	20	10	5	2	1
1	233	224	214	210	198	184	157	136
2	232	224	215	212	201	188	163	142
3	233	225	216	213	202	189	165	144
4	233	225	217	213	203	190	166	144
5	230	223	214	211	200	188	164	143
6	208	201	193	190	181	169	147	127
7	290	279	267	263	248	231	197	168
8	296	284	272	268	253	235	201	171
9	296	285	273	269	254	236	202	172
10	297	286	274	269	255	237	202	172
11	292	281	270	265	252	235	202	174
12	292	282	270	266	252	235	203	175
13	294	283	272	268	254	237	205	177
14	266	257	247	243	231	216	187	162
15	267	258	247	244	231	216	188	162
16	265	256	246	242	230	215	186	161
17	265	256	246	242	230	215	186	161
18	267	258	248	244	232	217	188	163
19	267	258	248	244	232	217	188	163
20	219	212	203	200	190	178	155	134
21	221	213	205	202	192	180	156	135
22	219	212	203	200	190	178	155	134
23	221	213	205	202	192	180	156	135
24	222	215	206	203	193	181	157	136
25	220	213	204	201	191	179	155	135
26	220	212	204	201	191	179	155	135
27	223	216	207	204	194	182	158	137
28	222	214	205	202	192	180	156	136
29	249	240	230	226	215	200	173	149
30	249	240	230	226	215	200	173	149
31	202	194	186	183	173	161	139	121
32	201	193	184	181	171	159	138	120
33	199	191	183	180	170	158	137	119
34	203	195	187	184	174	162	140	122
35	203	195	186	183	173	162	140	121
36	202	195	186	183	173	161	140	121
37	203	195	187	184	174	162	140	122
38	202	194	185	182	172	160	139	121
39	201	193	184	181	172	160	138	120
40	204	196	188	185	175	163	141	123
41	203	196	187	184	174	163	141	122
42	203	195	187	184	174	162	140	122
43	203	195	187	184	174	163	141	123
44	205	197	189	186	176	165	143	124
45	203	196	187	184	174	162	140	122
46	203	196	187	184	174	162	141	122
47	195	187	179	176	166	155	133	116
48	201	193	184	181	171	159	137	120
49	201	193	184	181	171	159	137	120
50	202	193	185	182	171	159	138	120
51	194	186	178	175	165	154	133	116
52	192	184	176	173	164	152	132	115
53	193	185	176	174	164	153	132	115

melting of glaciers and ice caps and the thermal expansion of seawater causes global sea levels to rise, threatening low-lying islands and coastal cities [4]. Additionally, it triggers extreme weather events in the mid-latitude regions [5–7], such as extreme snowfall in Europe [8] and smog in China [9]. As the Arctic sea ice decreases, Arctic shipping lanes (such as the Northwest Passage and the Northeast Passage) are becoming viable, providing new routing options for the global shipping industry [10]. This shortens the distance of some international routes, reduces transport costs, and improves transport efficiency [11,12]. Accurate single-point SIT data in the Arctic region can provide a real-time dynamic reference for route planning, helping ships avoid thick ice areas or dense floating ice zones and reducing the risk of ship damage and grounding [13]. Meanwhile, refined navigation and ice condition prediction can optimize speed and fuel consumption, reduce transportation costs, and enhance economic benefits. In addition, ice thickness prediction provides a scientific basis for deploying waterway infrastructure (such as ports and icebreaker scheduling) and helps promote the sustainable development of Arctic shipping.

Currently, there are several approaches to quantify the Arctic SIT. Remote sensing technology can determine SIT on a large scale [14–16]; however, the accuracy and spatial resolution of the resultant data are limited. Acoustic or electromagnetic induction measurements from underwater robots, submarines, aircraft, or ships cannot discern temporal variations in snow depth (SW) and SIT. The evolution of SIT can be obtained through numerical patterns of thermodynamic and dynamic processes based on known physical principles [17–19]. Still, the results are susceptible to substantial spatiotemporal variability due to the influence of physical parameterization schemes. On-site, manual observations and autonomous platform monitoring, such as drillings, ablation piles, and sea ice mass balance buoys (SIMBAs), are the most accurate methods for quantifying SW and ice thickness [20,21]. Collecting Arctic SIT data is resource intensive, time consuming, and in some cases, nearly impossible. In addition, the danger of offshore operations during sea ice melting intensified the difficulty of obtaining data. In 2019–2020, the Multidisciplinary drifting Observatory for the Study of Arctic Climate (MOSAiC) expedition gathered extensive, year-round data on Arctic sea ice and its connections with the atmosphere and ocean [22,23].

Deep learning (DL) [24] has been increasingly applied in sea ice prediction studies in recent years because of its inherent nonlinear fitting ability. For example, Chi and Liu et al. indicate that DL-based prediction models can be successfully utilized to fit, long term, the Arctic sea ice concentration (SIC) datasets and forecast monthly SICs throughout a year, although the prediction quality has slightly deteriorated due to accelerated sea ice melting rates in summer months [25,26]. Although many studies have validated the applicability of deep learning to predict Arctic SIC [27,28], research on predicting sea ice thickness (SIT) is still extremely limited. Only Song et al. [29] have systematically conducted relevant research in the public literature. The research of Song et al. verifies that DL models based on Concurrent Long Short-Term Memory (LSTM) [30] and fully convolutional U-net algorithms can provide monthly forecasts of SIT without considering complex physical processes. DL models demonstrate applicability in dealing with sea ice multi-scale spatiotemporal series forecasts, among which LSTM [31] has great potential for future application in the field of Arctic sea ice due to its significant advantage in handling complex series data. Nevertheless, the principal challenge remains the relationship between performance and the large amounts of data required.

Facing small-scale and imbalanced data, numerous studies have adopted data augmentation techniques to generate higher-quality samples, especially DL-based Generative Adversarial Networks (GANs) and Wasserstein GANs (WGANs) [32,33]. Currently, combining GANs (and their variants) with LSTM dealing with time series to process small numbers of complex time series has a very high potential, and the method has been used in areas such as satellite image prediction [34], intent recognition in collaborative tasks [35], and automatic recognition of human activities in aerial videos captured by unmanned aerial vehicles [36]. These studies have demonstrated that the use of hybrid data in network training can improve the accuracy and generality of prediction or classification in these domains compared to using only real training data.

In this study, we propose an innovative hybrid approach model for hindcasting the Arctic SIT, which uses as inputs the Arctic SIT and SW obtained by SIMBA analysis during MOSAiC and atmospheric forcing data from the European Centre for Medium-Range Weather Forecasts Reanalysis version 5 (ERA5) [37] for the same period. The core of this method lies in utilizing the powerful generation capability of WGAN to expand the feature set and processing complex time series through LSTM. We adopted a novel loss function for the model and conducted uncertainty estimation on the model. This study not only provides a new method for short-term prediction of the Arctic SIT but also provides innovative ideas for other sparse data-based physical feature identification studies.

The rest of the paper is organized as follows: Section 2 describes the composition of our model and evaluation methods; Section 3 describes the data sources and experimental setup; Section 4 describes the results of the experiments and the various evaluation results; and Section 5 presents the conclusions and perspectives.

2. Data and Methods

In this study, the primary focus is on proposing an innovative hybrid approach model for hindcasting the Arctic SIT and evaluating its accuracy and applicability. The dataset, model, and assessment methodology are described in detail below.

2.1. Dataset Description

In this study, the part of the Arctic drift sea ice during the MOSAiC survey was used as the research area to verify the proposed method's performance. Information on the study area and data is described below.

2.1.1. The SIT Dataset

The SIMBA used for on-site observation is a temperature chain autonomous ice mass balance buoy developed by SRSL, a Scottish Marine Science Society (SAMS) subsidiary. As shown in Figure 1, the float temperature chain has a total length of 4.8 m and is equipped with 241 digital chip temperature sensors with a spacing of 0.02 m. By carrying GPS and Iridium modules, the buoy can accurately monitor the temperature of ice and snow environments (SIMBA-ET), track the trajectory of sea ice movement, and regularly transmit data to a remote monitoring center. In addition, the SIMBA buoy integrates heating elements (heating resistors) to perform weak current heating on each sensor and the surrounding environment to determine the temperature change around the thermistor after heating (SIMBA-HT). As shown in Figure 2, during the MOSAiC survey, many SIMBA buoys were deployed near the observation station, with an overall observation period from October 2019 to August 2020 [38]. The SIMBA-ET sampling interval for all buoys is 6 h, and the SIMBA-HT sampling interval is 24 h.

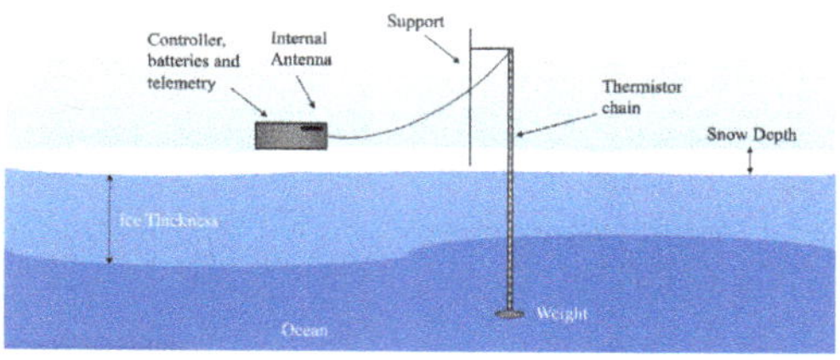

Figure 1. Schematic diagram of the snow and ice mass balance array (SIMBA). A sensor array is inserted into a 2-inch ice hole, linked to a Pelican Case with electronics, batteries, and an Iridium modem.

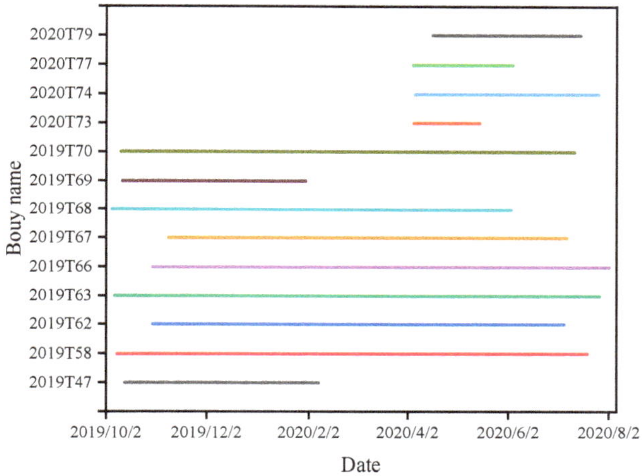

Figure 2. The operation periods of the buoys.

2.1.2. Atmospheric Forcing Data

ERA5 is the fifth-generation atmospheric reanalysis dataset developed by the European Centre for Medium-Range Weather Forecasts (ECMWF), covering the period from 1950 to the present with a time and spatial resolution of 1 h and $0.25° \times 0.25°$, respectively. It provides globally consistent meteorological and climate data covering the atmosphere, oceans, and land domains. This article matches the GPS positioning of buoys during the MOSAiC survey to the nearest point in the ERA5 data grid. It uses bilinear interpolation to extract atmospheric forcing field variables at the corresponding time.

As shown in Table 1, the following five atmospheric forcings obtained from ERA5 are part of the input network's features.

Table 1. Selected atmospheric forcing data for WGAN-LSTM model.

Variable	Source	Abbreviation or Calculation	Units
10 m wind field	ERA5	wind	m/s
Surface solar radiation downwards	ERA5	fsw	W/m^2
2 m air temperature	ERA5	t2m	°C
Total precipitation	ERA5	tp	mm/h
Proportion of total cloud cover	ERA5	tcc	%

2.1.3. Data Preprocessing

We upgraded the 6 h resolution SIT and SW data extracted by SIMBA buoys to 1 h resolution through linear interpolation, aligning them with ERA5 atmospheric forcing data (1 h resolution). Subsequently, sliding window technology is used to extract features from the interpolated data, with a window size of 24 h and a step size of 1 h. Next, this study selected the relevant data obtained from buoy 2020T74 (5 April to 26 July 2020) as a typical sample, during which the Arctic sea ice showed a seasonal transitional characteristic of slow growth followed by melting. The data are divided into training set, validation set, and testing set in an 8:1:1 ratio, while the remaining SIMBA data are used entirely to validate the model's generalization ability.

2.2. Model

2.2.1. Wasserstein Generative Adversarial Network

The recognition of DL models for sea ice prediction has shown promising improvements. However, these models require large amounts of high-quality, long time series historical data. DL models typically have a large number of parameters, making the models susceptible to overfitting if the amount of training data is insufficient or the quality of the data is poor [39]. Overfitted models perform well on the training set but make poor predictions on unseen data and have insufficient generalization [40].

GAN is a DL model proposed by Ian Goodfellow et al., which consists of two competing neural networks: generator and discriminator [32]. The generator produces synthetic data from a series of random noises, while the discriminator compares synthetic data with accurate data to determine whether the synthetic data are authentic. The loss function of GAN can be expressed as an equation related to Jensen–Shannon (JS) divergence. The JS divergence does not decrease as the two distributions approach each other, unless they completely overlap. Instead, it remains constant at log2 with high probability. This behavior poses a significant challenge for GAN training: when the generated distribution gets closer to the real distribution but does not fully overlap, the JS divergence fails to provide meaningful gradient signals. As a result, the original GAN algorithm often suffers from gradient vanishing, convergence issues, and instability during training. To address these limitations, WGAN introduces the Wasserstein distance as an alternative metric for measuring the distance between distributions:

$$W(p_{data} \parallel p_g) = \inf_{\gamma \in \prod S(p_{data}, p_g)} E_{(x,y)}[\parallel x - y \parallel] \tag{1}$$

where inf refers to the maximum lower bound, and $S\ (p_{data}, p_g)$ denotes all possible joint distributions in the distribution p_{data} and p_g. When the difference between two probability distributions is slight or does not overlap completely, compared to JS divergence, Wasserstein distance can still provide meaningful gradient information and smoother results for parameter updates in gradient descent methods. The WGAN algorithm has proven to be highly effective in generating accurate data; it can improve the stability of learning and get rid of problems such as pattern collapse.

2.2.2. Long Short-Term Memory Network

LSTM is a kind of Recurrent Neural Network (RNN) [41] that has proven very powerful at processing sequence data. It can solve standard RNNs' gradient disappearance and gradient explosion problems.

The three basic "gates" that make up the LSTM unit are the Input Gate, the Forget Gate, and the Output Gate. As shown in Figure 3, each gate of the LSTM cell is a neural network layer of a sigmoid activation function that controls the flow of information. The direction of the black arrow is the forward calculation process, and the red arrow's direction carries out the error's backpropagation. First, we must determine the network structure and loss function of LSTM. Second, each parameter should be initialized, and then, the model's accuracy calculated through the loss function. The criterion for updating the weights and bias terms is to minimize the loss of the specified objective function in the training samples. We must update the parameters by choosing the corresponding optimization algorithm if networks fail to reach the required accuracy. In this cycle, when the gradient reaches the required accuracy, we can determine the model's parameters, complete the LSTM model's modeling, and apply it to prediction or classification.

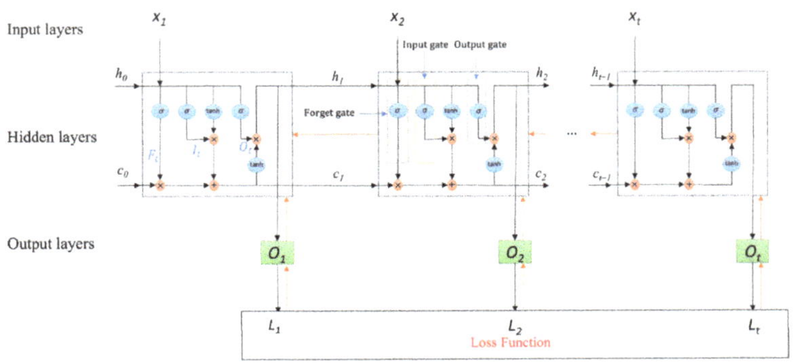

Figure 3. The structure of Long Short-Term Memory (LSTM). "×" represents matrix multiplication, "+" represents matrix addition, black arrows represent forward propagation, and red arrows represent backward propagation.

The computational expression within LSTM is as follows:

$$F_t = \sigma(W_{XF} \times X_t + W_{HF} \times H_{t-1} + b_F) \tag{2}$$

$$I_t = \sigma(W_{XI} \times X_t + W_{HI} \times H_{t-1} + b_I) \tag{3}$$

$$O_t = \sigma(W_{XO} \times X_t + W_{HO} \times H_{t-1} + b_O) \tag{4}$$

$$C_t = F_t \times C_{t-1} + I_t \times \tanh(W_{XC} \times X_t + W_{HC} \times H_{t-1} + b_C) \tag{5}$$

$$H_t = O_t \times \tanh(C_t) \tag{6}$$

where F_t, I_t, O_t, C_t, and H_t denote the outputs of the forgetting gate, input gate, output gate, cell state, and hidden layer, respectively; W and b denote the weight vector and bias vector, respectively; and tanh and σ denote the Tanh and Sigmoid functions for activating variables, respectively.

2.2.3. A New Loss Function: DISO

The mean square error (MSE) loss function is the default loss function for regression problems, defined as:

$$Loss_{MSE} = \frac{1}{n}\sum_{i=1}^{n} (y_i - \hat{y}_i)^2 \tag{7}$$

Mathematically, if the distribution of the target variable is Gaussian, it is the preferred loss function under the framework of maximum likelihood inference. Since the MSE squares the error, large outliers can have a significant impact on the overall error, resulting in large *MSE* values. If outliers are present in the data, it may be necessary to preprocess them or use another more robust error metric such as mean absolute error (*MAE*). Therefore it is important to choose a suitable statistical indicator as a loss function.

As opposed to a single statistical indicator, Hu et al. developed a new comprehensive index Distance between Indices of Simulation and Observation (DISO) [42], which quantitatively describes the overall performance of different models on the observed field. DISO combines different statistical measures, including correlation coefficient (*R*), *MAE*, and root mean square error (*RMSE*). In our case, the *DISO* loss function is defined as:

$$Loss_{DISO} = \sqrt{(R-1)^2 + MAE^2 + RMSE^2} \tag{8}$$

Overall, with the strong physical characteristic of the distance in three-dimensional spaces and the strict mathematical proof, this multidimensional evaluation capability can help the model optimize multiple objectives simultaneously during the training process and effectively solve the problem of conflicting evaluation results of different statistical indicators. Therefore, *DISO*, as a new comprehensive loss function for neural networks, will have great potential.

2.2.4. WGAN-LSTM

The WGAN-LSTM model combines the high-quality generative ability of WGAN with the ability of LSTM to process time series and learn long-term dependencies. As shown in Figure 4, the training process of this model starts with real datasets that contain multiple features, such as atmospheric forcing features, and key variables, such as SIT. After preprocessing the real dataset, it is divided into a training set, validation set, and test set. Based on the training set data, the WGAN network generates fake data that are relatively consistent with its distribution and splices them with the real training data to jointly train the LSTM. This hybrid dataset is used as the input to the LSTM network, which is used to train the model to recognize and learn long-term dependencies in the time series. After two LSTM networks, the model further processes and integrates features through a fully connected hidden layer to finally obtain the output.

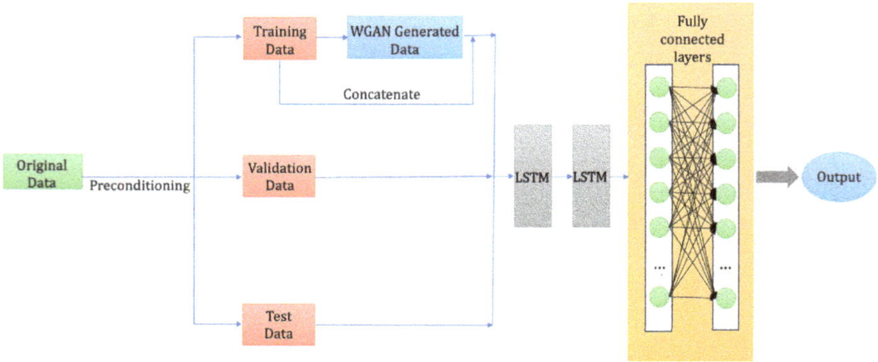

Figure 4. The proposed model for Arctic sea ice thickness (SIT) hindcast: Wasserstein GAN (WGAN)-LSTM.

We respectively employed two loss functions during training to optimize the LSTM network. First, we used the mean square error (MSE), a commonly used loss function, to measure the difference between the model's hindcast values and the observations. Secondly, we introduced a new comprehensive index DISO. Despite careful consideration of design and selection issues (e.g., hyperparameter tuning), the black-box nature of neural network architectures remains challenging. This black-box nature means that the internal working mechanism of the model is opaque and susceptible to small perturbations in the network weights, leading to inconsistent model test results.

In this study, we compared and qualitatively analyzed three models: LSTM, WGAN-LSTM (using MSE as the loss function on the LSTM section), and WGAN-LSTM-DISO (using DISO as the loss function on the LSTM section).

2.3. Performance Evaluation

The purpose of this study is to improve the accuracy and reliability of hindcast SIT. Thus, the performance of the proposed model needs to be evaluated considering two aspects. On the one hand, the accuracy of all models is evaluated based on the observed SIT at the same time. On the other hand, uncertainty analysis of the model will be performed to evaluate its reliability.

2.3.1. Accuracy Evaluation of SIT Hindcast

The statistical metrics are applied to quantitatively describe the accuracies of the models from different perspectives, such as the correlation coefficient (R) in measuring the strength and direction of the linear association between the simulated time series and the observed time series, mean absolute error (MAE) in measuring any persistent bias (underestimation with positive value and overestimation with negative value) for the observed time series, and RMSE in quantifying the averaged magnitude of the deviation [43].

In order to comprehensively assess the performance of the model, we use a variety of statistical metrics to measure the hindcast ability, including the center root mean square error (*RMSD*), *MAE*, *R*, and comprehensive index *DISO*. These metrics can reflect the hindcast accuracy and stability of the model from different perspectives. Their expressions are as follows:

$$RMSD = \sqrt{\frac{1}{n}\sum_{i=1}^{n}[(T_{O,i} - \overline{T}_O) - (T_{P,i} - \overline{T}_P)]^2} \tag{9}$$

$$MAE = \frac{1}{n}\sum_{i=1}^{n}|T_{O,i} - T_{P,i}| \tag{10}$$

$$R = \frac{\sum_{i=1}^{n}(T_{O,i} - \overline{T}_O)(T_{P,i} - \overline{T}_P)}{\sqrt{\sum_{i=1}^{n}(T_{O,i} - \overline{T}_O)^2}\sqrt{\sum_{i=0}^{n}(T_{P,i} - \overline{T}_P)^2}} \tag{11}$$

$$DISO = \sqrt{(R-1)^2 + MAE^2 + \frac{\sum_{i=1}^{n}(T_{O,i} - T_{P,i})^2}{n}} \tag{12}$$

where $T_O$and T_P denote the observed and forecast SIT, $\overline{T}_O$ and $\overline{T}_P$ denote the mean value of the observed and forecast SIT, and n denotes the forecast SIT sequence length.

2.3.2. Uncertainty Quantification

Model uncertainty is caused by uncertainty in the model structure and parameters, which reflects the model's fit to the data and generalization ability. Bayesian methods provide a robust framework for uncertainty analysis, especially in DL, and the Bayesian neural network (BNN) [44] is one of the main tools used to implement this framework.

In Bayesian neural networks, the weights and biases of the model are treated as random variables rather than fixed values. This approach allows the model to learn the probability distribution of the parameters during the training process, thus being able to quantify the uncertainty of the model.

Monte Carlo dropout (MC dropout) is an uncertainty estimation technique for DL that provides an effective means to visualize and quantify model uncertainty, which is crucial for improving model reliability and interpretability [45]. It is based on the dropout regularization method, which heavily discards a portion of the neurons during the training process, reducing the model's over-reliance on specific samples from the training data, thus reducing the risk of overfitting [46]. Dropout (and its variants) activation is still maintained during the neural networks testing phase to simulate the Bayesian distribution of the model parameters, which can be interpreted as a Bayesian approximation of the well-known probabilistic model: the Gaussian process [47].

In our problem, we apply dropout multiple times during the forward propagation of the model and compute multiple outputs to generate Monte Carlo samples. These whole samples are then used to quantify uncertainty by counting the mean and variance of the test results.

3. Experiment Settings

Our proposed model is developed in Python 3.7 using PyCharm 2023. 3. 4, and the torch 1. 13. 1 package is used on the backend. When using WGAN for data augmentation, the input features are atmospheric forcing data, SW, historical SIT, and the future predicted time point SIT, whose shape is 24 $\times$ 8. In the experiment, we generated an additional 5000 enhanced samples and concatenated them with real data to serve as the training data for LSTM. The remaining data from 2020T74 are used as the validation and test sets. The generators and discriminators of the WGAN model and the hyperparameters used in the LSTM model are shown in Table 2.

Table 2. Hyperparameters of all modules.

Module	Parameter	Values
WGAN generator	Noise vector dimension z	100
	Number of layers	five layers
	Activation function	LeakyReLU; output layer: tanh
	Number of neurons in each layer	(128, 256, 512, 1024, 192)
WGAN discriminator	Number of layers	three layers
	Activation function	LeakyReLU; output layer: tanh
	Number of neurons in each layer	(512, 256, 1)
WGAN model	Optimizer	RMSprop (lr = 0.00005)
	Batch size	64
	Number of epoch	1000
	Loss function	Wasserstein loss
LSTM model	Input frame resizing	24 × 7
	Batch size	256
	Number of LSTM layers	2
	Number of epoch	50
	Dropout rate	0.3
	Optimizer	AdamW

For model uncertainty estimation, we implemented dropout for the LSTM layer with a dropout rate of 0.3. The dropout rate for the fully connected layer after implementing the two-layer LSTM was chosen to be 0.5. In this experiment, we generated 100 MC samples for model uncertainty estimation by placing the data of each test set into the well-trained network model multiple times.

4. Results and Qualitative Analysis

4.1. Generated Features

We selected three moments during the WGAN training process, namely the 200th, the 500th, and the 1000th iterations, to evaluate the evolution of WGAN performance. We randomly select the same time point from the 5000 samples generated by WGAN, and pick out the generated features and corresponding generated ice thickness at that time point. Figure 5 shows the comparison between the scatter density maps of generated features and ice thickness with actual observations, and the loss function curve of the WGAN network, where the color scales from blue to red represent the increase in data point density.

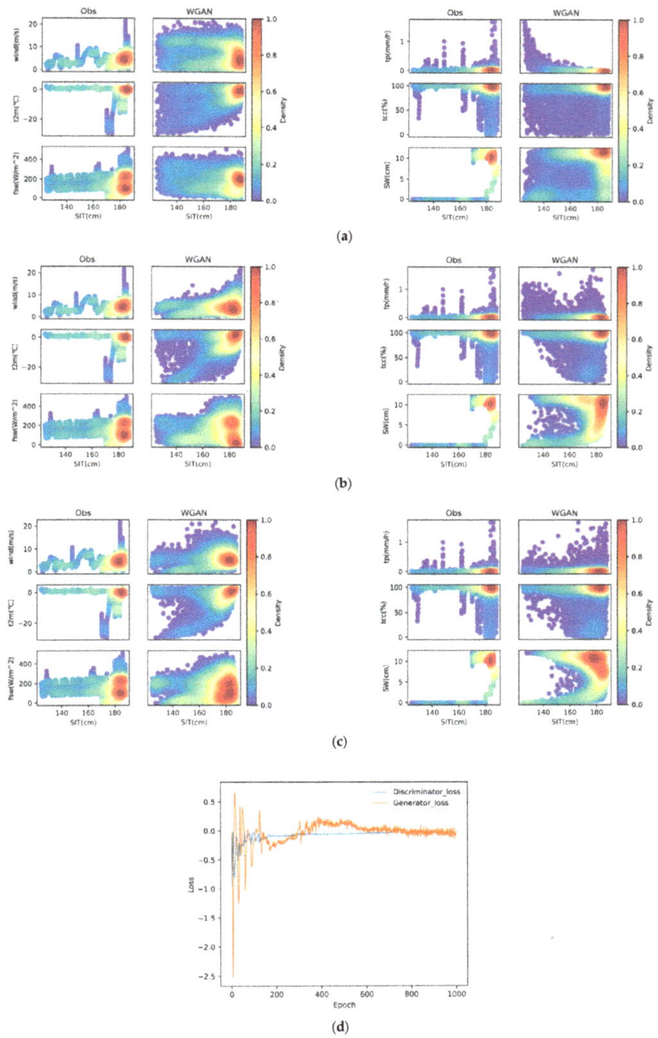

Figure 5. Scatter density plot between features and the corresponding SIT: (**a**) 200 iterations; (**b**) 500 iterations; (**c**) 1000 iterations; (**d**) loss function curve of the WGAN generator and discriminator.

We can see that as the number of iterations increases, the performance of the WGAN improves. As shown in Figure 5a, when the number of iterations reaches 200, the WGAN roughly captures the primary distribution of the observed features, although there are still some significant differences compared to the actual distribution. The area with a bluish color is relatively extensive, indicating that most of the data with lower density have been generated, in contrast to the high-density area of the observed data. However, as shown in Figure 5b, after 500 iterations, the color distribution of the scatter plot is more concentrated in the darker area, and the location of the high-density area is closer to the observed results. As the number of iterations increases, as shown in Figure 5c, the generated feature distribution is generally closer to the observed features and is better represented in some details. These changes indicate that the WGAN has gradually learned and improved its ability to simulate the observed distribution during training. It is worth noting that although the overall trend is close to the actual distribution, there are still some data points that deviate from the observed distribution. This phenomenon actually reflects the WGAN's ability to generalize, which can generate values that have not appeared in the observed samples. Figure 5d shows the loss curves of the WGAN model's generator and discriminator. It can be observed that the loss curves of the generator and discriminator start to stabilize and approach zero after about 800 iterations. The loss curve indicates that our model has achieved this balance and converged to a stable point, indicating that the model has received appropriate training. Combined with the visualization shown in Figure 5a–c, we can conclude that WGAN can effectively generate high-quality samples.

4.2. Model Hindcast Results

After network training, we use test set data for SIT hindcast. Figure 6 shows the observed SIT values and the SIT values exported by each model from 11:00 on 11 July 2020, to 11:00 on 26 July 2020. The SIT values exported by each model are the average values based on MC samples. The analysis results show that the three models' hindcast trends are consistent with the observed data, but the test deviation gradually increases over time. The main reason for this phenomenon is that most of our training data are in the ice growth period. As the melting degree increases, the model's overestimating ice thickness becomes more apparent. It is worth noting that the test results of LSTM and WGAN-LSTM on 23 July deviated from the observed data, showing an increase.

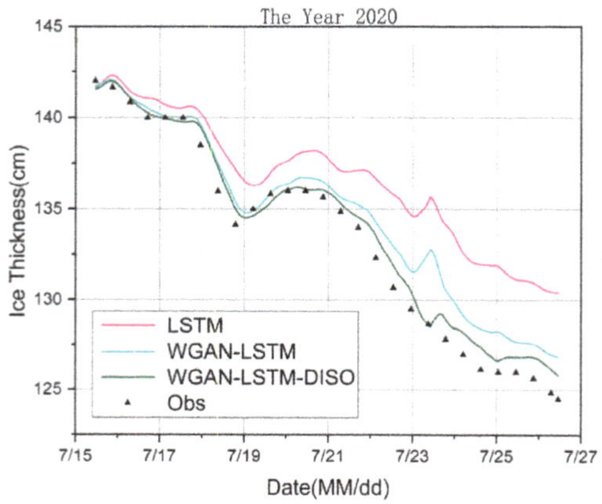

Figure 6. The time series of SIT exported by LSTM, WGAN-LSTM, and WGAN-LSTM-DISO, and measurement SIT. Obs represent measurement data (black diamond); take one point for every ten observations.

As shown in Figure 7, the tcc values of the training and validation sets (mostly during the ice growth period) are lower than those of the test set (during the ice melt period). Therefore, although the WGAN generated some training data, the network model still tended to learn the non-physical phenomenon of increased ice thickness when cloud cover decreased. The sudden decrease in the tcc value on 22 July 2020 caused the model to mistakenly believe that the ice thickness had increased, resulting in abnormal results. In contrast, the hindcast results of the WGAN-LSTM-DISO model were closest to the observed data, with a smoother test curve, effectively correcting the abnormal hindcast.

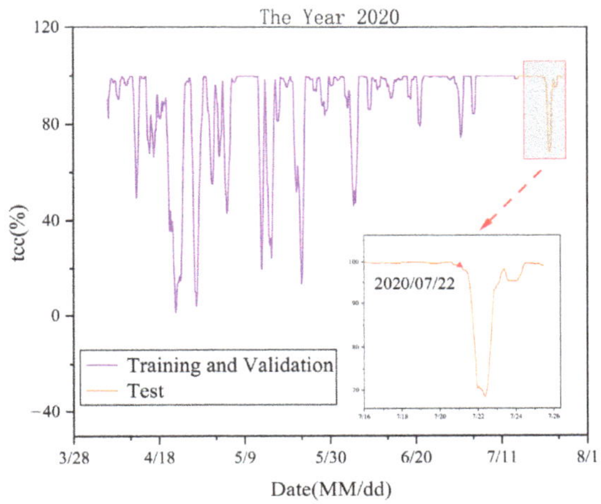

Figure 7. The proportion of total cloud cover (tcc) from 6 April to 26 July 2020. The red triangle presents the date when the tcc decreased on test data.

4.3. Model Evaluation

As shown in Table 3, the correlation coefficient R between the hindcasted and observed values of the three models is more excellent. At the same time, the RMSD and AE of LSTM, WGAN-LSTM, and WGAN-LSTM-DISO are reduced in turn. We can see that WGAN-LSTM can reduce the RMSD error of LSTM from 2.122 to 1.109, and the test RMSD error of WGAN-LSTM-DISO is further reduced to 0.686. However, it is worth noting that the MAE of all three models is greater than 0, indicating that all three models are overestimated, mainly due to the different periodic nature of the training and validation sets. Calculating the DISO difference between the two comparison models and the LSTM model and dividing it by the LSTM's DISO value obtains the performance improvement ratio, which quantitatively evaluates the degree of improvement in the model's overall performance. Compared with the LSTM model, the performance of WGAN-LSTM and WGAN-LSTM-DISO is improved by 51.9% and 75.2%, respectively. All of this shows that our data enhancement processing and loss function improvements can significantly improve the model's accuracy. This significant performance improvement shows that our data augmentation processing and loss function improvements are essential for improving model accuracy. Finally, we note that the standard deviation of each statistical metric for WGAN-LSTM-DISO is minimal, which means that the WGAN-LSTM-DISO model exhibits the most excellent stability across all evaluation metrics.

Table 3. The corresponding statistical metrics and their standard deviations between MC sample values and observed values.

Dataset	R	RMSD	MAE	DISO
Obs	1	0	0	0
LSTM	0.978 ± 0.010	2.212 ± 0.274	3.242 ± 2.119	0.044 ± 0.009
WGAN-LSTM	0.987 ± 0.005	1.109 ± 0.178	1.322 ± 1.020	0.021 ± 0.005
WGAN-LSTM-DISO	0.994 ± 0.002	0.686 ± 0.112	0.574 ± 0.572	0.011 ± 0.002

After a preliminary assessment of the performance of the three models, we further visualize the sample thickness contours predicted by the three models to analyze the effectiveness of the MC profile in capturing the uncertainty of the thickness hindcast. Figure 8 shows the mean and standard deviation of the model's hindcast at each time point for the complete dropout sample pool (MC samples are obtained every ten independent random trials). The error bars are generated based on two standard deviations around the mean, which can cover 99.7% of the samples under the Gaussian assumption. We can see that the average curve of the WGAN-LSTM-DISO model best matches the actual observations, and at any test time point, the sample distribution generated by this model accurately surrounds the observations. In contrast, the error bars of the LSTM and WGAN-LSTM models at the beginning of the validation set cover the basic facts. However, as time goes on, the actual observations begin to deviate from the sample distributions predicted by these two models and exceed the range of the error bars. In addition, the variance of the samples from the three models increases over time, making it more challenging for the model to predict the ice thickness at the test time point in the future.

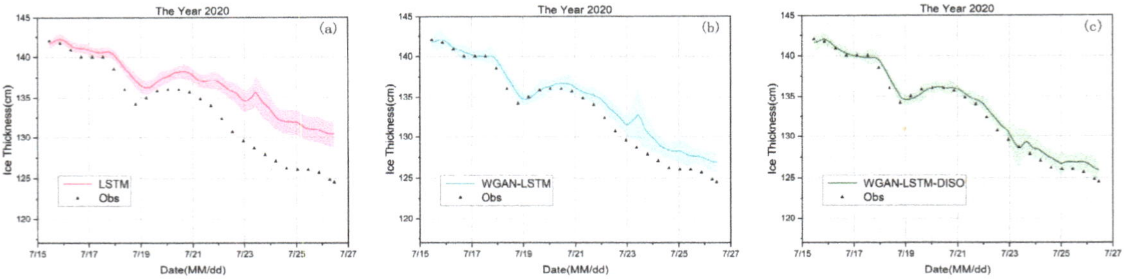

Figure 8. The means and standard deviations of the MC samples generated by LSTM, LSTM-DISO, and WGAN-LSTM-DISO in July 2020 (100 samples). From left to right: (**a**) the hindcast of LSTM; (**b**) the hindcast of GAN-LSTM; (**c**) the hindcast of GAN-LSTM-DISO.

Ideally, an effective uncertainty estimate for a model should ensure that the sample distribution it generates matches the actual observed value distribution at the test points. In other words, for the kth percentile of the sample generated by the ideal model, we expect k% of the actual observations to fall within that percentile range. To quantify this concept, we first fit a Gaussian distribution over the complete sample pool generated by the model and then estimate the two-tailed percentile of the observed underlying truth. Figure 9 shows the cumulative percentage of observations (y-axis) falling within a particular percentile of the model-predicted sample (x-axis). Ideal model performance is represented by the diagonal line y = x, where the proportion of observations falling within a percentile equals the percentile value. If the model curve lies below the diagonal line y = x, this indicates that the proportion of actual observations falling within a given percentile is lower than expected. This suggests that the model is overconfident, which could lead to over-reliance on the accuracy of the model predictions in practical applications.

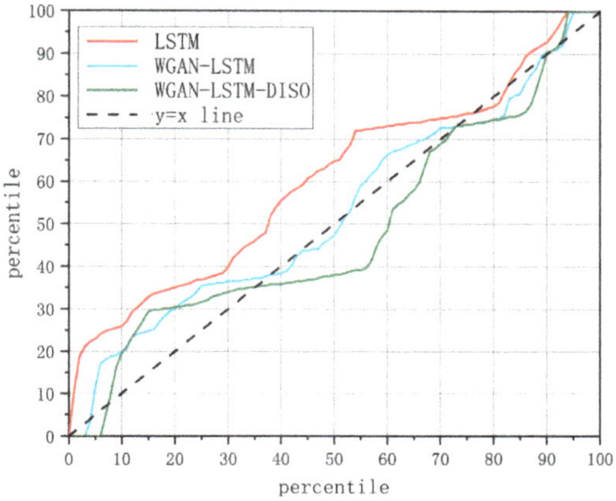

Figure 9. Cumulative percentage of observations within a certain percentile of samples of three models.

The WGAN-LSTM and WGAN-LSTM-DISO are partly below the diagonal line, producing slightly confident uncertainty estimates. The curve of the LSTM model is almost entirely above the diagonal, so its uncertainty estimate is much larger than expected in an ideal situation. When there is a lack of observational data information about the test date, it is generally preferable for the model to show a slightly less confident attitude. Overall, WGAN-LSTM (including WGAN-LSTM-MSE and WGAN-LSTM-DISO) produces good uncertainty estimates and effectively captures the characteristics of the distribution of actual observations. However, it is to our regret that the uncertainty of the WGAN-LSTM without DISO performs best. This is mainly due to the introduction of RMSE in DISO's design R; these indicators will assign higher weights to larger prediction errors. For example, RMSE amplifies the impact of errors, forcing the model to minimize such errors through extreme predictions. Similarly, when the weight of R is high, the model will enhance the linear correlation with the actual label through extreme predicted values.

4.4. Model Generalization Ability Evaluation and Physical Process

Based on the feature data extracted from 12 SIMBAs obtained by the MOSAiC program, this study systematically evaluated the generalization ability of the model proposed in this chapter in predicting Arctic SIT. These buoys are spatially distributed in the central Arctic Ocean, Fram Strait, and North Pole region, with a period from October 2019 to August 2020, including key phenological stages such as the Arctic sea ice growth period (November–April of the following year) and partial melting ice period (May–August). Figure 10 shows the time series of the predicted thickness and observed ice thickness of the models on different buoys. As shown in Figure 10a–i, the predicted ice thickness trends of the three models are consistent with the measured changes, and there is a significant deviation between the predicted results of the LSTM model and the measured SIT from October 2019 to January 2020, while it shows good agreement in other periods. This time-dependent error is mainly due to the limitations of the training data, as the model is only trained on data from April to June 2020, resulting in poor generalization ability of the LSTM model for samples far from the training period. In contrast, the testing performance of the WGAN-LSTM and WGAN-LSTM-DISO models on all buoys is significantly better than that of LSTM. However, from the intuitive comparison of the models in Figure 10, it can be seen that the visual difference in prediction results between the WGAN-LSTM and WGAN-LSTM-DISO models is not significant. This indicates that introducing the WGAN framework has mainly improved the models' performance, while the DISO module's improvement is relatively limited.

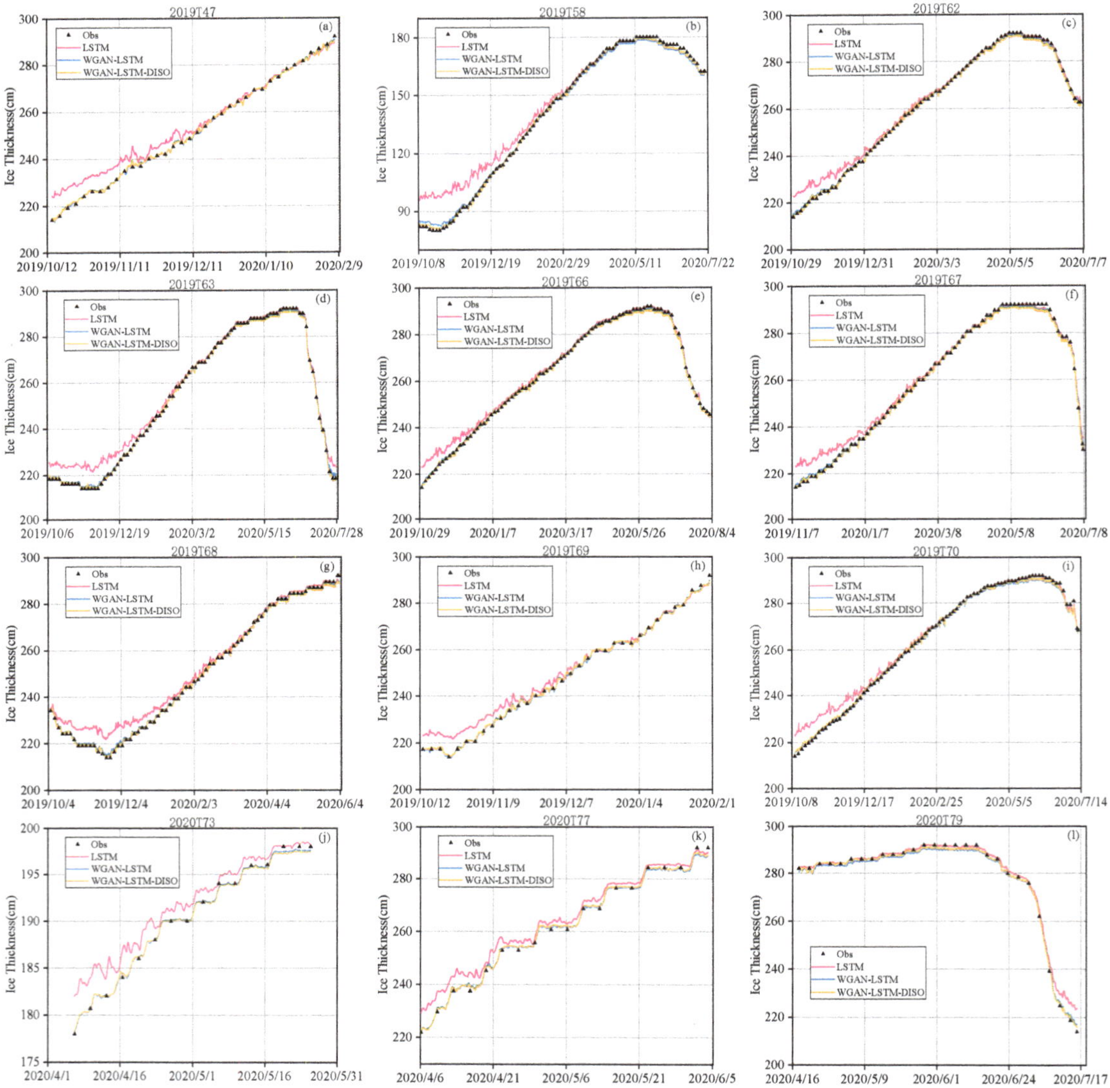

Figure 10. Results of the trained model on other buoys. (**a**–**l**) represent buoy 2019T47, buoy 2019T58, buoy 2019T62, buoy 2019T63, buoy 2019T66, buoy 2019T67, buoy 2019T68, buoy 2019T69, buoy 2019T70, buoy 2020T73, buoy 2020T77, buoy 2020T79.

In order to quantitatively analyze and demonstrate the generalization of the model, Table 4 shows the average error metrics of the model on all buoys. Compared to traditional LSTM models, the WGAN-LSTM model significantly improves prediction accuracy, with MAE reduced to 0.242, RMSD significantly reduced to 0.887, and the DISO index optimized to 0.005. Although the MAE value of WGAN-LSTM-DISO is slightly higher than that of WGAN-LSTM, its comprehensive indicator DISO has increased by 20% compared to the latter. The R values of all three models remain at a high level of 0.999, indicating that they can accurately reflect the trend of SIT changes. In summary, WGAN-LSTM-DISO exhibits optimal performance on buoy data from different regions and times, fully verifying its significant generalization ability.

Table 4. The mean error of models on all buoys.

Model	MAE	RMSD	R	DISO
LSTM	2.295	2.825	0.999	0.020
WGAN-LSTM	0.242	0.887	0.999	0.005
WGAN-LSTM-DISO	0.260	0.798	0.999	0.004

Due to air temperature being a key factor affecting Arctic SIT, this study analyzed the relationship between ice thickness predicted by the WGAN-LSTM-DISO model and measured air temperature during the freezing period (25 November 2019 to 1 January 2020) and melting period (1 June to 29 June 2020) based on observation data from buoy 2019T67. As shown in Figure 11, the model exhibits a negative correlation between ice thickness and air temperature increase under both ice conditions. During the frozen period, due to the insulation effect of snow cover, the heat exchange between the atmosphere and sea ice is suppressed, resulting in a slower response of ice thickness to air temperature changes. During the melting ice period, as snow melts and solar radiation increases, the heat absorbed by sea ice significantly increases, making ice thickness more sensitive to air temperature rise. This indicates that the model prediction results can accurately reflect the physical relationship between air temperature and ice thickness.

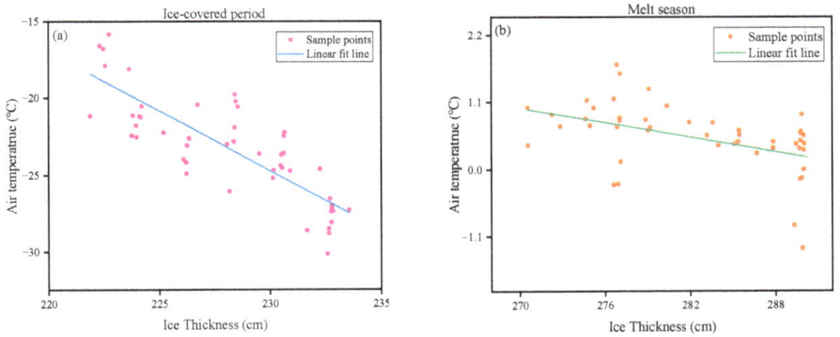

Figure 11. Scatter plot of actual air temperature observations and predicted ice thickness by the WGAN-LSTM-DISO model on buoy 2019T67. (**a**) represents part of the ice-covered period spanning 25 November 2019 to 1 January 2020; (**b**) represents part of the melt season spanning 1 June to 29 June 2020.

5. Conclusions

The acquisition of the Arctic sea ice data is limited by factors such as observation equipment and geographical environment, and the amount of data is relatively limited. In order to alleviate the difficulty of data volume that is too small and improve the accuracy of DL in predicting SIT, a novel hybrid model based on DL for hindcasting Arctic SIT was proposed: WGAN-LSTM. First, high-quality atmospheric forcing data and the corresponding SW and SIT were generated using WGAN and spliced with the original data. Second, the blended data were used as inputs to LSTM. We used different loss functions during training to obtain the SIT output at the next point. Finally, 100 MC samples generated by each model were used to evaluate the test results with model uncertainty analysis. The WGAN-LSTM model in this paper takes about 1 min to train once on iRIS xe. Through qualitative analysis, we found that the combined performance of the WGAN-LSTM and WGAN-LSTM-DISO models improved by 51.9% and 75.2%, respectively, compared to the LSTM model. Based on the test results of the models on other buoys, the two proposed models demonstrate good generalization performance.

The main contributions of this study are summarized as follows:

1. We propose a novel DL-based WGAN-LSTM method for small-scale Arctic SIT hindcasting, which alleviates the challenge of DL to address the insufficient amount of data for Arctic SIT prediction.
2. Compared with LSTM, the robustness and generalization ability of the WGAN-LSTM model is improved. The synthetic data generated by WGAN contains various possible sea ice change scenarios, which enables LSTM to better adapt to sea ice changes in different environments during the training process, and can more accurately capture the time-series features of SIT changes, thus improving the prediction accuracy and reducing the overfitting phenomenon.
3. Using a new comprehensive index DISO as the loss function, the anomalies appearing in the testing process are effectively corrected, but it can cause the model to become overconfident.
4. We demonstrate that the model prediction results can accurately reflect the physical relationship between air temperature and ice thickness.

However, the black-box nature of neural networks makes the model uncertain, and the predictions are less interpretable. We have only experimented on a small dataset, and the uncertainty and quality of the data seriously affect the model's performance, so the model's performance, when applied to a broader range of SIT predictions, remains to be explored. Sea ice changes show substantial variability and spatial–temporal heterogeneity due to the combined influence of complex physical mechanisms such as oceanic factors, atmospheric factors, and regional distribution. In the future, we can move toward combining deep learning models with expertise in the meteorological and oceanographic fields to improve the applicability and accuracy of the models through the selection and screening of input data, the constraints of physical mechanisms, and the processing of long time series.

Author Contributions: All authors contributed significantly to this manuscript. B.G.: writing—original draft preparation, software, drawings, and investigation; Y.L.: writing—review and editing, data curation, and resource acquisition; L.W.: writing—review and editing, resources, and funding acquisition; P.L.: funding acquisition; H.L.: supervision. All authors have read and agreed to the published version of the manuscript.

Funding: This study is supported by (1) the National Natural Science Foundation of China (No. 42320104004) and (2) the Fundamental Research Funds for the Central Universities (No. DUT24LK006).

Data Availability Statement: The data used in this study are cited in the references. The simulation results generated during this study are available from the corresponding author upon reasonable request.

Conflicts of Interest: The authors declare no conflicts of interest.

Abbreviations

The following abbreviations are used in this manuscript:

GAN	Generative adversarial network
WGAN	Wasserstein Generative adversarial network
LSTM	Long short-term memory
DL	Deep learning
SIT	Sea ice thickness
SW	Snow depth
SIMBA	Sea ice mass balance buoy
ERA5	European Centre for Medium-Range Weather Forecasts Reanalysis version 5
MOSAiC	Multidisciplinary drifting Observatory for the Study of Arctic Climate
weed	10 m wind field
fsw	Surface solar radiation downwards
t2m	2 m air temperature
tp	Total precipitation
tcc	Proportion of total cloud cover
MAE	Mean absolute error
RMSE	Root mean square error
RMSD	Center root mean square error
DISO	Distance between Indices of Simulation and Observation
MC	Monte Carlo
LeakyReLU	Leaky Rectified Linear Unit
RMSprop	Root Mean Square Propagation
AdamW	Adam with Weight Decay

References

1. Francis, J.A.; Chan, W.; Leathers, D.J.; Miller, J.R.; Veron, D.E. Winter Northern Hemisphere weather patterns remember summer Arctic sea-ice extent. *Geophys. Res. Lett.* **2009**, *36*, L07503. [CrossRef]
2. Serreze, M.C.; Barry, R.G. Processes and impacts of Arctic amplification: A research synthesis. *Glob. Planet. Change* **2011**, *77*, 85–96. [CrossRef]
3. Curry, J.A.; Schramm, J.L.; Ebert, E.E. Sea Ice-Albedo Climate Feedback Mechanism. *J. Clim.* **1995**, *8*, 240–247. [CrossRef]
4. Storlazzi, C.D.; Gingerich, S.B.; Van Dongeren, A.; Cheriton, O.M.; Swarzenski, P.W.; Quataert, E.; Voss, C.I.; Field, D.W.; Annamalai, H.; Piniak, G.A. Most atolls will be uninhabitable by the mid-21st century because of sea-level rise exacerbating wave-driven flooding. *Sci. Adv.* **2018**, *4*, eaap9741. [CrossRef]
5. Cohen, J.; Screen, J.A.; Furtado, J.C.; Barlow, M.; Whittleston, D.; Coumou, D.; Francis, J.; Dethloff, K.; Entekhabi, D.; Overland, J.; et al. Recent arctic amplification and extreme mid-latitude weather. *Nat. Geosci.* **2014**, *7*, 627–637. [CrossRef]
6. Huang, J.; Hitchcock, P.; Maycock, A.C.; Mckenna, C.M.; Tian, W. Northern hemisphere cold air outbreaks are more likely to be severe during weak polar vortex conditions. *Commun. Earth Environ.* **2021**, *2*, 147. [CrossRef]
7. Hou, Y.; Cai, W.; Holland, D.M.; Cheng, X.; Zhang, J.; Wang, L.; Johnson, N.C.; Xie, F.; Sun, W.; Yao, Y.; et al. A surface temperature dipole pattern between Eurasia and North America triggered by the Barents–Kara sea-ice retreat in boreal winter. *Environ. Res. Lett.* **2022**, *17*, 114047. [CrossRef]
8. Bailey, H.; Hubbard, A.; Klein, E.S.; Mustonen, K.R.; Welker, J.M. Arctic sea-ice loss fuels extreme European snowfall. *Nat. Geosci.* **2021**, *14*, 283–288. [CrossRef]
9. Zou, Y.; Wang, Y.; Zhang, Y.; Koo, J.H. Arctic sea ice, Eurasia snow, and extreme winter haze in China. *Sci. Adv.* **2017**, *3*, e160275. [CrossRef]
10. Smith, L.C.; Stephenson, S.R. New Trans-Arctic shipping routes navigable by midcentury. *Proc. Natl. Acad. Sci. USA* **2013**, *3*, E1191–E1195. [CrossRef]
11. Liu, M.; Kronbak, J. The potential economic viability of using the Northern Sea Route (NSR) as an alternative route between Asia and Europe. *J. Transp. Geogr.* **2010**, *8*, 434–444. [CrossRef]
12. Stephenson, S.R.; Smith, L.C.; Agnew, J.A. Divergent long-term trajectories of human access to the Arctic. *Nat. Clim. Change* **2011**, *1*, 156–160. [CrossRef]
13. Rigot- Müller, P.; Cheaitou, A.; Etienne, L.; Faury, O.; Fedi, L. The role of polarseaworthiness in shipping planning for infrastructure projects in the Arctic: The case of Yamal LNG plant. *Transp. Res. Part A Policy Pract.* **2022**, *155*, 330–353. [CrossRef]
14. Sallila, H.; Farrell, S.L.; McCurry, J.; Rinne, E. Assessment of contemporary satellite sea ice thickness products for Arctic sea ice. *Cryosphere* **2019**, *13*, 1187–1213. [CrossRef]

15. Gu, F.; Zhang, R.; Tian-Kunze, X.; Han, B.; Zhu, L.; Cui, T.; Yang, Q. Sea Ice Thickness Retrieval Based on GOCI Remote Sensing Data: A Case Study. *Remote Sens.* **2021**, *13*, 936. [CrossRef]
16. Landy, J.C.; Dawson, G.J.; Tsamados, M.; Bushuk, M.; Stroeve, J.C.; Howell, S.E.L.; Krumpen, T.; Babb, D.G.; Komarov, A.S.; Heorton, H.D.B.S. A year-round satellite sea-ice thickness record from CryoSat-2. *Nature* **2022**, *609*, 517–522. [CrossRef]
17. Wang, Y.; Yuan, X.; Bi, H.; Ren, Y.; Liang, Y.; Li, G.; Li, X. Understanding Arctic Sea Ice Thickness Predictability by a Markov Model. *J. Clim.* **2023**, *36*, 4879–4897. [CrossRef]
18. Hunke, E.C.; Hebert, D.A.; Lecomte, O. Level-ice melt ponds in the Los Alamos sea ice model, CICE. *Ocean Model.* **2013**, *71*, 26–42. [CrossRef]
19. Girard, L.; Amitrano, D.; Weiss, J. Failure as a critical phenomenon in a progressive damage model. *J. Stat. Mech. Theory Exp.* **2010**, *2010*, 577–611. [CrossRef]
20. Rösel, A.; Itkin, P.; King, J.; Divine, D.; Wang, C.; Granskog, M.A.; Krumpen, T.; Gerland, S. Thin sea ice, thick snow, and widespread negative freeboard observed during N-ICE2015 north of Svalbard. *J. Geophys. Res. Oceans* **2018**, *123*, 1156–1176. [CrossRef]
21. Jackson, K.; Wilkinson, J.; Maksym, T.; Meldrum, D.; Beckers, J.; Haas, C.; Mackenzie, D. A Novel and Low-Cost Sea Ice Mass Balance Buoy. *J. Atmos. Ocean. Technology.* **2013**, *30*, 2676–2688. [CrossRef]
22. Nicolaus, M.; Perovich, D.K.; Spreen, G.; Granskog, M.A.; Albedyll, L.V.; Angelopoulos, M.; Anhaus, P.; Arndt, S.; Belter, H.J.; Bessonov, V.; et al. Overview of the MOSAiC expedition: Snow and sea ice. *Elem. Sci. Anthr.* **2022**, *10*, 000046. [CrossRef]
23. Rabe, B.; Heuzé, C.; Regnery, J.; Aksenov, Y.; Allerholt, J.; Athanase, M.; Bai, Y.; Basque, C.; Bauch, D.; Baumann, T.M.; et al. Overview of the MOSAiC expedition: Physical oceanography. *Elem. Sci. Anthr.* **2022**, *10*, 00062. [CrossRef]
24. Lecun, Y.; Bengio, Y.; Hinton, G. Deep learning. *Nature* **2015**, *521*, 436–444. [CrossRef]
25. Chi, J.; Kim, H.C. Prediction of Arctic Sea Ice Concentration Using a Fully Data Driven Deep Neural Network. *Remote Sens.* **2017**, *9*, 1305. [CrossRef]
26. Liu, L. A review of deep learning for cryospheric studies. In *Deep Learning for the Earth Science: A Comprehensive Approach to Remote Sensing, Climate Science, and Geosciences*; John Wiley & Sons: Hoboken, NJ, USA, 2021; pp. 258–268.
27. Anderson, T.R.; Hosking, J.S.; Pérez-Ortiz, M.; Paige, B.; Elliott, A.; Russell, C.; Law, S.; Jones, D.C.; Wilkinson, J.; Phillips, T. Seasonal Arctic sea ice forecasting with probabilistic deep learning. *Nat. Commun.* **2021**, *12*, 5124. [CrossRef]
28. Ren, Y.; Zhang, W. A data-driven deep learning model for weekly sea ice concentration prediction of the pan-arctic during the melting season. *IEEE Trans. Geosci. Remote Sens.* **2022**, *60*, 4304819. [CrossRef]
29. Song, C.T.; Zhu, J.; Li, X.C. Assessments of Data-Driven Deep Learning Models on One-Month Predictions of Pan-Arctic Sea Ice Thickness. *Adv. Atmos. Sci.* **2024**, *41*, 1379–1390. [CrossRef]
30. Shi, X.; Chen, Z.; Wang, H.; Yeung, D.Y.; Wong, W.K.; Woo, W.C. Convolutional LSTM Network: A Machine Learning Approach for Precipitation Nowcasting. In Proceedings of the 29th Annual Conference on Neural Information Processing Systems (NIPS), Montreal, QC, Canada, 7–12 December 2015.
31. Hochreiter, S.; Schmidhuber, J. Long short-term memory. *Neural Comput.* **1997**, *9*, 1735–1780. [CrossRef]
32. Goodfellow, I.; Pouget-Abadie, J.; Mirza, M.; Xu, B.; Warde-Farley, D.; Ozair, S.; Courville, A.; Bengio, Y. Generative Adversarial Nets. In Proceedings of the 28th Conference on Neural Information Processing Systems (NIPS), Montreal, QC, Canada, 8–13 December 2014.
33. Arjovsky, M.; Chintala, S.; Bottou, L. Wasserstein Generative Adversarial Networks. In Proceedings of the 34th International Conference on Machine Learning, Sydney, Australia, 6–11 August 2017.
34. Xu, Z.; Du, J.; Wang, J.J.; Jiang, C.X.; Ren, Y. Satellite Image Prediction Relying on GAN and LSTM Neural Networks. In Proceedings of the IEEE International Conference on Communications (IEEE ICC), Shanghai, China, 20–24 May 2019.
35. Mavsar, M.; Morimoto, J.; Ude, A. GAN-Based Semi-Supervised Training of LSTM Nets for Intention Recognition in Cooperative Tasks. *IEEE Robot. Autom. Lett.* **2024**, *9*, 263–270. [CrossRef]
36. Bousmina, A.; Selmi, M.; Ben Rhaiem, M.A.; Farah, I.R. A Hybrid Approach Based on GAN and CNN-LSTM for Aerial Activity Recognition. *Remote Sens.* **2023**, *15*, 3626. [CrossRef]
37. Hersbach, H.; Bell, B.; Berrisford, P.; Hirahara, S.; Horányi, A.; Muñoz-Sabater, J.; Nicolas, J.; Peubey, C.; Radu, R.; Schepers, D.; et al. The ERA5 global reanalysis. *Q. J. R. Meteorol. Soc.* **2020**, *146*, 1999–2049. [CrossRef]
38. Lei, R.B.; Cheng, B.; Hoppmann, M.; Zhang, F.Y.; Zuo, G.Y.; Hutchings, J.K.; Lin, L.; Lan, M.S.; Wang, H.Z.; Regnery, J.; et al. Seasonality and timing of sea ice mass balance and heat fluxes in the Arctic transpolar drift during 2019–2020. *Elem. Sci. Anthr.* **2022**, *10*, 000089. [CrossRef]
39. Shorten, C.; Khoshgoftaar, T.M. A survey on Image Data Augmentation for Deep Learning. *J. Big Data* **2019**, *6*, 60. [CrossRef]
40. Schmidhuber, J. Deep learning in neural networks: An overview. *Neural Netw.* **2015**, *61*, 85–117. [CrossRef] [PubMed]
41. Elman, J.L. Finding Structure in Time. *Cogn. Sci.* **1990**, *14*, 179–211. [CrossRef]

42. Hu, Z.; Chen, X.; Zhou, Q.; Chen, D.; Li, J. DISO: A rethink of Taylor diagram. *Int. J. Climatol.* **2019**, *39*, 2825–2832. [CrossRef]
43. Hu, Z.; Hu, Q.; Zhang, C.; Chen, X.; Li, Q. Evaluation of Reanalysis, Spatially-interpolated and Satellite Remotely-sensed Precipitation data sets in Central Asia. *J. Geophys. Res.-At.* **2016**, *121*, 5648–5663. [CrossRef]
44. Kononenko, I. Bayesian neural networks. *Biol. Cybern.* **1989**, *61*, 361–370. [CrossRef]
45. Gal, Y.; Ghahramani, Z. Dropout as a Bayesian Approximation: Representing Model Uncertainty in Deep Learning. In Proceedings of the 33rd International Conference on Machine Learning, New York, NY, USA, 20–22 June 2016.
46. Srivastava, N.; Hinton, G.; Krizhevsky, A.; Sutskever, I.; Salakhutdinov, R. Dropout: A Simple Way to Prevent Neural Networks from Overfitting. *J. Mach. Learn. Res.* **2014**, *15*, 1929–1958.
47. Rasmussen, C.E.; Williams, C.K.I. Gaussian Processes for Machine Learning. *Int. J. Neural Syst.* **2004**, *14*, 69–106.

water

MDPI

Article

Statistical Optimization and Analysis on the Spatial Distributions of Ice Ridge Keel in the Northwestern Weddell Sea, Antarctica

Bing Tan [1,*], Yanming Chang [1], Chunchun Gao [2,*], Ting Wang [1], Peng Lu [3], Yingzhe Fan [1] and Qingkai Wang [3]

[1] School of Mathematics and Statistics, Nanyang Normal University, Nanyang 473061, China; ymchang@nynu.edu.cn (Y.C.); tingwang526@nynu.edu.cn (T.W.); yzfan@nynu.edu.cn (Y.F.)
[2] College of Geodesy and Geomatics, Shandong University of Science and Technology, Qingdao 266590, China
[3] State Key Laboratory of Coastal and Offshore Engineering, Dalian University of Technology, Dalian 116024, China; lupeng@dlut.edu.cn (P.L.); wangqingkai@dlut.edu.cn (Q.W.)
* Correspondence: tanbing@nynu.edu.cn (B.T.); gaochunchun@sdust.edu.cn (C.G.); Tel.: +86-0377-63510029 (B.T.); +86-0532-80681159 (C.G.)

Abstract: Statistical optimization methods serve as fundamental tools for studying sea-ice-related phenomena in the polar regions. To comprehensively analyze the spatial distributions of ice ridge keels, including the draft and spacing distributions, a statistical optimization model was developed with the aim of determining the optimal keel cutoff draft, which differentiates ice ridge keels from sea ice bottom roughness. By treating the keel cutoff draft as the identified variable and minimizing the relative errors between the theoretical and measured keel spatial distributions, the developed model aimed to seek the optimal keel cutoff draft and provide a precise method for this differentiation and to explore the impact of the ridging intensity, defined as the ratio of the mean ridge sail height to spacing, on the spatial distributions of the ice ridge keels. The utilized data were obtained from observations of sea ice bottom undulations in the Northwestern Weddell Sea during the winter of 2006; these observations were conducted using helicopter-borne electromagnetic induction (EM-bird). Through rigorous analysis, the optimal keel cutoff draft was determined to be 3.8 m, and this value was subsequently employed to effectively differentiate ridge keels from other roughness features on the sea ice bottom. Then, building upon our previous research that clustered measured profiles into three distinct regimes (Region 1, Region 2, and Region 3, respectively), a detailed statistical analysis was carried out to evaluate the influence of the ridging intensity on the spatial distributions of the ice ridge keels for all three regimes. Notably, the results closely matched the predictions of the statistical optimization model: Wadhams'80 function (a negative exponential function) exhibited an excellent fit with the measured distributions of the keel draft, and a lognormal function proved to effectively describe the keel spacing distributions in all three regimes. Furthermore, it was discovered that the relationship between the mean ridge keel draft and frequency (number of keels per kilometer) could be accurately modeled by a logarithmic function with a correlation coefficient of 0.698, despite considerable data scatter. This study yields several significant results with far-reaching implications. The determination of the optimal keel cutoff draft and the successful modeling of the relationship between the keel draft and frequency represent key achievements. These findings provide a solid theoretical foundation for analyzing the correlations between the morphologies of the sea ice surface and bottom. Such theoretical insights are crucial for improving remote sensing algorithms for ice thickness inversion from satellite elevation data, enhancing the accuracy of sea ice thickness estimations.

Keywords: mathematical modeling; antarctica; keel spatial distribution; optimization; statistical analyses

1. Introduction

Climate warming has triggered a series of profound environmental changes in the polar regions. The declining polar sea ice and consequent rising sea levels are not only reshaping polar ecosystems and natural environments but also redrawing the geostrategic landscape of the Arctic and Antarctic, which have attracted increasing global attention [1–6].

The complex seasonal and interannual variations in sea ice represent defining characteristics of polar seas and are closely intertwined with the global climate system and ecosystems. They significantly modulate the momentum, heat, and mass exchanges among the atmosphere, sea ice, and ocean, contributing to phenomena such as rising sea levels, accelerated climate warming, and more frequent extreme weather events [7–11]. Understanding the morphology and distribution of sea ice is essential for improving parameterization schemes for sea ice thermodynamics and optimizing remote sensing algorithms for ice thickness estimation by combining them with sea ice roughness data, which are the direct thermodynamic outcomes of interactions among the atmosphere, sea ice, and ocean [12–15].

Ice ridges, characterized by their sails (above-surface protrusions) and keels (below-bottom extensions) and formed by dynamic deformation induced by wind, waves, and ocean currents, as well as the crushing and overlapping of collisions within brash ice at the edges of ice floes, are prominent morphological features of sea ice and key elements in the sea ice structure [14]. A predefined cutoff height or draft is commonly used to distinguish ice ridge sails/keels from other sea ice surface/bottom undulations: features with height/draft values below this cutoff are categorized as non-ridged ice surface/bottom undulations, while those above are identified as ice ridge sails/keels [16,17]. This cutoff parameter is crucial for estimating the volume and thickness of sea ice, understanding the exchanges of heat, energy, and momentum between the atmosphere and ocean, and developing large-scale sea ice thermodynamic models [18,19]. Consequently, the study of ice ridge sail and keel morphologies and distributions has become an increasing area of research [13,18–23].

Despite its importance, large-scale, continuous, and high-precision field observations and quantitative analyses of sea ice bottom draft remain scarce, owing to the significant challenges posed by harsh and complex polar environments, the random and intricate spatial distribution of sea ice, and the limitations of observational equipment in terms of adaptability and resolution. The current methods for inferring sea ice bottom morphology and spatial distribution from ice surface data obtained via satellite altimetry (e.g., IceBridge, CryoSat-2, and SMOS) predominantly rely on empirical approaches, which often employ preassumed parameters regarding snow and ice morphology and distribution [14,24,25]. However, unrealistic theoretical assumptions and the omission of key parameters compromise the accuracy of representing real-situation conditions and explaining the variations in polar sea ice morphology and distribution. This limitation hinders the refinement of polar marine climate models.

In contrast, mathematical models and methods offer a powerful alternative. They can effectively describe the complex processes of sea ice freezing and melting, its random spatiotemporal distribution, and the interactions between sea ice and external driving forces such as wind, currents, and waves. Additionally, these mathematical models and methods can generate data that are difficult to obtain through field measurements or laboratory simulations, making them invaluable tools in sea ice research [7,17].

Against this backdrop, utilizing a dataset of sea ice bottom undulations collected in the Northwestern Weddell Sea during the winter of 2006 and building on our previous research on ice ridge sail morphology and classification [17,22], this study aims to determine the optimal keel cutoff draft and analyze the morphology and distribution of ice ridge keels by integrating data analysis, mathematical modeling, statistical optimization, and parameter identification techniques. The ultimate goal is to provide a novel and objective theoretical basis and experimental validation for the inversion of polar sea ice thickness using satellite or other altimetry techniques derived from surface elevation data.

This paper is organized as follows. Section 2 details the geographical region, field conditions, and data collection and processing methods. Section 3 develops a nonlinear-constrained statistical optimization model for determining the optimal keel cutoff draft, where the performance function is defined as the sum of relative errors between measured and theoretical keel spatial distributions, and the cutoff draft serves as the parameter to be identified. Section 4 presents a comprehensive analysis of the influence of the ridging intensity (the ratio of mean ridge sail height to spacing) on the keel spatial distributions, integrating previous clustering results [22], and explores the relationship between keel draft and frequency (number of ridge keels per kilometer), as well as that between the sail height and frequency based on earlier research [17]. Finally, the conclusions are presented in the last section.

2. Investigated Region and Data

2.1. Geographical Region and Field Conditions

From 24 August to 29 October 2006, the Alfred Wegener Institute for Polar and Marine Research carried out the Winter Weddell Outflow Study in the northwest Weddell Sea of Antarctica. During this study, a helicopter-borne electromagnetic induction bird (EM-bird) system was employed to collect data on the sea ice roughness.

The measurement area spanned a latitude ranging from 60° S to 66° S and a longitude ranging from 40° W to 60° W. A total of 17 helicopter flights were conducted, comprehensively covering the diverse ice conditions and types encountered throughout the expedition (Figure 1).

Considering the field conditions and the sea ice formation mechanism [19], the investigated area was partitioned into three distinct sub-regions: the northern, central, and southern sub-regions, respectively.

The northern sub-region, located within the marginal ice zone between approximately 60° S and 62° S, was characterized by relatively weak dynamic forces, which led to the minimal deformation of both the sea ice surface and bottom. Consequently, ice ridge sails and keels were scarce and smaller, mainly due to limited ice floe overlap. In fact, some ice bottom profiles in this sub-region showed the complete absence of ice ridge keels.

In contrast, the central sub-region, ranging from around 62° S to 63.5° S, experienced a significant increase in sea ice thickness. This growth was driven by the dynamic interactions between first- and second-year ices. Ice that endured the summer underwent further freezing and development during the subsequent winter, giving rise to a greater abundance of larger ice ridge sails and keels.

The southern sub-region was predominantly dominated by first-year ice formed in the Larsen Polynya. Environmental forces induced icebergs and glacial ice with distinct velocities, generating a remarkable diversity of ice ridge sails and keels. The outflowing branch of the Weddell Gyre, notable for its intense ice deformation processes, featured the most prominent and densely packed ridges along the shelf ice edge. Meanwhile, the Larsen Polynya exhibited an almost complete absence of ridge keels on the ice bottom and a low ridge frequency on the ice surface [17]. Here, ice ridges were smaller in size and fewer in number, and were primarily formed by broken level ice.

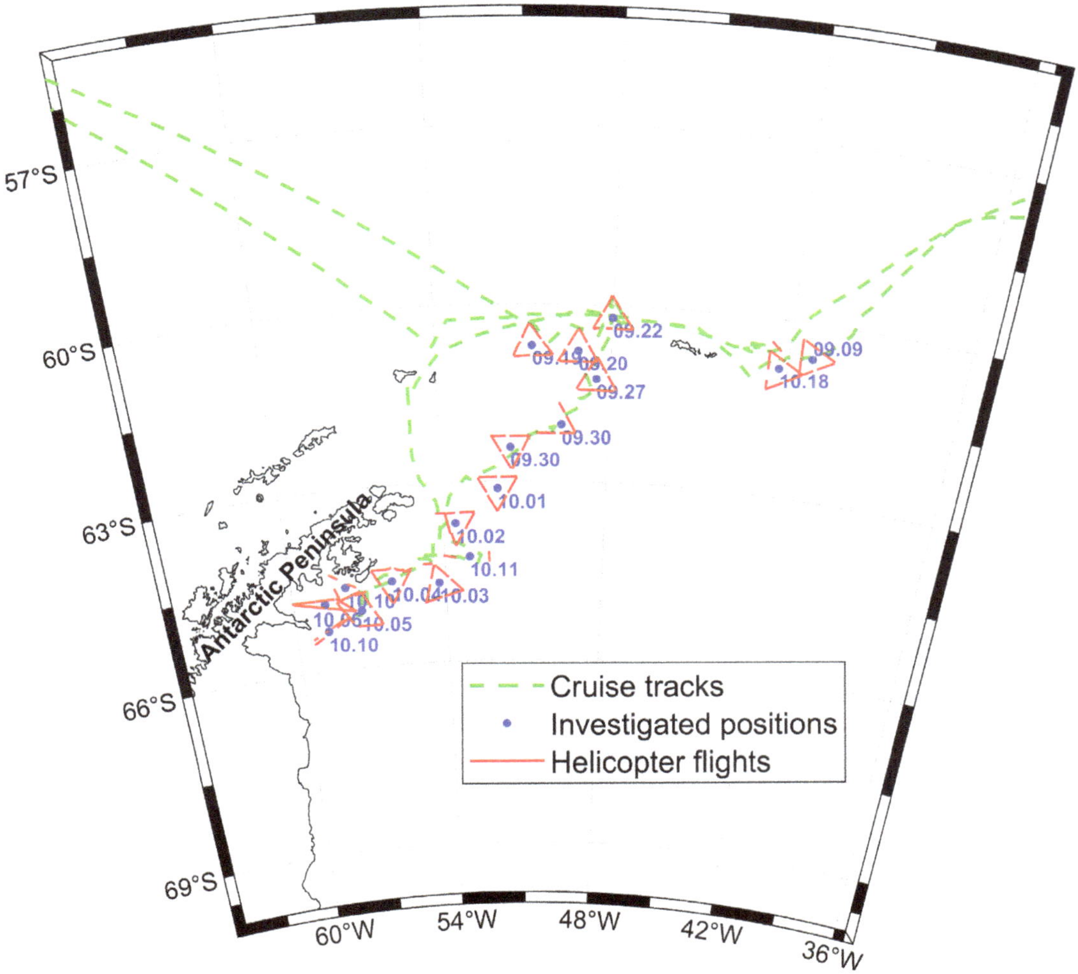

Figure 1. Cruise tracks (green lines) and helicopter flights (red lines) during field investigation in Northwestern Weddell Sea; the locations of the observations along the helicopter flights are denoted by blue circles with the appropriate date (the month and day of 2006).

2.2. Data Collection and Processing

A total distance of 2988.5 km was surveyed, comprising 94 profiles of sea ice surface and bottom undulations. These profiles varied in length, ranging from 6.3 to 56.8 km.

The EM-bird system used in the survey was trailed 20 m below the helicopter and operated at an altitude of 10–20 m above the sea ice surface [17] (Figure 2). The Riegl LD90 laser altimeter, mounted on the nose of the EM-bird, was responsible for randomly measuring sea ice surface elevations. This altimeter achieved an accuracy of 2.5 cm, emitting infrared pulses at a wavelength of 905 nm with a sampling frequency of 100 Hz.

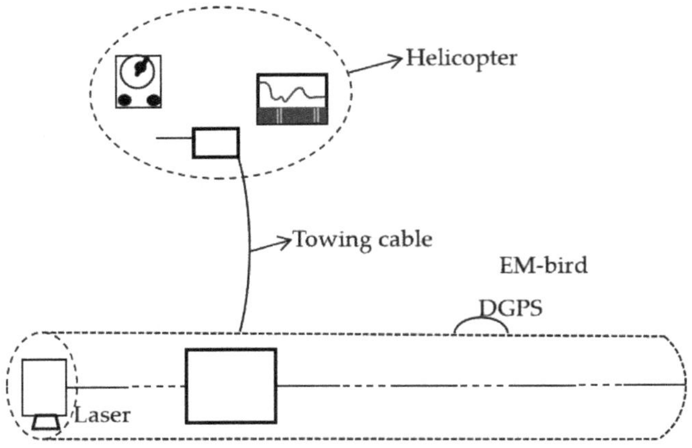

Figure 2. Sketch of the EM-bird system employed during field investigation in Northwestern Weddell Sea.

Regarding the EM-bird itself, it was 3.5 m in length, 0.35 m in diameter, and weighed 105 kg. It collected sea ice bottom undulation data at a sampling frequency of 10 Hz. The system offered an accuracy of more than 0.1 m when measuring level ice; however, this accuracy diminished when measuring rough or deformed ice bottoms.

The helicopter maintained a flight speed of approximately 40 m/s. This speed resulted in distinct spatial sampling distances for the measurement instruments: the laser altimeter had a sampling distance of 0.3–0.4 m between adjacent data points, while the EM-bird's sampling distance ranged from 3 m to 4 m.

A positioning system (GPS) fixed on the helicopter deck recorded the altitude of the helicopter relative to a reference point at intervals of 0.1 s. Meanwhile, an onboard computer facilitated the real-time storage and processing of all collected data, ensuring efficient data management during the survey.

The accuracy of sea ice surface and bottom undulation measurements is highly susceptible to variations in the flight altitude and speed of the helicopter. To mitigate the influence of the helicopter's motion state on the elevation data, a three-step automatic filtering method, as described in [17], is indispensable. This method combines three key data sources: the flight altitude of the helicopter relative to the water line (h), the distance from the helicopter to the sea ice surface measured by the laser altimeter (h_s), and the distance from the helicopter to the ice bottom measured by the EM-bird (h_b).

For corresponding measurement points in the two datasets, the sea ice surface elevation and sea ice bottom draft can be accurately calculated as $h - h_s$ and $h_b - h$, respectively. And the corresponding morphologies of the sea ice profile are illustrated in Figure 3, by assuming the ice ridge keel cross-sections as isosceles triangles.

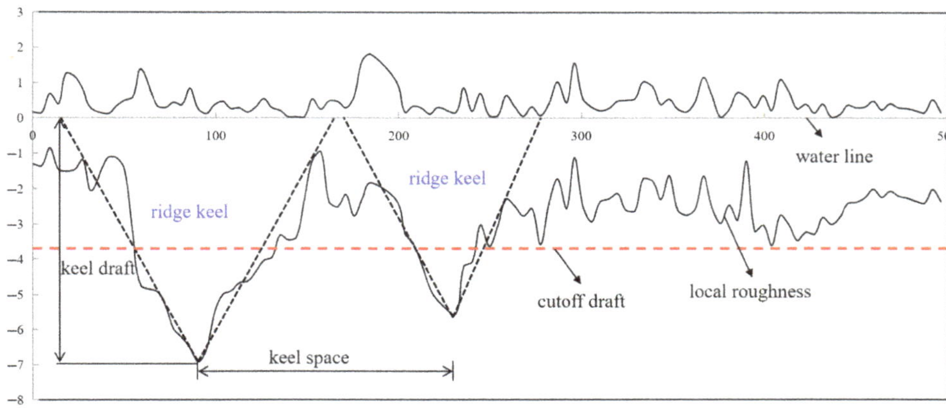

Figure 3. Diagrammatic representation of sea ice bottom morphologies. The black isosceles triangles with dotted lines represent the approximate cross-sections of ridge keels, while the black curves denote the measured data of sea ice roughness.

3. Optimal Model of the Cutoff Draft

The optimal identification of the keel cutoff draft is widely recognized as a pivotal parameter for accurately distinguishing ridge keels from other ice bottom undulations. This parameter significantly influences the correlation between sea ice surface and bottom characteristics, as well as the effective estimation of ice–water dynamic interactions, mirroring the critical impact of the cutoff height on ice ridge sails [17]. Therefore, determining the keel cutoff draft represents a fundamental step in analyzing ridge keel morphology and understanding the relationship between sea ice surface and bottom features.

Traditionally, the keel cutoff draft has been determined by integrating the Rayleigh criterion [26] with field observation experience [27,28]. Nevertheless, owing to the absence of a defined upper limit for the threshold value, the keel cutoff draft selected by the Rayleigh criterion depended generally on the considerate data set and the researcher's subjective, empirical judgment. An excessively large cutoff value would tend to overlook some smaller ridge keels. These smaller keels, although less prominent, can still play a significant role in the overall dynamics of sea ice, including its interaction with ocean currents, heat exchange, and the mechanical properties of the ice pack. Moreover, the lack of an upper bound constraint for the Rayleigh criterion is associated with an arbitrary cutoff parameter and inconsistent results in different previous studies, undermining the reliability and comparability of studies related to sea ice morphology, highlighting the need for a more refined and data-driven approach to determining the keel cutoff draft.

In this section, we will develop a nonlinear statistical optimization model. Subsequently, the optimal keel cutoff draft obtained from this model will be used to objectively differentiate local undulations and ridge keels on the sea ice bottom.

3.1. Classical Theoretical Distributions of the Keel Draft and Spacing

Let $f_d = f_d\ (d; d_{c-k}, \Theta)$ and $f_s = f_s\ (s; d_{c-k}, \Psi)$ denote the theoretical probability density functions for the keel draft and spacing, respectively. Both functions are Lipschitz continuous. Here, $d \in D{:} = [d_{c-k}, d_{\max}]$ denotes the ridge keel draft, where $d_{\max} \in R^+$ (R^+ is the set of positive real numbers) is the maximum keel draft and $d_{c-k} \in U(d_{c-k})$ is the keel cutoff draft, with $U(d_{c-k})$ being the admissible parameter set for this parameter. Additionally, $s \in S{:} = [s_{\min}, s_{\max}]$ denotes the keel spacing, with $s_{\min}$ and $s_{\max}$ being the minimum and maximum spacings, respectively, both $s_{\min}$ and $s_{\max}$ belonging to R^+ and satisfying $s_{\min} < s_{\max}$, Θ, and Ψ representing the parameter sets closely associated with the cutoff draft.

Two classical theoretical functions have been proposed to characterize the ridge keel draft distribution, one of which was proposed by Hibler et al. [29]

$$f_d(d; d_{c-k}, \lambda_1) = \frac{2\sqrt{\lambda_1/\pi}\exp(-\lambda_1 d^2)}{erfc(d_{c-k}\sqrt{\lambda_1})},\ d > d_{c-k}, \tag{1}$$

where $erfc(x) = \frac{2}{\pi}\int_x^{+\infty} e^{-t^2}dt$ and λ_1 is a relevant parameter, with the mean keel draft <d> determined by the following:

$$< d >= \frac{\exp(-\lambda_1 {d_{c-k}}^2)}{\sqrt{\lambda_1 \pi} erfc(d_{c-k}\sqrt{\lambda_1})}. \tag{2}$$

Another theoretical function was proposed by Wadhams [30], who showed that the following negative exponential function can accurately model the distribution of keel drafts and proved it to be a powerful tool for characterizing the statistical behavior of sea ice keel drafts.

$$f_d(d; d_{c-k}, \lambda_2) = \lambda_2 \exp(-\lambda_2(d - d_{c-k})),\ d > d_{c-k} \tag{3}$$

where λ_2 is a relevant parameter with a mean keel draft <d> determined by the following:

$$< d >= d_{c-k} + {\lambda_2}^{-1} \tag{4}$$

In subsequent sections, Equation (1) will be referred to as the Hiblers'72 function, and Equation (3) as the Wadhams'80 function for simplicity.

Building on the assumption of spatial independence, Hibler et al. [29] employed a homogeneous Poisson process to model the occurrence of ridge keels and derived a negative exponential distribution density function that effectively describes the distribution of ridge keel spacing.

$$f_s(s; d_{c-k}, \lambda_3) = \lambda_3 \exp(-\lambda_3 s),\ \ d > d_{c-k}, \tag{5}$$

where λ_3 is the distribution shape parameter related to the mean spacing <s> by the following:

$$\lambda_3 =< s >^{-1}. \tag{6}$$

Wadhams and Davy [31] showed that a lognormal function provides a superior fit for the distribution of ridge keel spacing compared to other models.

$$f_s(s; d_{c-k}, \theta, \mu, \sigma) = \frac{1}{\sqrt{2\pi}\sigma(s-\theta)} \cdot \exp\left(-\frac{(\ln(s-\theta)-\mu)^2}{2\sigma^2}\right)\ s > \theta,\ d > d_{c-k}, \tag{7}$$

where θ is the shift parameter, and μ and σ are the mean and standard error of ln(s–θ), respectively. They are all related to the mean spacing <s> by the following:

$$< s >= \theta + \exp(\mu + \frac{\sigma^2}{2}). \tag{8}$$

3.2. Statistical Optimization Model of the Cutoff Draft

As the following two equations illustrate, let $E_d(d_{c-k})$ and $E_s(d_{c-k})$ represent the relative errors, respectively, between the measured ridge keel draft distribution and its corresponding theoretical function, and between the measured ridge keel spacing distribution and its associated theoretical function.

$$E_d(d_{c-k}) = \frac{\sum_{i=1}^{n} |f_d(d_i, d_{c-k}; \lambda_2) - f_{d-i}|}{\sum_{i=1}^{n} |f_{d-i}|}, \; d_i > d_{c-k}, \tag{9}$$

$$E_s(d_{c-k}) = \frac{\sum_{j=1}^{m} |f_s(s_j, d_{c-k}; \theta, \mu, \sigma) - f_{s-j}|}{\sum_{j=1}^{m} |f_{s-j}|}, \tag{10}$$

where d_i is the measured ridge keel draft, s_j is the measured ridge keel spacing, f_{d-i} is the probability density of the measured ridge keel draft, f_{s-j} is the probability density of the measured ridge keel spacing, and i = 1, 2, . . ., n, j = 1, 2, . . ., m, for the given cutoff draft $d_{c-k} \in U(\cdot)$.

The admissible parameter set for the keel cutoff draft $U(\cdot) = \{d_{c-k} \mid 3.5 \text{ m} \leq d_{c-k} \leq 4.2 \text{ m}\}$ is limited by two critical constraints: the Rayleigh criterion and the necessity of adequately encompassing the measured keel drafts. Mathematically, this set possesses distinct properties. Specifically, $U(\cdot)$ is nonempty, bounded, closed, and convex, indicating the existence of valid limited parameter values within it.

To find out the best fits for the ridge keel spatial distributions, set $d_{c-k,0}$ = 3.5 m as the initial cutoff draft and Δd_{c-k} = 0.1 m as the step increment. The analytical results indicate that, for any keel cutoff draft within the admissible parameter set $U(\cdot)$, the Wadhams'80 function provided the optimal fit for the observed keel draft distribution. In contrast, the lognormal function proved to best fit the measured keel spacing distribution. This suggests that these two respective functions effectively capture the statistical characteristics of keel drafts and spacings across the admissible range of the cutoff draft parameter.

We set

$$J_{d-s}(d_{c-k}) = E_{d-w}(d_{c-k}) + E_{s-l}(d_{c-k}), \tag{11}$$

$$E_{d-w}(d_{c-k}) = \frac{\sum_{i=1}^{n} |f_{d-w}(d_i, d_{c-k}; \lambda_2) - f_{d-i}|}{\sum_{i=1}^{n} |f_{d-i}|}, \; d_i > d_{c-k}, \tag{12}$$

$$E_{s-l}(d_{c-k}) = \frac{\sum_{j=1}^{m} |f_{s-l}(s_j, d_{c-k}; \theta, \mu, \sigma) - f_{s-j}|}{\sum_{j=1}^{m} |f_{s-j}|}, \tag{13}$$

where $J_{d-s}(d_{c-k})$ denotes the performance function, $E_{d-w}(d_{c-k})$ is the relative error between the measured keel draft distribution and Wadhams'80 function, $E_{s-l}(d_{c-k})$ is the relative error between the measured keel spacing distribution and lognormal distribution, $f_{d-w}(d_i, d_{c-k}; \lambda_2)$ denotes the probability density function of the Wadhams'80 function for the keel draft (Equation (3), i = 1, 2, . . ., n), and $f_{s-l}(s_j, d_{c-k}; \theta, \mu, \sigma)$ is the lognormal distribution for keel spacing (Equation (7), j = 1, 2, . . ., m), respectively.

A statistical optimization model of the cutoff draft is then established as follows:

$$\begin{cases} \min J_{d-s}(d_{c-k}) \\ s.t. \;\; f_{d-w}(d_i, d_{c-k}; \lambda_2) \in M(U_{ad}(d_{c-k})), \; i = 1, \cdots, n \\ \qquad f_{s-l}(s_j, d_{c-k}; \theta, \mu, \sigma) \in N(U_{ad}(d_{c-k})), \;\; j = 1, \cdots, m \\ \qquad d_{c-k} \in U_{ad}(d_{c-k}) \end{cases}, \tag{14}$$

where

$M(U(d_{c-k})) = \{f_{d-w}(d_i, d_{c-k}; \lambda_2) \mid f_{d-w}(d_i, d_{c-k}; \lambda_2)$ is the corresponding value of Wadhams'80 probability density function at the keel cutoff draft $d_{c-k} \in U(d_{c-k})$ (refer to Equation (3)), i = 1, 2, . . ., $n\}$.

$N(U(d_{c-k})) = \{f_{s-l}(s_j, d_{c-k}; \theta, \mu, \sigma) \mid f_{s-l}(s_j, d_{c-k}; \theta, \mu, \sigma)$ is the corresponding value of the lognormal probability density function at the keel cutoff draft $d_{c-k} \in U(d_{c-k})$ (refer to Equation (7)), $j = 1, 2, \ldots, m\}$.

Figure 4 shows how the performance function $J_{d-s}(d_{c-k})$, the relative errors $E_{d-w}(d_{c-k})$ and $E_{s-l}(d_{c-k})$ vary when the keel cutoff draft increases. Evidently, when the cutoff draft attains a value of 3.8 m, $E_{d-w}(d_{c-k})$, $E_{s-l}(d_{c-k})$, and $J_{d-s}(d_{c-k})$ each reach their respective minimum values, so the value of 3.8 m is judiciously selected as the optimal keel cutoff draft. Consequently, the statistical optimization model (Equation (14)) offers a precise and data-driven criterion for distinguishing ridge keels from other ice bottom features.

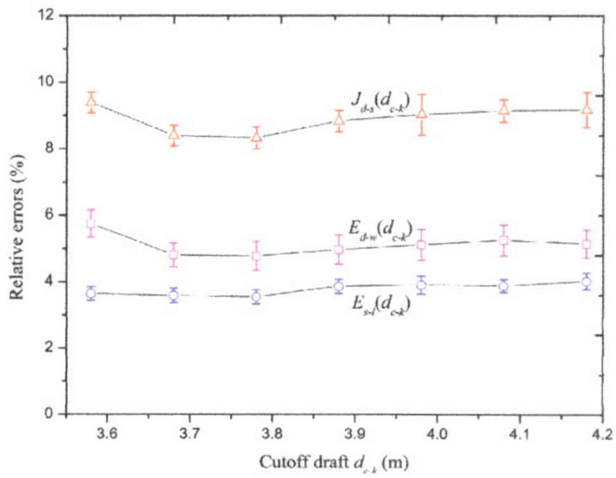

Figure 4. Variations in the relative errors $E_{d-w}(d_{c-k})$, $E_{s-l}(d_{c-k})$ ((Equations (12) and (13), respectively), and performance function $J_{d-s}(d_{c-k})$ (Equation (11)) for different cutoff drafts d_{c-k}.

4. Results

4.1. Keel Draft and Spacing Distributions in Different Regimes

Considering the influences of geographical location and growth environment, Tan et al. [22] developed an intelligent algorithm that employed ridging intensity R_i as the clustering index to cluster the measured profiles into three distinct regimes based on an increasing ridging intensity, labeled as Region 1 ($R_i \leq 0.01$), Region 2 ($0.01 < R_i \leq 0.026$), and Region 3 ($R_i > 0.026$), respectively. The results showed that Region 1 corresponds to the northern and part of the southern sub-regions, Region 2 to the remaining southern sub-region, and Region 3 to the central sub-region.

The clustering results from reference [22] are fundamental for all subsequent statistical analyses and discussions, which aim to explore how the ridging intensity affects the ridge keel spatial distributions, thereby providing a crucial framework for understanding the complex relationships between sea ice morphological characteristics and environmental factors.

4.1.1. Distribution of Keel Draft

To thoroughly compare the observed and theoretical distributions of ridge keel draft, Figure 5 illustrates the probability density functions (PDFs) of keel draft for the three regions (Region 1, Region 2, and Region 3). In Region 1 (Figure 5a), the data exhibit only slightly perceptible scatter, while Regions 2 (Figure 5b) and 3 (Figure 5c) show minor scatter at the higher ends. Notably, Wadhams'80 function demonstrates a strong fit with the measured keel draft distribution for each of the three regions.

In contrast, for all three regions, the Hiblers'72 function systematically overestimates intermediate keel draft values while consistently underestimating the observed data at both the lower and higher tails. Although the deviations at the high tails gradually decrease from Region 1 to Region 3, Figure 5 clearly highlights the poor fits between the Hiblers'72 function and the observed keel draft distributions for all regions.

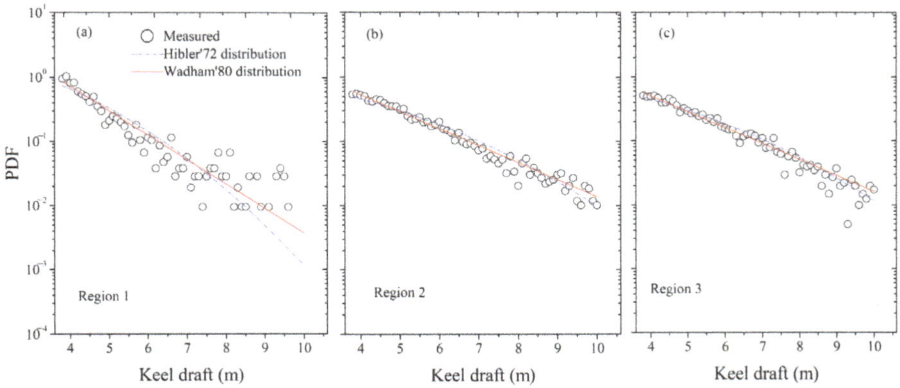

Figure 5. PDFs of ridge keel draft for each of the three regions; the Wadhams'80 distribution is shown by red solid lines, whereas the Hiblers'72 distribution is indicated by blue dash dotted lines. (**a**) Region 1; (**b**) Region 2; (**c**) Region 3.

To compare the fitted results of Hiblers'72 and Wadhams'80 functions to the keel draft distributions for the three regions, linear regression analysis was performed between the theoretical and the real measured values. Simultaneously, to quantify the deviation between the theoretical and measured keel draft distributions, we define the following mean absolute error *MAED*:

$$MAED = \frac{\sum_{i=1}^{n} |f_d(d_i, d_{c-k}; \Theta) - f_{d-i}|}{n}, \; d_i > d_{c-k}, \tag{15}$$

where the variables and functions are defined in accordance with Equation (9).

The analysis results are presented in Table 1. For Region 2 (moderate ridging intensity), Wadhams'80 function exhibits the strongest correlation with the measured keel draft distribution, with a linear correlation coefficient of 0.984 and a *MAED* value of 0.076. Both Region 1 and Region 3 show linear correlation coefficients exceeding 0.900; however, Region 1 has a *MAED* value of 0.194—an order of magnitude higher than other two, indicating a distinct error. The corresponding Hiblers'72 function demonstrates similar trends in the three regions but with consistently smaller linear correlation coefficients and larger *MAED* values than those of Wadhams'80 function. These results further confirm that Wadhams'80 function provides a better fit to the observed keel draft distribution compared to Hiblers'72 for each of the three regions.

Table 1. Linear correlation coefficients and *MAED* values between the theoretical and measured keel draft distributions.

Cluster	Hiblers'72 Distribution		Wadhams'80 Distribution	
	r	*MAED*	r	*MAED*
Region 1	0.872	0.258	0.908	0.194
Region 2	0.976	0.089	0.984	0.076
Region 3	0.970	0.085	0.973	0.070

4.1.2. Distribution of Keel Spacing

To compare the theoretical distributions with the observed keel spacing data, Figure 6 presents the probability density functions (PDFs) of keel spacing for the three regimes. Although there is a slight overestimation at the lower tails of the observed data, the lognormal function effectively describes the keel spacing distribution in each region. In contrast, the exponential function systematically overestimates moderate ridge spacings while underestimating observed keel spacings at both the lower and higher tails in all regions. Notably, the deviation of the measured data tails decreases gradually from Region 1 to Region 3, aligning with a reduction in the range of overestimation.

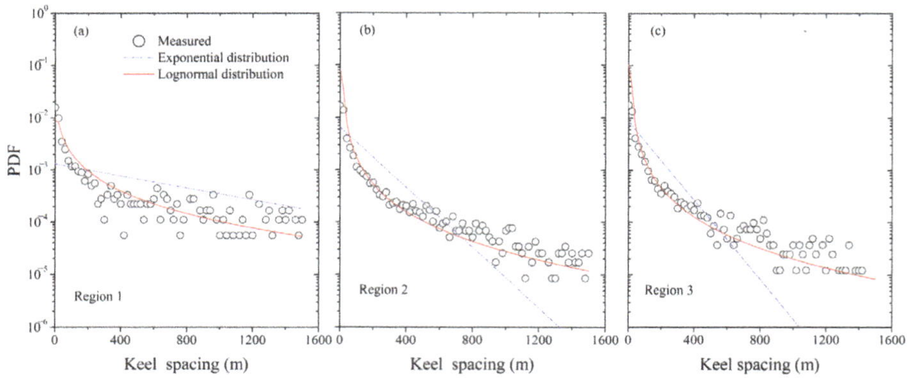

Figure 6. Ridge keel spacing PDFs for each of the three regions; the lognormal distribution is shown by red solid lines, whereas the exponential distribution is indicated by blue dash dotted lines. (**a**) Region 1; (**b**) Region 2; (**c**) Region 3.

As shown in Table 2, which lists the correlation coefficients between the two theoretical functions and the measured data, the linear correlations between the lognormal function and the observed keel spacing distributions are significantly stronger than those of the exponential function for all regions.

Table 2. Linear correlation coefficients and *MAES* values between the theoretical and measured keel spacing distributions.

Cluster	Exponential Distribution		Lognormal Distribution	
	r	*MAES*	r	*MAES*
Region 1	0.723	0.472	0.866	0.279
Region 2	0.904	0.613	0.968	0.183
Region 3	0.888	0.875	0.964	0.196

Similarly, to quantify the deviation between the measured keel spacing distribution and theoretical function, we conducted a linear regression analysis, and *MAES* represents the mean absolute error between the measured keel spacing distribution and theoretical function, defined as follows:

$$MAES = \frac{\sum_{j=1}^{m} \left| f_s(s_j, d_{c-k}; \Psi) - f_{s-j} \right|}{m} \tag{16}$$

where the variables and functions are defined in accordance with Equation (10).

The analysis results are shown in Table 2. In Region 1, where ice ridging intensities are the lowest, the lognormal function exhibits the weakest correlation with the measured keel spacing distributions, with a linear correlation coefficient of only 0.866 and a maximum *MAES* of 0.279. In contrast, Regions 2 and 3 show stronger correlations: the linear correlation coefficients exceed 0.960 and *MAES* values decrease to below 0.200. Notably, Region 2 (representing moderate ice ridging intensities) achieves the largest linear correlation coefficient (0.968) and the lowest *MAES* (0.183).

For the exponential function, the results show lower linear correlations and significantly larger *MAES* values than those of the lognormal function. Additionally, the trend in *MAES* variation for the exponential function contradicts the correlation coefficient for all three regions. These findings indicate that the lognormal function is more effective for predicting keel spacing distributions, strongly supporting the results presented in Figure 6.

4.2. Correlation Between the Draft and Frequency of Ridge Keel

To comprehensively characterize the morphological and spatial distribution of ice ridge keels, this study analyzes their horizontal and vertical features, quantified by the keel frequency (number of keels per kilometer) and keel draft, respectively. Figure 7a illustrates the relationship between the mean keel draft and frequency, with data in the investigated area considered as a whole.

Notably, despite higher data dispersion compared to the relationship between the mean sail height and frequency (Figure 7b), a significant logarithmic correlation (correlation coefficient $r = 0.698$) exists between the keel draft and frequency. This correlation reflects the complex, irregular nature of the keel structure and distribution, which exhibit greater morphological variability than the ridge sails. Logarithmic regression models further confirm that both the mean keel draft and sail height increase logarithmically with frequency. However, the increment ratios of the keel draft and sail height to the corresponding frequencies decrease as the frequency increases, indicating the non-linear evolutionary patterns of the ice ridge.

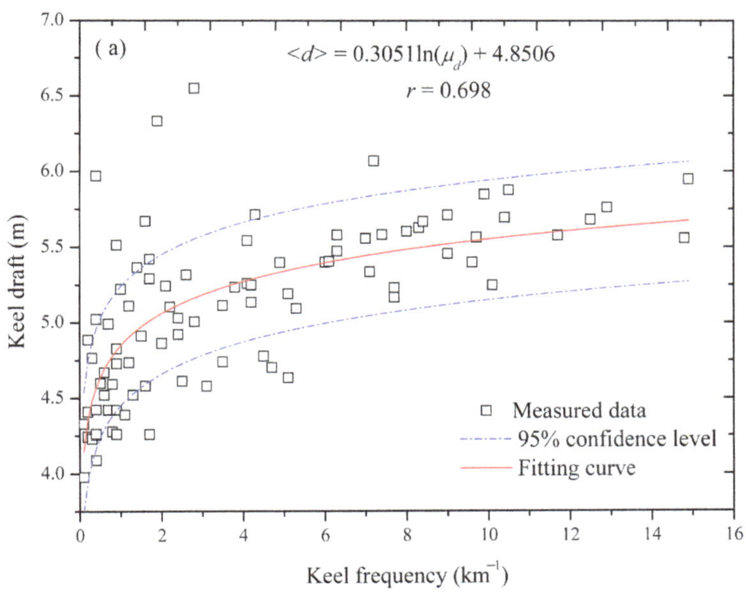

Figure 7. *Cont.*

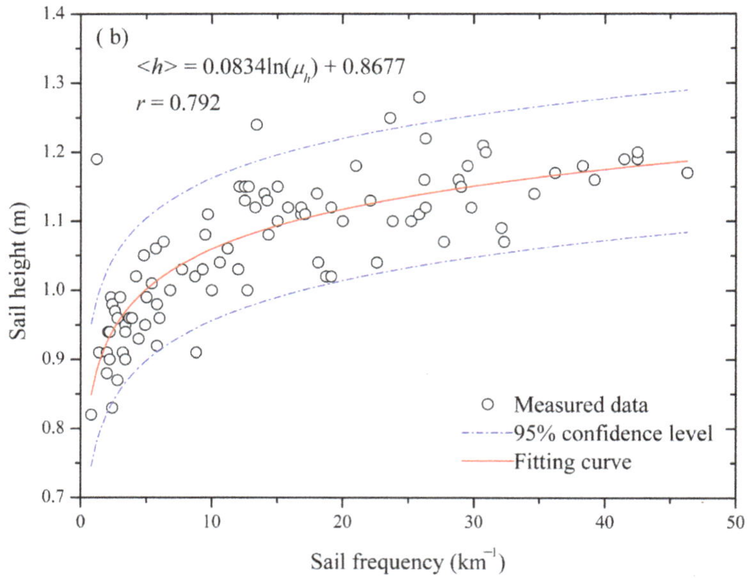

Figure 7. Comparison of the relationship between the mean keel drafts and frequencies, as well as the mean sail heights and frequencies, where r is the correlation coefficient, and μ_h and μ_d are the ridge sail and keel frequencies, respectively. The regression fits are shown by the red solid lines, while the 95% confidence interval bounds for the regression models are shown by the blue dash dotted lines. (**a**) the relationship between the mean keel draft and frequency; (**b**) the relationship between the mean sail height and frequency.

An analysis of the observed data in Figure 7 reveals a region-dependent response: for a given increase in the ridge frequency, there are more pronounced increases in both draft and spacing in the lower ridging intensity regions compared to the higher ridging intensity regions. The logarithmic regression further highlights that the mean keel draft exhibits substantially less variability than the mean keel spacing. These consistent findings validate the use of logarithmic relationships as a robust mathematical framework for modeling the morphological evolution of ice ridge keels.

5. Conclusions

During the 2006 Winter Weddell Outflow Study in the northwest Weddell Sea, Antarctica, a helicopter-borne EM-bird system was utilized to comprehensively measure sea ice bottom roughness, effectively covering the diverse ice types and regimes within the investigated region. A total of 94 sea ice bottom roughness profiles were collected, with lengths ranging from 6.3 to 56.8 km, aggregating to a cumulative length of 2988.5 km. This study conducted a detailed analysis of the field environment, the underlying formation mechanisms, and the distinct characteristics of ice ridges in the three different investigated sub-regions: the northern, central, and southern sub-regions, respectively. By processing the sea ice bottom roughness data, the distributions of ridge keel draft and spacing, as well as the correlation between keel draft and frequency, were systematically analyzed by statistical methods. The key findings are summarized as follows:

1. A statistical optimization model with nonlinear constraints was developed to accurately determine the optimal keel cutoff draft by minimizing the relative errors between the theoretical and measured ridge keel spatial distributions. The analysis

revealed that, irrespective of the cutoff draft value, Wadhams'80 function (a negative exponential function) consistently provided the best fit for the measured keel draft distribution. Conversely, a lognormal function effectively characterized the corresponding keel spacing distribution. Furthermore, an optimal cutoff draft value of 3.8 m was obtained by combining the statistical optimization model, serving as a definitive discriminant threshold for differentiating ridge keels from other sea ice bottom roughness features in the study region.

2. Under the optimal cutoff draft of 3.8 m, the influences of ridging intensity on the ridge special distributions were systematically analyzed based on the clustering results of our previous research [22], which clustered the measured profiles into three distinct regimes according to the increasing ridging intensity: Region 1 ($R_i \leq 0.01$), Region 2 ($0.01 < R_i \leq 0.026$), and Region 3 ($R_i > 0.026$), respectively. For any region, Wadhams'80 function and the lognormal function demonstrated the best fit for keel draft and spacing distributions, respectively, achieving larger linear correlation coefficients and smaller mean absolute errors (*MAED* for keel draft, and *MAES* for keel spacing). In contrast, the Hiblers'72 model and exponential function exhibited significant deviations from the observed data. Notably, Wadhams'80 and lognormal functions are applicable across a broader spectrum of the ice ridging intensity, enabling more accurate predictions of keel draft and spacing distributions in complex, variable sea ice environments.
3. A logarithmic regression analysis, considering both the vertical and horizontal characteristics of the ridge keels, demonstrated a significant logarithmic increase in the mean keel draft with frequency. This relationship was statistically validated by a correlation coefficient of $r = 0.698$ at a 95% confidence level. Notably, while the increment ratios of keel draft to frequency exhibited a decreasing trend with increasing keel frequency, the mean keel draft generally increased, consistent with the measured distributions of keel morphology and spacing.

Author Contributions: Conceptualization, B.T. and C.G.; validation, Y.C., P.L. and T.W.; data curation, Q.W. and Y.F.; writing—original draft preparation, B.T. and C.G.; writing—review and editing, P.L., Q.W., Y.C. and T.W.; project administration, B.T.; funding acquisition, B.T. All authors have read and agreed to the published version of the manuscript.

Funding: This research was funded by the National Natural Science Foundation of China (41876219, 42074094), Key Science Foundation for Universities of Henan Province (24A110008), Senior-end Foreign Expert Introduction Project of Henan Province (HNGD2025034), and Cultivation Project of the National Natural Science Foundation of China.

Data Availability Statement: The data used to support the findings of this study are available for the corresponding authors upon request.

Acknowledgments: Xiang Fu and Yupeng Leng at Dalian University of Technology are thanked for English examination.

Conflicts of Interest: The authors declare no conflicts of interest. The funders had no role in the design of the study; in the collection, analyses, or interpretation of data; in the writing of the manuscript; or in the decision to publish the results.

References

1. Comiso, C.J.; Meier, N.W.; Gersten, R. Variability and trends in the Arctic Sea ice cover: Results from different techniques. *J. Geophys. Res. Oceans* **2017**, *122*, 6883–6900. [CrossRef]
2. Wang, Z.; Li, Z.; Zeng, J.; Liang, S.; Zhang, P.; Tang, F.; Chen, S.; Ma, X. Spatial and temporal variations of Arctic sea ice from 2002 to 2017. *Earth Space Sci.* **2020**, *7*, e2020EA001278. [CrossRef]

3. Parkinson, C.L.; DiGirolamo, N.E. Sea ice extents continue to set new records: Arctic, Antarctic, and global results. *Remote Sens. Environ.* **2021**, *267*, 112753. [CrossRef]
4. Kumar, P.; Verma, A.; Gajbhiye, D.; Chandra, V.; Goswami, A.; Dutta, S. Impact of Changing Climate Over Polar Ice Sheet: A Case Study from Larsemann Hills, East Antarctica. In *Climate Change and Environmental Impacts: Past, Present and Future Perspective*; Springer International Publishing: Cham, Switzerland, 2023; pp. 189–204. [CrossRef]
5. Rantanen, M.; Karpechko, A.Y.; Lipponen, A.; Nordling, K.; Vihma, T.; Hyvärinen, O.; Ruosteenoja, K.; Vihma, T.; Laaksonen, A. The Arctic has warmed nearly four times faster than the globe since 1979. *Commun. Earth Environ.* **2022**, *3*, 168. [CrossRef]
6. Yumashev, D.; Hope, C.; Schaefer, K.; Riemann-Campe, K.; Iglesias-Suarez, F.; Jafarov, E.; Burke, E.J.; Young, P.J.; Shorbany, Y.E.; Whiteman, G. Climate policy implications of nonlinear decline of Arctic land permafrost and other cryosphere elements. *Nat. Commun.* **2019**, *10*, 1900. [CrossRef]
7. Regan, H.C.; Lique, C.; Armitage, T.W.K. The Beaufort Gyre extent, shape, and location between 2003 and 2014 from satellite observations. *J. Geophys. Res. Oceans* **2019**, *124*, 844–862. [CrossRef]
8. Bacon, S. Arctic sea ice, ocean, and climate evolution. *Science* **2023**, *381*, 946–947. [CrossRef]
9. Cohen, J.; Zhang, X.; Francis, J.; Jung, T.; Kwok, R.; Overland, J.; Ballinger, T.J.; Bhatt, U.S.; Chen, H.W.; Coumou, D.; et al. Divergent consensuses on Arctic amplification influence on midlatitude severe winter weather. *Nat. Clim. Change* **2020**, *10*, 20–29. [CrossRef]
10. Koenigk, T.; Key, J.; Vihma, T. Climate change in the Arctic. In *Physics and Chemistry of the Arctic Atmosphere*; Kokhanovsky, A., Tomasi, C., Eds.; Springer International Publishing: Cham, Switzerland, 2020; pp. 673–705. [CrossRef]
11. Perovich, D.; Raphael, I.; Moore, R.; Clemens-Sewall, D.; Lei, R.; Sledd, A.; Polashenski, C. Sea ice heat and mass balance measurements from four autonomous buoys during the MOSAiC drift campaign. *Elem. Sci. Anthr.* **2023**, *11*, 00017. [CrossRef]
12. Yang, Q.; Losa, S.N.; Losch, M.; Tian-kunze, X.; Nerger, L.; Liu, J.; Kaleschke, L.; Zhang, Z. Assimilating SMOS sea ice thickness into a coupled ice-ocean model using a local SEIK filter. *J. Geophys. Res. Oceans* **2014**, *119*, 6680–6692. [CrossRef]
13. Lei, R.; Gui, D.; Heil, P.; Hutchings, J.K.; Ding, M. Comparisons of sea ice motion and deformation, and their responses to ice conditions and cyclonic activity in the western Arctic Ocean between two summers. *Cold Reg. Sci. Technol.* **2020**, *170*, 102925. [CrossRef]
14. Martin, T. Comparison of different ridge formation models of Arctic sea ice with observations from laser profiling. *Ann. Glaciol.* **2006**, *44*, 403–410. [CrossRef]
15. Chen, F.; Xu, R.; Lao, Q.; Chen, C.; Liu, S.; Cai, S.; Xu, J.; Jin, G. Characteristics and mechanisms of ocean fronts-induced decomposition of particulate organic matter and its implication for marine carbon burial. *J. Geophys. Res. Oceans* **2025**, *130*, e2024JC021804. [CrossRef]
16. Leppäranta, M. *The Drift of Sea Ice*; Springer Science & Business Media: Berlin/Heidelberg, Germany, 2011; 347p. [CrossRef]
17. Tan, B.; Li, Z.; Lu, P.; Haas, C.; Nicolaus, M. Morphology of sea ice pressure ridges in the Northwestern Weddell Sea in winter. *J. Geophys. Res. Ocean.* **2012**, *117*, C06024. [CrossRef]
18. Bashmachnikov, I.; Machín, F.; Mendonça, A.; Martins, A. In situ and remote sensing signature of meddies east of the mid-Atlantic ridge. *J. Geophys. Res. Oceans* **2009**, *114*, C05018. [CrossRef]
19. Haas, C.; Friedrich, A.; Li, Z.; Nicolaus, M.; Pfaffling, A.; Toyota, T. Regional variability of sea ice properties and thickness in the northwestern Weddell Sea obtained by in-situ and satellite measurements. In *The Expedition of the Research Vessel "Polarstern" to the Antarctic in 2006 (ANT-XXX III/7)*; Lemke, P., Ed.; Reports on Polar and Marine Research; Helmholtz Gemeinschaft: Bonn, Germany, 2009; Volume 586, pp. 36–74.
20. Wadhams, P. New predictions of extreme keel depths and scour frequencies for the beaufort sea using ice thickness statistics. *Cold Reg. Sci. Technol.* **2012**, *76–77*, 77–82. [CrossRef]
21. Ekeberg, O.; Høyland, K.; Hansen, E. Ice ridge keel geometry and shape derived from one year of upward looking sonar data in the Fram Strait. *Cold Reg. Sci. Technol.* **2015**, *109*, 78–86. [CrossRef]
22. Tan, B.; Wang, T.; Wang, L.; Feng, E.; Lu, P.; Li, Z. Clustering and analyzing the pressure ridge morphology in northwestern Weddell Sea of Antarctic via an improved k-means clustering algorithm. *Pac. J. Optim.* **2016**, *12*, 623–633.
23. Lei, R.; Tian-Kunze, X.; Li, B.; Heil, P.; Wang, J.; Zeng, J.; Tian, Z. Characterization of summer Arctic sea ice morphology in the 135°–175°W sector using multi-scale methods. *Cold Reg. Sci. Technol.* **2017**, *133*, 108–120. [CrossRef]
24. Farrell, S.L.; Kurtz, N.; Connor, L.N.; Elder, B.C.; Leuschen, C.; Markus, T.; Mcadoo, D.C.; Panzer, B.; Richter-Menge, J.; Sonntag, J.G. A first assessment of IceBridge snow and ice thickness data over Arctic sea ice. *IEEE Trans. Geosci. Remote Sens.* **2012**, *50*, 2098–2111. [CrossRef]
25. Kern, S.; Ozsoy-Çiçek, B.; Worby, A. Antarctic sea-ice thickness retrieval from ICESat: Inter-comparison of different approaches. *Remote Sens.* **2016**, *8*, 538. [CrossRef]

26. Lowry, R.T.; Wadhams, P. On the Statistical Distribution of Pressure Ridges in Sea Ice. *J. Geophys. Res. Oceans* **1979**, *84*, 2487–2494. [CrossRef]
27. Tin, T.; Jeffries, M.O. Morphology of Deformed First-year Sea Ice Features in the Southern Ocean. *Cold Reg. Sci. Technol.* **2003**, *36*, 141–163. [CrossRef]
28. Salganik, E.; Lange, B.A.; Itkin, P.; Divine, D.; Katlein, C.; Nicolaus, M.; Hoppmann, M.; Neckel, N.; Ricker, R.; Høyland, K.V.; et al. Different mechanisms of Arctic first-year sea-ice ridge consolidation observed during the MOSAiC expedition. *Elem. Sci. Anthr.* **2023**, *11*, 8. [CrossRef]
29. Hibler, W.D., III; Weeks, W.F.; Mock, S.J. Statistical aspects of sea ice ridge distributions. *J. Geophys. Res.* **1972**, *77*, 5954–5970. [CrossRef]
30. Wadhams, P. A comparison of sonar and laser profiles along corresponding tracks in the Arctic Ocean. In *Sea Ice Processes and Models, Proceedings of the Arctic Ice Dynamics Joint Experiment International Commission on Snow and Ice Symposium*; Pritchard, R.S., Ed.; University of Washington Press: Seattle, WA, USA, 1980; pp. 283–299.
31. Wadhams, P.; Davy, T. On the spacing and draft distribution for pressure ridge keels. *J. Geophys. Res. Oceans* **1986**, *91*, 10697–10708. [CrossRef]

water

MDPI

Article

Numerical Simulation of Regular Wave and Ice Floe Interaction Using Coupled Eulerian–Lagrangian Method

Chaoge Yu [1,2] and Yukui Tian [1,2,*]

1 China Ship Scientific Research Center, Wuxi 214082, China; yucg@cssrc.com.cn
2 National Key Laboratory of Hydrodynamics, Wuxi 214082, China
* Correspondence: tianyukui@cssrc.com.cn

Abstract: Wave propagation is impacted by the presence of ice floes. The influence of waves, on the other hand, causes ice floes to overlap and accumulate. In this paper, the interaction of ice floes and regular waves was simulated using the Finite Element Method. Firstly, natural ice floe fields were generated using the Python 3.10 programing language, with floe size distribution and randomness taken into consideration. Then, using the velocity inlet boundary wave generation method, regular simple harmonic waves were produced. The process where ice floes couple with waves was simulated with the Coupled Eulerian–Lagrangian (CEL) approach. Variations in wave height after passing through the ice floe field were investigated, and further research was conducted on the movement and fragmentation characteristics of ice floes. Simulations employing the Coupled Eulerian–Lagrangian (CEL) approach reveal that (1) ice floe motion exhibits periodic characteristics synchronized with incident wave periods; (2) wave height attenuation increases by 62–80% with rising ice concentration (70–90%); and (3) fragmentation predominantly occurs at wave trough phases due to flexural stress concentration. These findings quantitatively characterize wave–ice energy transfer mechanisms critical for polar navigation safety assessments.

Keywords: ice floes; propagation characteristics; regular waves; numerical simulation; Coupled Eulerian–Lagrangian approach

1. Introduction

Global warming has accelerated the melting of sea ice in the Arctic, and sea ice cover has been decreasing year by year, with changes in sea ice extent observed since satellite observations began in 1979, with summer and autumn being the most significant, showing a reduction of about 50 percent. The feasibility of Arctic navigation by ships has increased significantly. The marginal ice zone is located between an ice sheet and open water, which is the area where ice floes are concentrated, and it is also the key area for polar ship navigation. According to an analysis by Li [1], in recent decades, the effective wave height and wave period in most areas of the Arctic Ocean in summer have increased, which means that the number of waves carrying more energy have increased. The interaction mechanism between wave and ice floes is the technical basis of ship–wave–ice floe coupling analysis and is an important part of evaluating the navigation safety of ships in the marginal ice zone. The interaction between waves and ice floes is an extremely complex fluid–structure interaction process, and research work on this involves the macroscopic physical morphology and mechanical properties of sea ice, as well as complex wave theory. Therefore, the coupling relationship between waves and ice floes is worthy of in-depth study.

In the early days, scholars mainly applied theoretical models of the interaction between waves and ice floes, such as the mass loading model proposed by Peters [2], the elastic sheet bending model proposed by Wadhams [3], and, subsequently, a series of models accounting for the viscosity of seawater. Luo [4] believes that theoretical models have their own advantages and are suitable for different scenarios; for example, the mass load model is mainly suitable for discontinuous lotus-leaf-like ice regions, the elastic sheet model is suitable for continuous ice sheets, the viscous layer model is mainly suitable for grease ice regions, and the viscoelastic model can be applied to the entire marginal ice zone with appropriate coefficients. Because the correlation coefficient of the viscoelastic model is difficult to determine by experimental measurements and other means, it is difficult to apply it directly.

In terms of experimental approaches, due to the complexity of the interaction between waves and ice floes, field observations and model experiments provide valuable insights. Specifically, field observations are often based on buoys, submersibles, and navigation but, due to the constraints of a harsh environment and poor accessibility, field observations are relatively few, and the time and space coverage is low. Typically, only the basic data of waves are obtained from field observations, and measurements of wave–ice floe interactions are rarely carried out. For example, Dean [4] used a ship-borne wave recorder to record the attenuation of waves in the ice margin region, while Rogers [5] observed the attenuation of waves in the floating ice region in the Beaufort Sea. Furthermore, indoor model tests are demanding and require both sea ice simulation and wave-making capabilities, and the Hamburg Ice Pool is one of the few ice pools with wave-making capabilities. Notably, with the support of the joint project LS-WICE, a series of wave–ice interaction model tests were carried out in HSVA [6,7], which involved the analysis of wave propagation and attenuation characteristics in sea ice regions, the motion characteristics of ice floes in waves, the breakage characteristics of level ice under wave action, and the structural ice load under the combined action of waves and ice floes. Due to the limitation of experimental conditions, Wang [8] used paraffin non-frozen model ice to simulate the size distribution of ice floes in real-world conditions, measured the motion response of ice floes in regular waves of different wavelengths, summarized their longitudinal and vertical motion responses, and compared them with the motion response laws of isolated ice floes in waves, and the results showed that the longitudinal motion of the ice floes in the waves showed two motion states: irregular motion in the short-wave interval and regular motion in the long-wave range.

Regarding numerical simulations, Meylan [9] studied the coupling effect of different shapes of sea ice under wave action and found that waves are most obviously affected by the bending stiffness of ice floes but have no obvious relationship with the geometry of ice floes. Wang [10] studied the interaction between waves and arbitrarily shaped elastic ice floes based on the coupled boundary element method, and the calculation results were consistent with Meylan's conclusion. Wu [11] carried out numerical simulation studies of waves and single ice floes based on STAR-CCM+ and comprehensively analyzed the effects of thickness, size, and shape of ice floes on their motion characteristics in waves. Ni [12] used the VOF model to study the interaction characteristics of two-dimensional Stokes waves with single and multiple ice floes and analyzed the motion and force characteristics of ice floes under wave action, as well as the propagation characteristics of waves. Wang [13] studied the interaction between regular waves and ice floes based on the S-ALE method in LS-DYNA, focusing on the longitudinal drift motion of ice floes under wave action, analyzing the influence of the ratio of wavelength to the size of ice floes on the longitudinal drift motion characteristics of ice floes, and validating the numerical model based on model experiments. Lu [14] combined FLUENT with EDEM, simulating ice floes as discrete elements and waves using the VOF method. Wu [15] simulated the level ice failure under wave action based on the DEM-SPH coupling model, examining the effects of wave height, wave period, and ice thickness on ice failure mode. Zhang [16] developed a novel two-way SPH method incorporating a density-ratio-sensitive fluid–ice interaction model, an ice particle separation algorithm for fracture, and unequal time-step treatment. This method efficiently simulated wave-induced breakup of ice floes, was validated against regular waves experiments, and applied to solitary and focused waves. He [17] established a one-way CFD-DEM coupling model using FVM for waves and bonded DEM particles for ice. This model simulated ice sheet breakup under regular waves, was validated against HSVA experiments, and revealed nonlinear floe size distributions dependent on wave height and ice thickness. Zhang [18] developed an SPH model with an improved fluid–ice interface treatment scheme, using ice particles as dummy particles to simulate wave-induced kinematic responses and flexural deformations of ice floes, validated against experimental data for surge/heave motions and bending patterns. Ni [19] synthesizes research on ice deformation and fracture induced by flexural-gravity waves (FGWs) from both above-ice and underwater moving loads. This synthesis covered diverse ice conditions (intact sheets, cracks, leads, and channels) and explored factors like load velocity, ice properties, and boundary effects.

In summary, methodologies for studying wave–ice floe interactions include theoretical models, experimental observations, and numerical simulations. Among these, numerical simulation holds potential, enabling effective characterization of wave propagation and ice floe motion dynamics under wave action. Building on these advances, in this paper, the interaction between ice floes and waves is simulated based on the coupled Eulerian–Lagrangian method, with a focus on the motion and fragmentation characteristics of ice floes in waves, as well as the propagation and attenuation characteristics of waves within ice fields.

2. Basic Theory

2.1. CEL and Hydrodynamic Models

This work utilizes the explicit dynamics module of a finite element software platform, an industry-standard framework for simulating dynamic phenomena involving nonlinear processes such as high-speed collisions. Since the dynamical equations are in differential format, there is no need to perform equilibrium iterations. Computational convergence is inherently ensured when the sufficiently small time steps are employed. The equation of state that can describe fluid dynamics is introduced into ABAQUS/Explicit and, combined with the CEL method, enables robust simulation of fluid–structure interaction (FSI). In recent years, the CEL method has been extensively applied to FSI problems with significant success. For instance, Majlesi [20] employed the CEL method to simulate the interaction between waves and bridge piers and the simulation results were in good agreement with the experiments, and Kim [21] used the CEL method to simulate the tank fall study to analyze the impact of the fluid on the oil tank during the tank fall.

The CEL method employs distinct kinematic descriptions, the Eulerian formulation for fluids and the Lagrangian formulation for solids. In this work, waves are modeled within a Eulerian domain, while ice floes are represented using a Lagrangian mesh. Interaction between these domains is governed by the principles of the immersed boundary method. A dedicated fluid–structure interaction (FSI) contact algorithm defines the influence of the Lagrangian body (ice floe) motion on the Eulerian domain (fluid), coupling Eulerian elements with Lagrangian elements for simultaneous solution. Figure 1 illutrates the Lagrangian body and Eulerian body within the velocity field. Crucially, the Lagrangian mesh deforms and displaces congruently with the material it represents (the ice floe). Conversely, the Eulerian mesh remains spatially fixed; material deformation and flow within this domain are tracked via the evolving volume fraction of material within each cell, calculated using the VOF method.

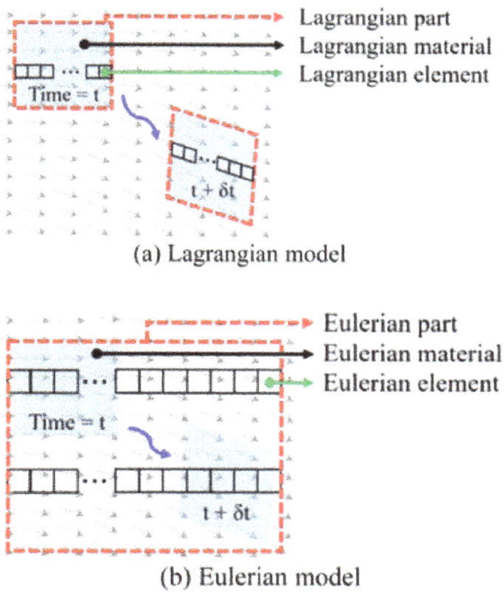

Figure 1. Schematic representation of Lagrangian and Eulerian frameworks [21].

The Eulerian framework, characterized by its non-deforming mesh, circumvents mesh distortion issues inherent in Lagrangian methods. This makes it particularly suitable for simulating large material deformations. Consequently, wave simulation in this study is implemented within the Eulerian domain. The governing equations for the fluid are given by the conservation laws:

$$\frac{\partial \rho}{\partial t} + \nabla \cdot (\rho v) = 0 \tag{1}$$

$$\rho \frac{\partial v}{\partial t} + \rho v \cdot \nabla (v) = \nabla \cdot \sigma + f \tag{2}$$

$$\rho \frac{\partial e}{\partial t} + \rho v \cdot \nabla (e) = \sigma : \nabla (v) + f \cdot v \tag{3}$$

where ρ is the density, f is the density of the external force, v is the velocity vector, and e is the specific internal energy. The water body is described by the Mie–Grüneisen equation of state in the form of a linear U_s-U_p Hugoniot.

$$U_s = c_0 + \Psi_0 U_p \tag{4}$$

$$p = \frac{\rho_0 c_0^2 \mu}{(1 - \Psi_0 \mu)^2} \left(1 - \frac{\Gamma_0 \mu}{2}\right) + \Gamma_0 \rho_0 E_m \tag{5}$$

where p is pressure, E_m is specific energy, U_s is linear shock wave velocity, U_p is particle velocity, c_0 is speed of sound, Ψ_0 and Γ_0 are reference material constants, and μ is the viscosity coefficient.

2.2. Wave Generation Method

In this study, the fluid domain is modeled using the Eulerian approach. Periodic waves are generated by prescribing a time-dependent velocity boundary condition along the Eulerian domain boundary. Consistent with wave theory, these periodic waves are fully characterized by three independent parameters, wave height *H*, period *T*, and water depth *h*, as shown in Figure 2. Referring to the research conclusions of Montoya [22], it is possible to generate ideal Stokes second-order waves through Equations (6)–(11). When the wave height is too large relative to the water depth, the wave is prone to breakage, so the relevant parameters of the wave need to meet Equations (12) and (13).

$$\xi(t) = -\xi_0\left[\cos(wt) + \frac{a}{2hn_1}\left(\frac{3}{4\sinh^2(kh)} - \frac{n_1}{2}\right)\sin(2\omega t)\right] \tag{6}$$

$$v(t) = -\omega\xi_0\left[-\sin(\omega t) + \frac{a}{hn_1}\left(\frac{3}{4\sinh^2(kh)} - \frac{n_1}{2}\right)\cos(2\omega t)\right] \tag{7}$$

$$\xi_0 = \frac{an_1}{\tanh(kh)} \tag{8}$$

$$\omega^2 = gk\tanh kh \tag{9}$$

$$n_1 = \frac{1}{2}\left(1 + \frac{2kh}{\sinh 2kh}\right) \tag{10}$$

$$n(x) = \frac{H}{2}\cos(kx - \omega t) + \frac{\pi H^2}{8L}\frac{\cosh(kh)}{\sinh^3(kh)}(\cosh(2kh) + 2)\cos 2(kx - \omega t) \tag{11}$$

$$U_r = \frac{H\gamma^2}{h^3} \tag{12}$$

$$U_r < \frac{8}{3}\pi^2 \tag{13}$$

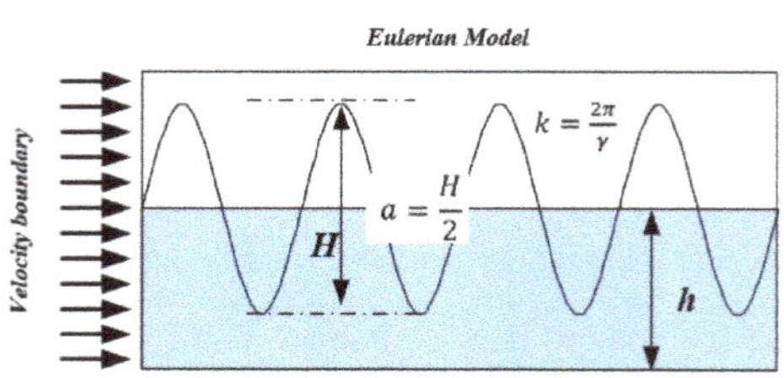

Figure 2. Periodic wave schematic.

In the formula, $\xi(t)$ references the displacement boundary that should be applied since the Euler body only supports the definition of velocity boundary, $v(t)$ can be used to generate an ideal wave, k is the wave number, ω is circular frequency, $n(x)$ is theoretical wave free surface, γ is wave length, and the other parameters are process physical quantities.

2.3. Generation of the Ice Floe and Material Model of Sea Ice

The temporal and spatial distribution of Arctic ice floes is quite random, and the distribution rules of ice floes in different sea areas vary greatly and have pronounced seasonal/interannual characteristics. The distribution data of ice floes are primarily derived from satellite imagery, aerial surveys, and ship navigation photography equipment. Limited to factors such as resolution and measurement range, there is no unified function expression of the distribution law of ice floe at present. Consequently, this study generates ice floe fields assuming a normal size distribution. ABAQUS has a flexible secondary development interface, which can be implemented through the scripting language Python. At present, the generation methods of ice floe mainly include the polygon overlap method, Vino cutting method, and image recognition method. Among them, the polygon overlap method can generate arbitrary shaped ice floe blocks, and it is convenient to adjust the size distribution law and roundness of ice floe blocks, but the time complexity is high when the program is designed, and the overlap probability of polygons is too high when the density is high, so the generation efficiency of the high-density ice floe field is low. The image recognition method requires the input of a high-quality target image, and accurate contour segmentation becomes challenging with low image quality or contacting floes. Therefore, this paper generates an ice floe based on the Wino polygon function in Python, firstly by adjusting the arrangement of the Vino seeds, changing the size distribution of the ice floe, then scaling the ice floe in equal proportions to make the ice floe density reach the specified value, and finally extracting the corner coordinates of the ice floe to construct the ice floe through the ABAQUS built-in function. As shown in Figure 3, 90%, 70%, and 50% concentration ice floes were generated, and the area of each ice floe was statistically fitted, and it was found that its size distribution was normally distributed, which met the expectations of the program design.

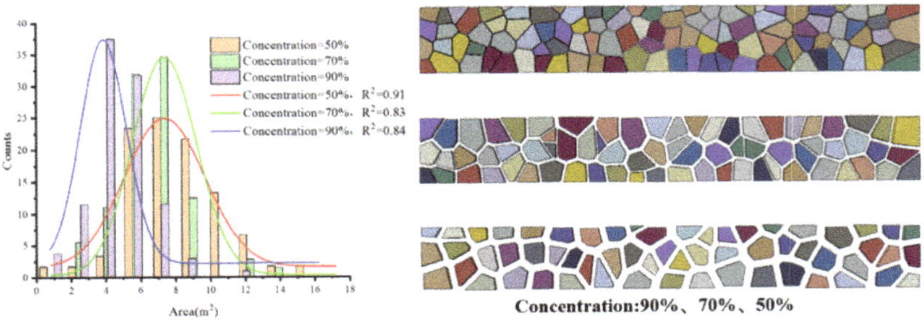

Figure 3. Generated ice fields at 90%, 70%, and 50% concentrations.

Ice floes may break up under the action of waves, which is closely related to the size of the ice floes, etc. With the development of computer technology and algorithms, it has become common to use FEM to analyze material fracture failure, and the most common method for simulating failure is the element deletion model. Before the damage starts, the mechanical behavior of the material obeys the preset constitutive model and, when the damage starts, the stiffness of the material degrades until the bearing capacity is completely lost and the element is deleted, which is macroscopically manifested as a crack. In summary, when using the element deletion model, it is necessary to set the constitutive model, the damage initiation criterion, and the element failure criterion. In this paper, the pre-damage initiation model follows the Drucker Prager plastic model, which exhibits strain-hardening behavior after yielding, and the material begins to damage and stiffness degrades when the equivalent plastic strain reaches a certain value. The Drucker Prager plastic model introduces the effect of hydrostatic pressure into the Mises yield criterion.

When this criterion is applied to yield, there are usually three yield surface forms, namely linear form, hyperbolic form, etc.; as shown in Figure 4a, a simple linear form is selected, which provides a yield surface in the partial plane that cannot be circular to match the different yield values of triaxial tension and compression, the associated inelastic flow in the partial plane, and the separate expansion and friction angles. d' is the cohesive term of Drucker Prager's model, p is the hydrostatic pressure, β is the friction angle, and t is the Mises equivalent stress.

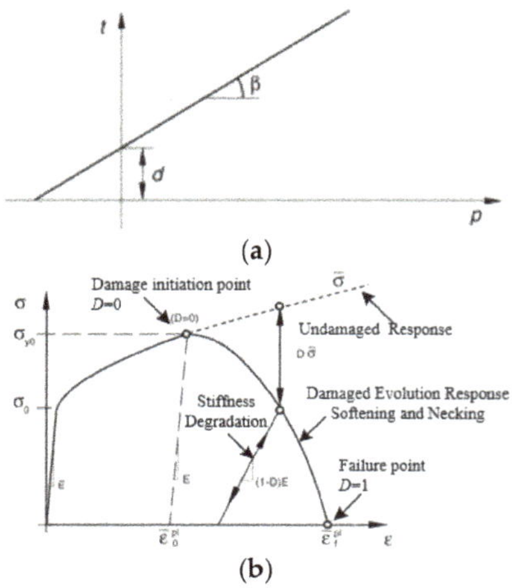

Figure 4. The constitution of sea ice materials. (**a**) Drucker Prager plastic; (**b**) ductile damage.

3. Numerical Simulation and Analysis

Following the methodology established in Section 2, this section first carries out the verification of the wave generation method and then simulates the interaction between the wave and the ice floe based on the CEL method and then analyzes the propagation and attenuation characteristics of the wave in the ice floe area, as well as the movement and failure characteristics of the ice floe under waves action, as shown in Figure 5.

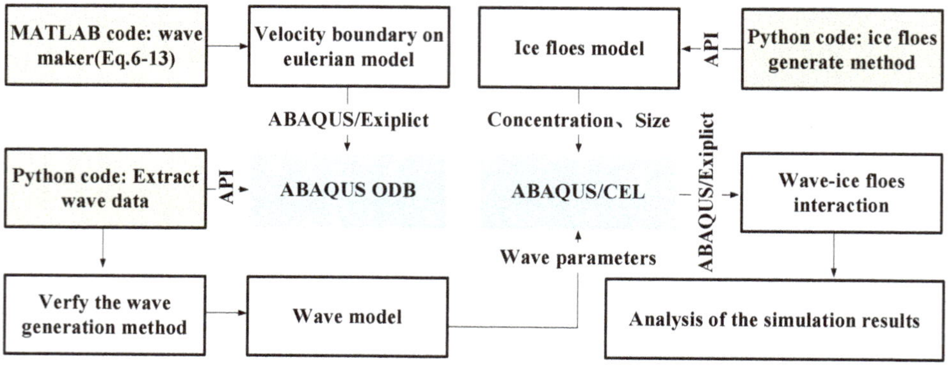

Figure 5. Framework of the numerical simulation.

3.1. Wave Generation and Verification

According to the wave generation theory in Section 2.2, the simulation prescribed a wave height $H = 1$, period $T = 5$ s, and initial water depth $h = 8$ m. The time-dependent velocity boundary condition, generated using MATLAB R2016b, was applied to specific boundaries of the Eulerian domain as illustrated in Figure 6. The velocity boundary condition and Euler boundary condition are set on the YZ plane, wherein the non-reflection boundary condition can reduce the reflection of the fluid and ensure the wave surface height. The Euler boundary condition is set on the upper surface of XY to simulate the atmosphere, and the velocity in three directions is set to 0 on the lower surface of XY to simulate the no-slip boundary condition. Since the mesh of the Eulerian body does not change with the movement and deformation of the water body, the size of the Euler body should be large enough to cover the area where the fluid may flow; in this case, the Euler body is 200 m long, 10 m wide, and 9.5 m deep.

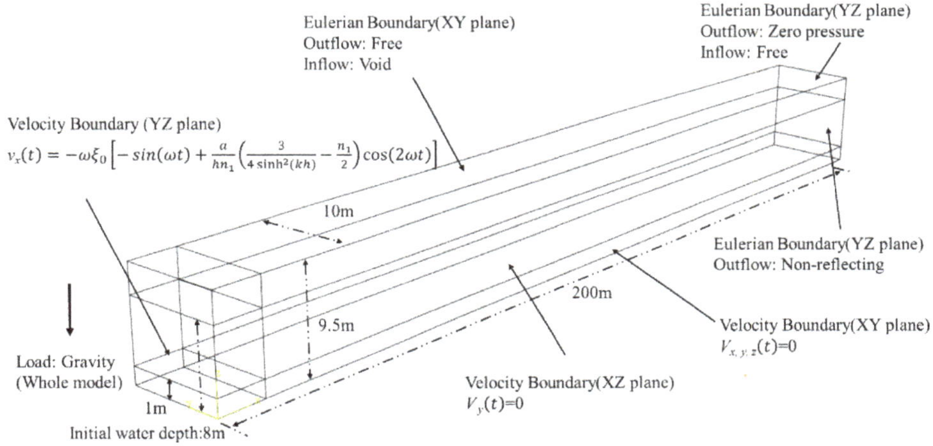

Figure 6. Eulerian domain dimensions and boundary.

Wave generation and propagation over 60 s were simulated by explicit analysis, and the material parameters of the water are listed in Table 1. As shown in Figure 7a, the Abaqus/ODB file is redeveloped using the Python language, enabling fluid free-surface visualization. As demonstrated in Figure 7b, the simulated wave profile exhibits close agreement with theoretical Stokes second-order characteristics, validating the numerical implementation. Based on the research of Montoya [22], the meshes near the velocity boundary region, the wave generation region, and the region near the bottom of the Euler body should be encrypted and the grid size should be consistent with that shown in Figure 8 and Table 2 to ensure the accuracy of the wave data generated.

Table 1. Material parameters of sea ice and water.

Properties of Sea ice			
Density (kg/m^3)	900	Dilation angle (°)	12.0
Young's modulus (MPa)	500	Fracture strain	1×10^{-10}
Poisson's ratio	0.33	Flow stress ratio	1.0
Friction angle (°)	36.0	Fracture energy (J/m^2)	5.0
Properties of Water			
Density (kg/m^3)	1000		
Speed of sound (m/s)	1500		
Viscosity (Pa·s)	0.001		
Material constant $\Psi_0 \backslash \Gamma_0$	0.0		

Table 2. Mesh sizes of Eulerian domain.

Domain	**Mesh Size (m)**		
	Δx	**Δy**	**Δz**
①	0.25	0.10	1.0
②	0.50	0.10	
③	0.50	0.10	
④	0.25	0.30	
⑤	0.50	0.30	

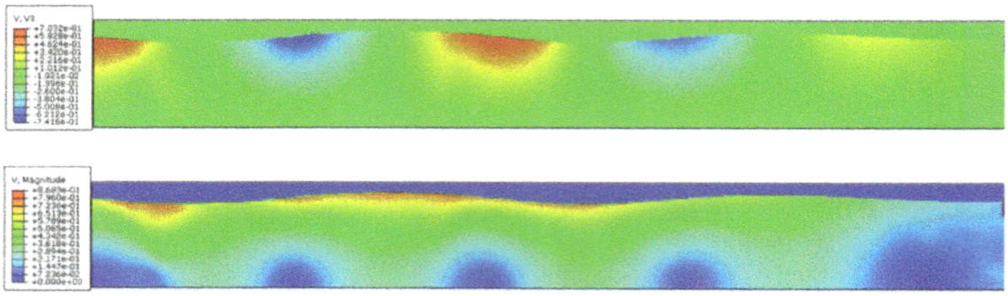

(**a**)

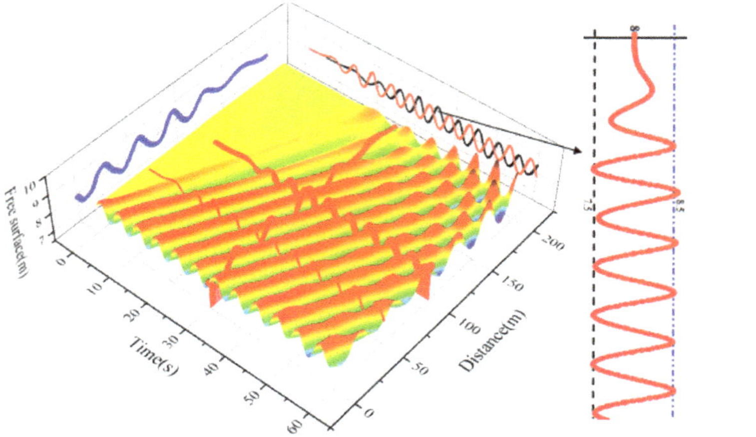

(**b**)

Figure 7. Numerical simulation of wave generation. (**a**) Velocity distribution; (**b**) free surface elevation.

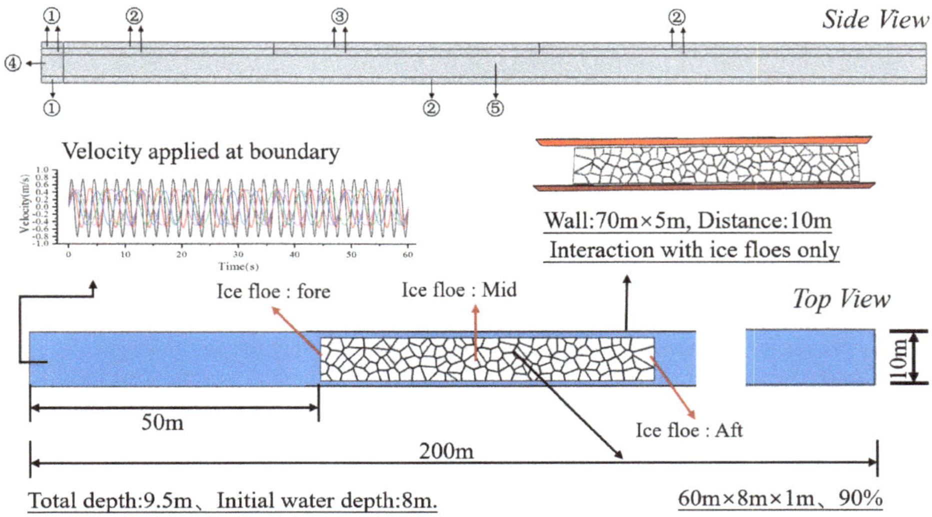

Figure 8. Numerical modeling of wave–ice floe interaction (side view; top view).

3.2. Numerical Models and Working Conditions

In this section, the CEL approach is used to study the interaction between waves and ice floes. The numerical model configuration (Figure 8) maintains consistent Eulerian domain dimensions and boundary conditions with the validated wave tank (Figure 6). Mesh size (Table 2) was determined following the study of Montoya [22], with mesh independence confirmed through systematic refinement studies, and the mesh near the velocity boundary, the area of wave generation, and the bottom region of the Euler body are appropriately encrypted. The ice floe is arranged on the left side of the Euler body, 50 m away from the boundary, and steel plates are arranged on both sides of the ice floe to restrict the movement of the ice floe in the Y direction and prevent it from escaping the Eulerian domain. The material parameters for water and sea ice are listed in Table 1.

This investigation employs numerical simulation to characterize wave–ice floe interactions, specifically examining wave propagation/attenuation through ice fields and ice floe kinematic/fracture responses under wave loading, while evaluating the influence of sea ice concentration and wave period. Three distinct ice concentration configurations were implemented: 100% concentration representing continuous level ice, alongside 90% and 70% concentrations. The numerical conditions are listed in Table 3.

Table 3. Numerical conditions.

Case	Ice Floe Concentration	Wave Period (s)	Wave Height (m)	Wave Length (m)
1	70%	5.0	0.8	34.89
2	70%	5.0	1.0	34.89
3	70%	5.0	1.2	34.89
4	70%	4.5	1.0	29.57
5	70%	4.75	1.0	32.24
6	90%	5.0	1.0	34.89
7	100%	5.0	1.0	34.89

3.3. Analysis of Results

Initial analysis focused on ice floe kinematics and fracture dynamics under wave loading. At 90% concentration, longitudinal floe displacement exhibited phase-correlated motion with wave propagation, while vertical movement synchronized with free-surface elevation. This behavior is visualized in Figure 9a,b through displacement magnitude contour plots, where color gradients correspond to wave phase (crest/trough). Larger ice features underwent significant flexural deformation prior to fracture, as demonstrated by the bending failure sequence in Figure 9c,d. In contrast, smaller floes exhibited no fracture due to insufficient wave-induced stresses relative to their structural capacity.

Three ice floes (fore, mid, and aft, as shown in Figure 8) within the ice floe field were analyzed for the multi-directional motion characteristics. During simulations, the gravity step is between 0 and 5 s, with only gravity load applied during this step, and wave step is between 5 and 60 s. As shown in Figure 10, it is worth noting that, because the initial position of the ice floe is above the water surface, the vertical displacement occurs due to the action of gravity during the simulation process, so the vertical displacement is basically less than 0 in Figure 10d. The movement of ice floes under the action of waves has obvious periodic characteristics, and the period is approximately equal to the wave period, which is consistent with the phenomenon when a wave interacts with a single ice floe. The rotational motion of the ice floe is between ± 0.1 rad, and the longitudinal ice floe moves reciprocating with the waves, but, on the whole, it moves forward and, at 90% concentration, due to the small spacing between multiple ice floes, the probability of collision is larger, so the longitudinal displacement has no obvious reciprocating characteristics.

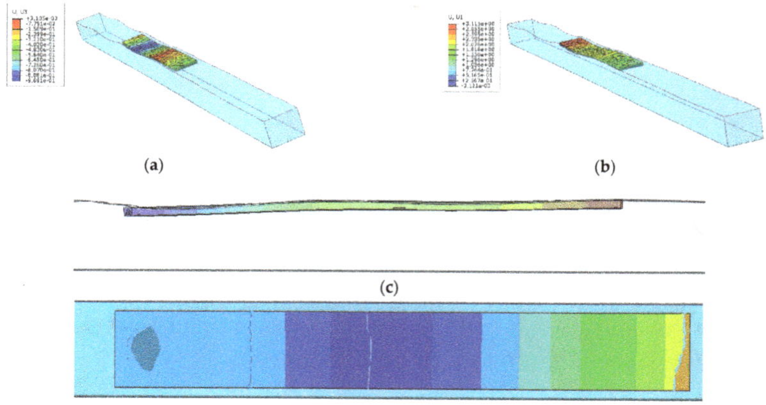

Figure 9. Motion and fracture of ice floes under regular waves. (**a**) Time = 20 s, Case 6; (**b**) time = 30 s, Case 6; (**c**) ice bending, Case 7; (**d**) ice bending fracture, Case 7.

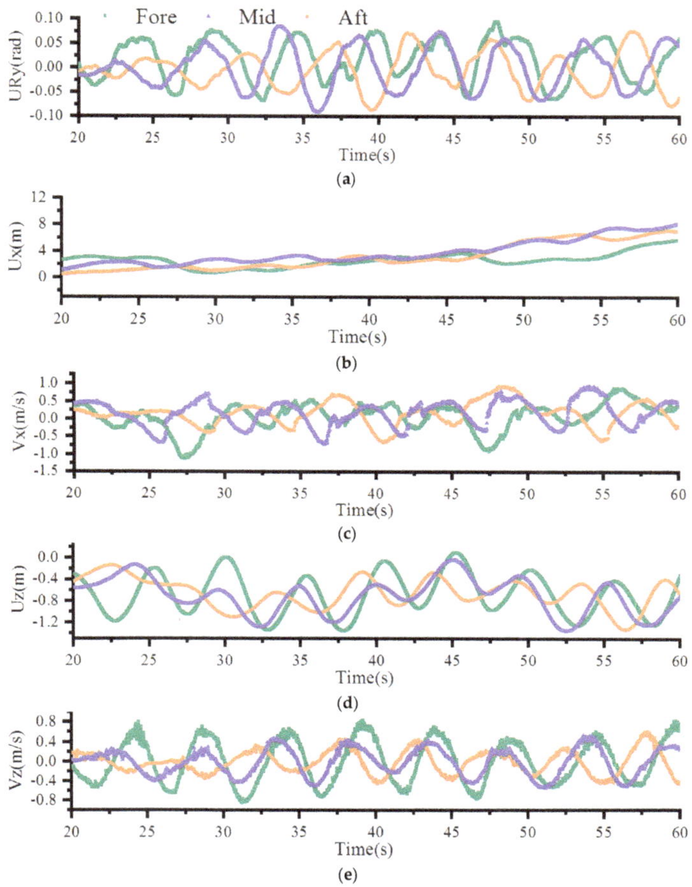

Figure 10. The motion displacement and velocity of typical ice floes (Case 6). (**a**) Rotational motion; (**b**) longitudinal displacement; (**c**) longitudinal velocity; (**d**) vertical displacement; (**e**) vertical velocity.

At 70% concentration, the longitudinal motion of the ice floe shows obvious reciprocating characteristics, as shown in Figure 11a. Compared with the longitudinal displacement, the vertical displacement of ice floe is more periodic, but it is also significantly correlated with the collision between ice floes, as shown in Figure 11b, the vertical displacement at low density is more regular, and the vertical displacement amplitude is basically the same and basically consistent with the height of the wave. Comparing the vertical velocity and vertical displacement, there is also a certain phase difference between the two, as shown in Figure 11c,e, consistent with viscous damping effects in the ice–water system. Wave-induced attenuation is evident in Figure 11c, where the aft-positioned ice floe shows reduced vertical displacement amplitude, suggesting energy dissipation during wave propagation through the ice field.

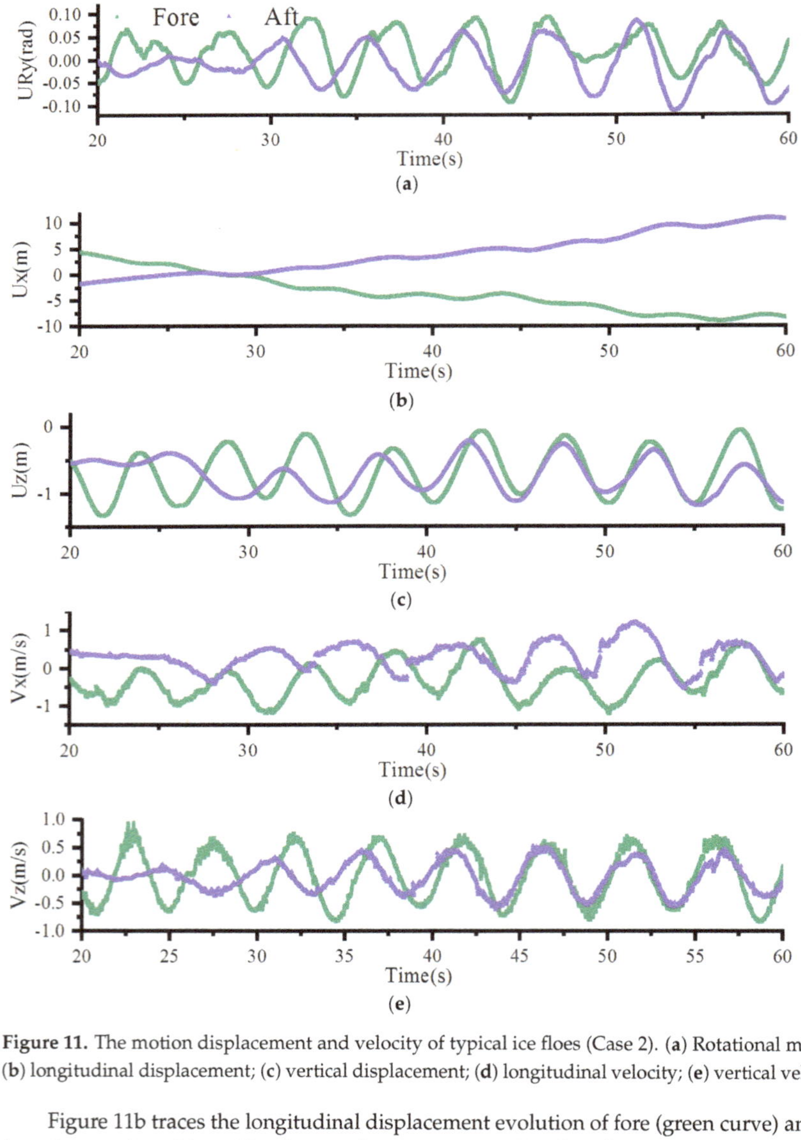

Figure 11. The motion displacement and velocity of typical ice floes (Case 2). (**a**) Rotational motion; (**b**) longitudinal displacement; (**c**) vertical displacement; (**d**) longitudinal velocity; (**e**) vertical velocity.

Figure 11b traces the longitudinal displacement evolution of fore (green curve) and aft (purple curve) positioned ice floes under wave action. The fore-floe exhibits a pronounced phase opposition to the aft-floe. This anti-phase relationship suggests wave-induced differential forcing modulated by floe spacing and hydrodynamic coupling. Figure 12 provides granular insight into the collision mechanics. The longitudinal velocity of fore-floe (Vx, red curve) demonstrates pre-collision harmonic oscillations. At t = 18.3 s, Vx undergoes a discontinuous transition from 0.5 m/s to −1.0 m/s, consistent with impulsive force dynamics during floe contact. The inset illustrates contact mechanics, where stress concentration (red zones) at the floe edges aligns with velocity gradient discontinuities, validating the momentum transfer mechanism.

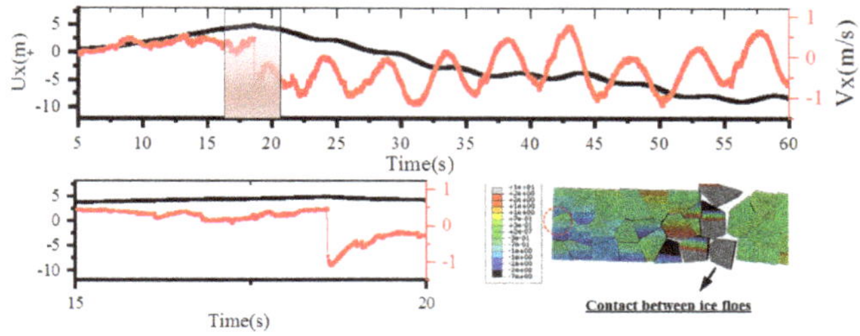

Figure 12. The motion displacement and velocity of ice floes at fore in Case 2.

Rotational dynamics of aft-positioned ice floes across varying wave conditions (Cases 1–5) reveal distinct oscillation regimes (Figure 13). All cases show angular displacements within ±0.15 rad, with Case 4 (purple) demonstrating the largest amplitude (±0.18 rad) and Case 2 (black) the smallest (±0.08 rad). The rotational frequency aligns with wave periodicity. Notably, Case 4 exhibits a transitional regime (T = 4.75 s), where rotational motion transitions from periodic to chaotic behavior at t = 40 s and t = 50 s, suggesting incipient floe–floe contact events. Longitudinal displacement exhibits monotonic positive drift, reaching maximum at t = 60 s, indicative of persistent Stokes drift under short-period waves.

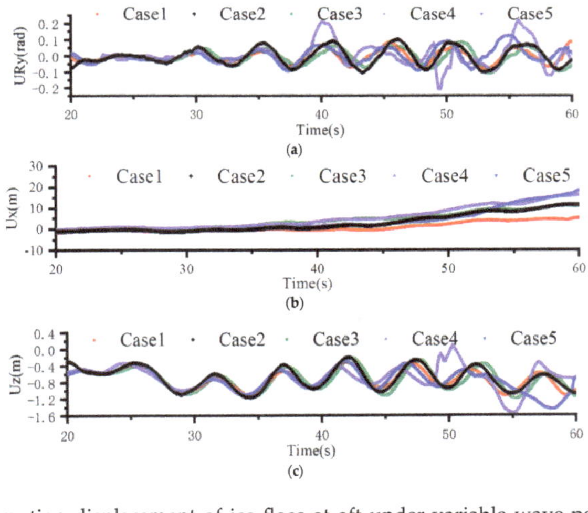

Figure 13. The motion displacement of ice floes at aft under variable wave period and height (70% concentration ice floes). (**a**) Rotational motion; (**b**) longitudinal displacement; (**c**) vertical displacement.

Longitudinal ice floe displacement statistics across wave conditions (Cases 1-5) are quantified in Figure 14. From Figure 14a, it can be observed that, when the wave period is relatively short, the longitudinal displacement of the floating ice increases significantly. Figure 9b shows that, as the wave propagation distance increases, the standard deviation of the vertical displacement of the ice floe gradually decreases, suggesting that the waves attenuate as the propagation distance increases.

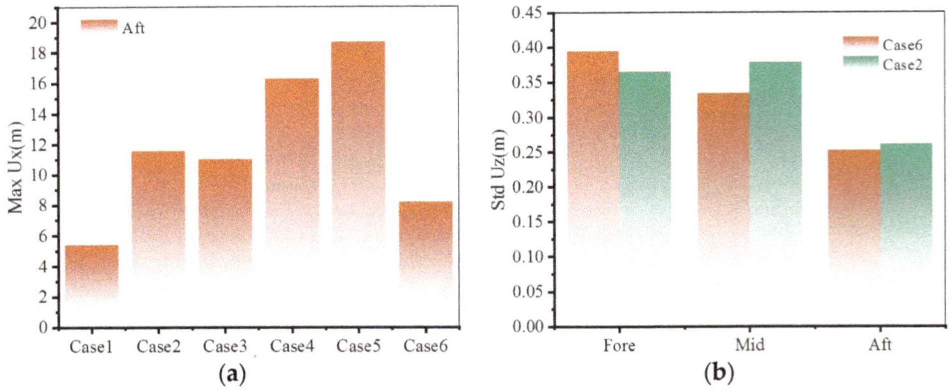

Figure 14. Comparison of the ice floe displacements at aft of different conditions. (**a**) Maximum of longitudinal displacement; (**b**) standard deviation of longitudinal displacement.

It is obvious that the ice floe will also have a certain impact on the propagation of waves, and the conversion of wave energy into the kinetic energy of the ice floe will inevitably cause its own attenuation, as shown in Figure 15. It can be clearly seen from the figure that the wave has a significant attenuation after passing through the ice floe area. In addition, as shown in Figure 15b, the waves are strongly affected by the reflection of the level ice, and the regularity of the waves in front of the level ice gradually weakens over time.

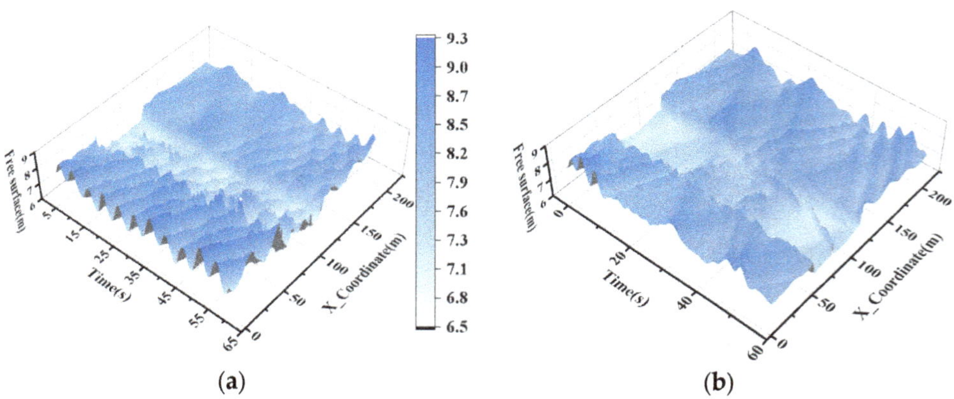

Figure 15. Typical free surface elevation of wave passthrough ice floes. (**a**) Case 3; (**b**) Case 7.

As shown in Figure 16, the ice floe or level ice absorbs a significant amount of wave energy. The characteristics of the waves are represented by the standard deviation of the free surface height between 30 s and 60 s, with wave attenuation of more than 50% in the presence of sea ice (Case 2, Case 7 about 62%, and Case 6 about 80%). Comparing Case 2 and Case 6, it can be observed that the increase in sea ice concentration leads to an increase in the attenuation of waves. When waves pass through level ice, their attenuation is less than when passing through 90% concentration ice floes. This may be due to the boundary conditions of the level ice being fixed, not moving with the wave motion.

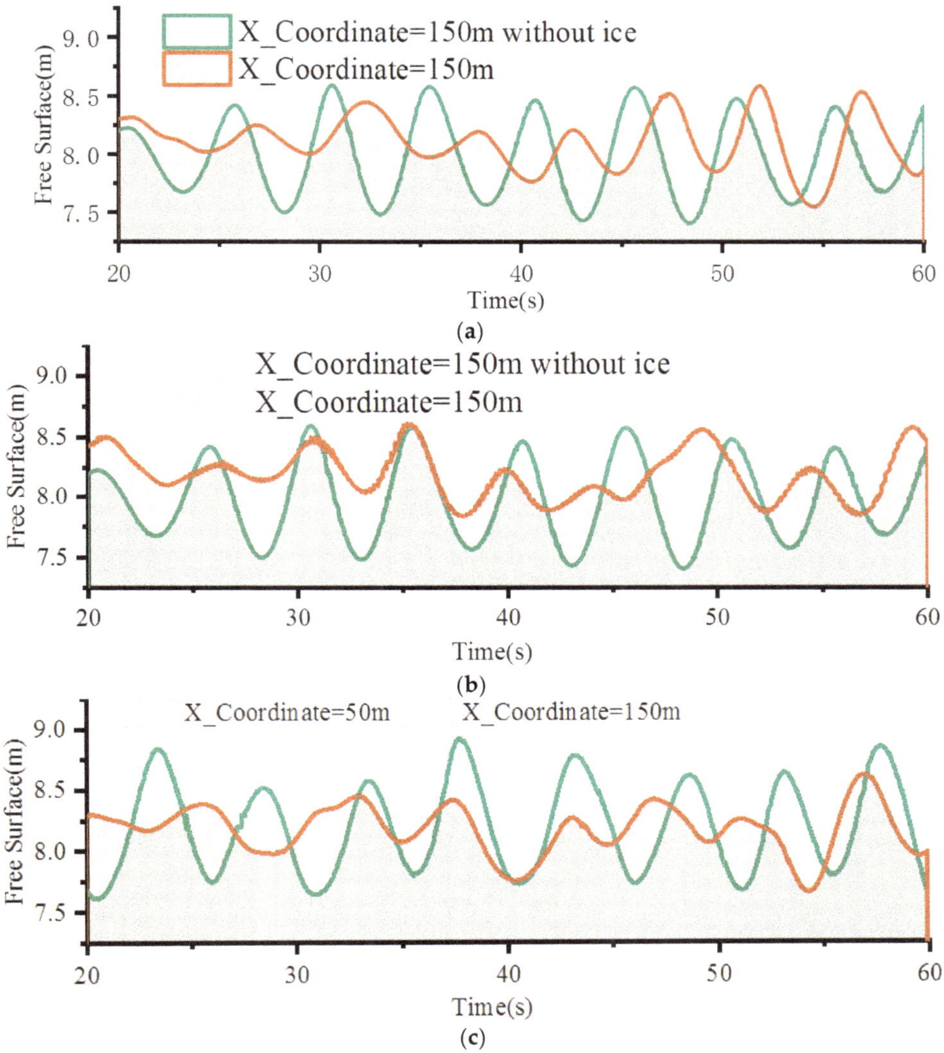

Figure 16. Influence of ice floes on free surface of wave (70% concentration ice floes). (**a**) Case 2; (**b**) Case 7; (**c**) Case 6.

Figure 17 illustrates the changes in the wave free surface under different conditions with the same ice concentration. By analyzing the wave attenuation levels under the same wave height but different wave periods, it was found that the attenuation level is greater at a period T = 4.75 s at 68%, while, at T = 4.5 s, it is the lowest at 59%. This can be explained from Figure 14a, where the longitudinal displacement of the ice floe is maximized at T = 4.75 s, indicating that the ice floe absorbs more energy from the waves.

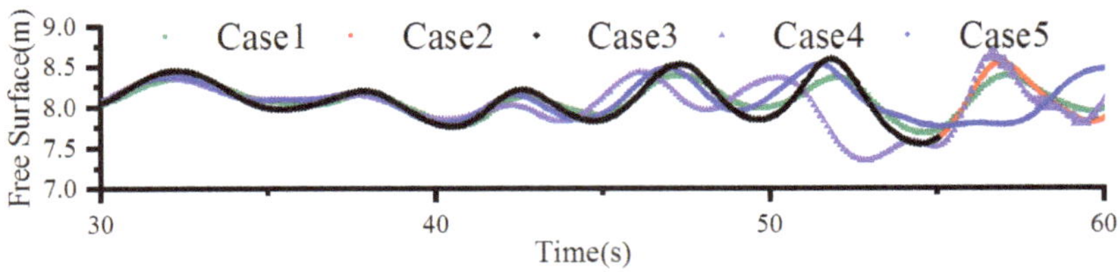

Figure 17. Attenuation of wave passthrough ice floes.

4. Conclusions

This paper establishes a methodology for generating regular waves and random ice floes compatible with ABAQUS. Utilizing the Coupled Eulerian–Lagrangian (CEL) approach, we conduct numerical simulations of interactions between regular waves, ice floes, and level ice. The motion and fragmentation characteristics of ice floes under wave action, alongside wave attenuation through ice-covered waters, are analyzed. Key findings include:

(1) Based on the velocity inlet boundary wave-generating method, a relatively ideal regular wave can be generated.

(2) The ice floes move periodically in both the longitudinal and vertical directions under the action of waves, and the period is consistent with the waves, but there is a certain phase difference. The longitudinal and vertical motion characteristics of ice floe are affected by the interaction between ice floes, and the movement of ice floes under low-ice-floe concentration is more regular, and the amplitude is more consistent in different cycles.

(3) The conversion of wave energy into kinetic energy of ice floe causes its own attenuation; the increase in sea ice concentration leads to an increase in the attenuation of waves.

Author Contributions: Conceptualization, Y.T. and C.Y.; methodology, C.Y.; investigation, C.Y.; data processing and analyzing, C.Y.; writing—original draft preparation, C.Y.; writing—review and editing, Y.T.; supervision.; project administration, Y.T. All authors have read and agreed to the published version of the manuscript.

Funding: This research was supported by the National Natural Science Foundation of China (Grant Nos. 52192690 and 52192693).

Data Availability Statement: The original contributions presented in this study are included in the article. Further inquiries can be directed to the corresponding author.

Conflicts of Interest: The authors declare no conflicts of interest.

References

1. Li, S.; Dou, T.; Xiao, C. Research progress for the changes of Arctic Ocean surface wave with diminishing sea ice. *Trans. Atmos. Sci.* **2021**, *44*, 118–127. (In Chinese)
2. Peters, A.S. The effect of a floating mat on water waves. *Commun. Pure Appl. Math.* **1950**, *3*, 319–354. [CrossRef]
3. Wadhams, P. The seasonal ice zone. In *The Geophysics of Sea Ice*; Untersteiner, N., Ed.; Springer: Boston, MA, USA, 1986; pp. 825–991. [CrossRef]
4. Luo, W.; Guo, C.; Su, Y.; Zhang, H. The study and progress of ship-wave-sea ice interaction in marginal ice zone. *Shipbuild. China* **2017**, *58*, 12–20. (In Chinese)
5. Rogers, W.E.; Thomson, J.; Shen, H.H.; Doble, M.J.; Wadhams, P.; Cheng, S. Dissipation of wind waves by pancake and frazil ice in the autumn Beaufort Sea. *J. Geophys. Res. Ocean.* **2016**, *121*, 7991–8007. [CrossRef]

6. Cheng, S.; Tsarau, A.; Li, H.; Herman, A.; Evers, K.U.; Shen, H. Loads on Structure and Waves in Ice (LS-WICE) project, Part 1: Wave attenuation and dispersion in broken ice fields. In Proceedings of the 24th International Conference on Port and Ocean Engineering under Arctic Conditions (POAC), Busan, Republic of Korea, 11–16 June 2017.
7. Herman, A.; Tsarau, A.; Evers, K.U.; Li, H.; Shen, H. Loads on Structure and Waves in Ice (LS-WICE) project, Part 2: Sea ice breaking by waves. In Proceedings of the 24th International Conference on Port and Ocean Engineering under Arctic Conditions (POAC), Busan, Republic of Korea, 11–16 June 2017.
8. Wang, C.; Wang, J.; Wang, C.; Song, M. Experimental study on longitudinal motion law of ice floes under wave action. *J. Huazhong Univ. Sci. Technol. (Nat. Sci. Ed.)* **2022**, *50*, 143–148. (In Chinese)
9. Meylan, M.H. Wave response of an ice floe of arbitrary geometry. *J. Geophys. Res. Ocean.* **2002**, *107*, 5-1–5-12. [CrossRef]
10. Wang, C.D.; Meylan, M.H. A higher-order-coupled boundary element and finite element method for the wave forcing of a floating elastic plate. *J. Fluids Struct.* **2004**, *19*, 557–572. [CrossRef]
11. Wu, T.; Luo, W.; Jiang, D.; Deng, R.; Huang, S. Numerical study on wave-ice interaction in the marginal ice zone. *J. Mar. Sci. Eng.* **2020**, *9*, 4. [CrossRef]
12. Ni, B.; Liu, C.; Hu, W.; Han, D. Numerical simulation on the interaction between two-dimensional Stokes wave and multiple ice floes. *Chin. J. Hydrodyn.* **2019**, *34*, 141–148. (In Chinese)
13. Wang, C.; Wang, J.; Wang, C.; Wang, Z.; Zhang, Y. Numerical study on wave–ice floe interaction in regular waves. *J. Mar. Sci. Eng.* **2023**, *11*, 2235. [CrossRef]
14. Lu, T.C.; Zou, Z.J. Numerical simulation of ice-wave interaction by coupling DEM-CFD. In Proceedings of the 38th International Conference on Ocean, Offshore and Arctic Engineering (OMAE 2019), Glasgow, UK, 9–14 June 2019; American Society of Mechanical Engineers: New York, NY, USA, 2019; Volume 8, p. V008T07A015. [CrossRef]
15. Wu, J.; Liu, L.; Ji, S. Analysis of wave attenuation and ice breaking characteristics based on DEM-SPH coupling method. *Chin. J. Hydrodyn.* **2023**, *38*, 176–186. (In Chinese)
16. Zhang, N.; Ma, Q.; Zheng, X.; Yan, S. A two-way coupling method for simulating wave-induced breakup of ice floes based on SPH. *Eng. Fract. Mech.* **2023**, *282*, 109148. [CrossRef]
17. He, K.; Ni, B.; Xu, X.; Wei, H.; Xue, Y. Numerical simulation on the breakup of an ice sheet induced by regular incident waves. *Appl. Ocean Res.* **2023**, *120*, 103024. [CrossRef]
18. Zhang, N.; Zheng, X.; Ma, Q. Study on wave-induced kinematic responses and flexures of ice floe by smoothed particle hydrodynamics. *Comput. Fluids* **2019**, *189*, 46–59. [CrossRef]
19. Ni, B.; Xiong, H.; Han, D.; Zeng, L.; Sun, L.; Tan, H. A review of ice deformation and breaking under flexural-gravity waves induced by moving loads. *J. Mar. Sci. Appl.* 2024, *advance online publication*. [CrossRef]
20. Majlesi, A.; Shahriar, A.; Nasouri, R.; Montoya, A.; Du, A.; Testik, F.Y.; Matamoros, A. The evaluation of explicit parameters on Eulerian-Lagrangian simulation of wave impact on coastal bridges. *Coast. Eng.* **2024**, *191*, 104540. [CrossRef]
21. Kim, D.H.; Kim, S.; Kim, S.W. Numerical analysis of drop impact-induced damage of a composite fuel tank assembly on a helicopter considering liquid sloshing. *Compos. Struct.* **2019**, *229*, 111438. [CrossRef]
22. Montoya, A.; Matamoros, A.; Testik, F.; Nasouri, R.; Shahriar, A.; Majlesi, A. *Structural Vulnerability of Coastal Bridges Under extreme Hurricane Conditions*; Report No. 18STTSA04; Transportation Consortium of South-Central States: Kunming, China, 2019.

water

MDPI

Article

In Situ Tests on the Flexural Strength and Effective Elastic Modulus of Brackish Ice During Different Ice Periods

Shaopeng Ji [1,2], Yubo Liu [3,*], Qingkai Wang [3], Peng Lu [3] and Shunqi Yuan [3]

1 State Key Laboratory of Structural Analysis, Optimization and CAE Software for Industrial Equipment, Dalian University of Technology, Dalian 116024, China; www.jsp@163.com
2 China Ship Scientific Research Center, National Key Laboratory of Science and Technology on Hydrodynamics, Wuxi 214082, China
3 State Key Laboratory of Coastal and Offshore Engineering, Dalian University of Technology, Dalian 116024, China; wangqingkai@dlut.edu.cn (Q.W.); lupeng@dlut.edu.cn (P.L.); Yuansq_Rex101@mail.dlut.edu.cn (S.Y.)
* Correspondence: liuyubokk@mail.dlut.edu.cn

Abstract: Global warming reduces the thickness and duration of seasonal lake ice, increasing the risk of ice cover failure. To investigate the bending behavior of ice cover, six groups of full-scale cantilever beam tests were conducted on a brackish water lake during the winter of 2023–2024, covering the following three ice periods: growth, stable, and melt. A total of 16 upward-loaded beams and 24 downward-loaded beams were tested. The results showed that the flexural strength of brackish ice was 374.21 ± 99.93 kPa, and the effective elastic modulus was 2.77 ± 0.93 GPa. The square root of bulk porosity, fitted with an exponential function, is the optimal predictor of flexural performance. Both flexural strength and effective elastic modulus systematically decreased with increasing porosity, and empirical regression formulas were established. On average, downward-loaded flexural strength was approximately 17.3% to 38.8% higher than upward-loaded strength, whereas elastic modulus showed no significant difference between the two loading directions. Flexural mechanical properties during the melt period reduced significantly, with a strength and modulus about 33.0% to 61.1% lower than those in the growth and stable periods. Comparisons with existing datasets demonstrate that the mechanical properties of brackish ice are lower than those of freshwater ice but higher than those of sea ice. This study provides new in situ data on the full-scale flexural mechanical properties of brackish ice and offers an important basis for assessing ice loads in lakes and estuarine environments under climate change.

Keywords: brackish ice; flexural strength; effective elastic modulus; cantilever beam; porosity

1. Introduction

Global warming has exerted significant impacts on the cryosphere [1–3]. For lake ice, both observations and modelings have demonstrated that seasonal-ice-covered lakes exhibit reduced ice duration and thickness [4,5]. In particular, the increasing frequency of winter extreme weather events introduces additional uncertainties to ice conditions [6]. As ice temperature rises, ice cover becomes prone to fracturing under the effects of wave [7]. Such ice break-up can substantially compromise the safety of ice-related activities.

Flexural failure is one of the primary modes of ice fracture. The flexural strength of lake ice serves as a fundamental mechanical parameter that governs the load-bearing capacity and fracture resistance of ice cover [8]. Experiments for determining the flexural strength and effective elastic modulus of ice mainly employ two approaches: simple beam tests and cantilever beam tests, both based on linear elasticity theory [9]. Simple beam tests (three-point and four-point bending) are typically performed on small-scale samples extracted from the ice cover, with tests conducted in situ or under controlled laboratory conditions. However, most of these tests rely on "dry" specimens preserved at uniform subzero temperatures, which fail to reproduce the through-thickness temperature gradients and brine connectivity typical of natural floating ice. To address this limitation, Wei et al. [10–12] developed a laboratory freezing model capable of reproducing realistic thermal gradients and liquid brine inclusions, and investigated the fracture toughness and elastic modulus of columnar saline ice under such conditions. Their experiments demonstrated that floating saline ice with vertical temperature gradients and brine interconnectivity exhibited an elastic modulus 21% lower than that of dry, isothermal specimens. Moreover, the fracture toughness and flexural strength of the floating, relatively warm samples were only about 30% of those measured in cold, dry specimens. These findings highlight the significant influence of thermal and brine structures on the mechanical behavior of ice and emphasize the importance of reproducing realistic environmental conditions. However, even laboratory experiments that simulate thermal gradients cannot fully capture the scale effects, crystal textures, and structural heterogeneity inherent to natural ice covers. To capture the mechanical behavior of ice under true in situ conditions, full-scale field methods have been developed [13–16]. Marchenko et al. [14] and Voermans et al. [15] estimated the effective elastic modulus of sea and lake ice in situ by analyzing the vibration of fixed-end floating ice beams and the low-frequency dispersion of surface waves, respectively. These methods are non-destructive and dynamic, allowing for assessment of the overall elastic response of natural ice covers. Complementary to these techniques, destructive full-scale tests, such as cantilever beam tests, can directly measure both flexural strength and effective elastic modulus. Karulina et al. [16] conducted more than seventy full-scale cantilever beam tests on both sea ice and freshwater ice. Their results showed that the flexural strength of freshwater ice was significantly higher than that of sea ice, and that the loading direction had little influence on the measured strength. In addition to the above methods, some studies have also adopted disc center-loading tests [17,18]; however, this approach is limited to the determination of flexural strength only. Overall, these methods differ not only in environmental realism but also in scale, which exerts a major influence on measured strength. Aly et al. [9] systematically quantified this scale effect through regression analysis of a multi-scale experimental database, establishing a relationship between flexural strength and beam volume. Their results confirmed that larger specimens systematically exhibit lower apparent strength, primarily due to the increased probability of internal flaws and structural heterogeneity.

The flexural strength and effective elastic modulus of ice are dependent on multiple factors, including intrinsic material properties (e.g., temperature, brine volume) and extrinsic testing conditions (e.g., loading configuration, sample size). Timco and O'Brien established a negative exponential relationship between sea ice flexural strength and the square root of brine volume through analysis of considerable data points [19]. Lainey and Tinawi found that both flexural strength and effective elastic modulus of sea ice increased as the temperature decreased [20]. Han et al. further identified distinct strain rate dependencies in the effective elastic modulus of freshwater ice. They observed that the effective elastic modulus showed no significant correlation with strain rate at temperatures above -2 °C, but increased with increasing strain rate at temperatures below -5 °C [21]. Moreover, cyclic loading has been shown to enhance the flexural strength of both freshwater and saline ice [22,23]. In addition to cyclic effects, the internal pore composition also strongly influences ice mechanics. Recent experiments on gas-enriched sea ice showed that both flexural strength and strain modulus decrease with increasing porosity, but brine inclusions cause greater weakening than gas [24]. Gas-dominated ice retained a higher strength and stiffness at comparable porosity levels, indicating that pore composition and connectivity, rather than total porosity alone, govern the bending behavior of natural ice. Despite these advances, most previous studies have focused on freshwater or sea ice, with little systematic attention given to the brackish ice commonly found in estuaries or semi-enclosed lakes. The physical properties of brackish ice (e.g., salinity, density, and porosity) lie between those of freshwater ice and sea ice, yet its mechanical behavior remains poorly understood. Moreover, while many empirical relationships between ice mechanical parameters and physical properties have been established, few studies have systematically explored the optimal predictors of full-scale flexural performance, particularly when considering loading direction and layered ice properties.

To address these gaps, this study conducted full-scale cantilever beam tests on brackish ice at Hanzhang Lake, northern China, during the winter of 2023–2024. This method was chosen for its ability to capture the in situ flexural behavior of the ice cover at full scale while simultaneously determining both flexural strength and effective elastic modulus, thus ensuring direct comparability with previous sea and freshwater ice datasets. In combination with observations of ice physical properties, the tests allowed a systematic evaluation of the relationships between physical conditions and flexural behavior. Special attention was given to the influence of loading direction and ice period, and the results were further compared with existing freshwater ice and sea ice datasets. This approach provides new insights into the flexural mechanical properties of brackish ice and their controlling factors.

2. Materials and Methods

2.1. Research Area

The field experiments were conducted at Hanzhang Lake (Figure 1), a natural lake situated in the northern temperate zone in China. The lake experiences complete freezing during winter, with the ice cover typically persisting from early December to mid-March. Typically, the mean air temperature during the freezing period is −4.4 °C, and the average winter precipitation is 35.7 mm. Notably, the snow accumulation duration accounts for only 20% of the total freezing period. Hanzhang Lake is located near the Bohai Sea estuary and is connected to the sea through a sluice gate, resulting in a water salinity range of 5–7 ppt. During winter, the sluice gate remains closed, isolating the lake from external inflows or outflows. Consequently, ice formation is primarily governed by thermodynamic processes, leading to a flat ice surface devoid of ridges. During the winter of 2023–2024, continuous records of ice thickness and air temperature were obtained at a fixed observation site. The monitoring results indicated a typical three-stage evolution. Figure 1d shows the variations in air temperature and ice thickness during the ice periods. From 9 December 2023 to 25 January 2024, the thickness steadily increased from less than 0.10 m to approximately 0.30 m, with a corresponding mean air temperature of −6.3 °C; this stage was defined as the growth period. From 26 January to 28 February, the thickness further increased to a peak of about 0.40 m and remained relatively stable, with a mean air temperature of −4.8 °C; this stage was defined as the stable period. From 29 February to 15 March, the air temperature rose to around 1.1 °C, the ice thickness gradually decreased to 0.23 m, with surface meltwater and porosity becoming evident. Subsequently, between 16 March and 18 March, the ice cover rapidly disintegrated and disappeared. This stage was defined as the melt period. Based on this ice period evolution, six groups of full-scale cantilever beam tests were conducted across the three ice periods.

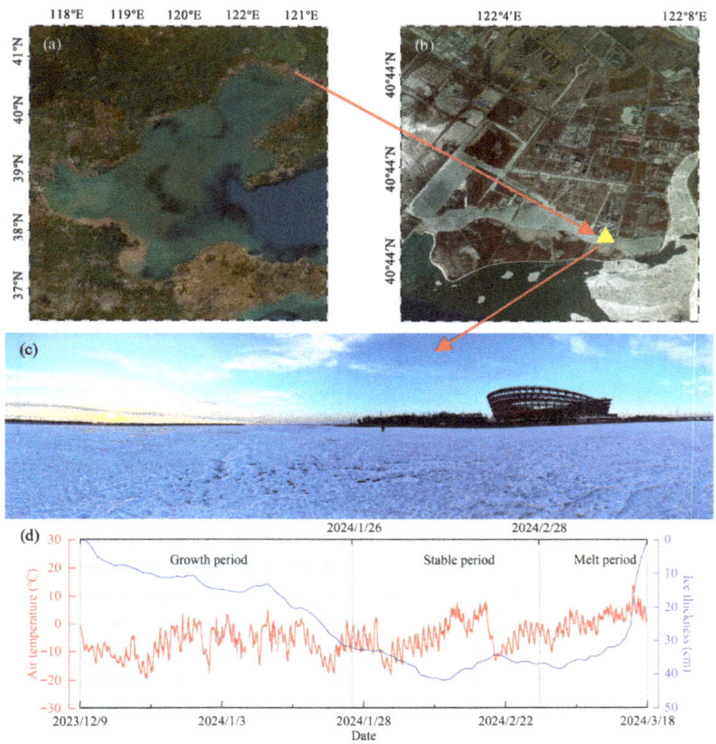

Figure 1. Location and temperature of the study area: (**a**) Bohai Sea; (**b**) Hanzhang Lake; (**c**) Field view of the experimental field; (**d**) line graph of ice thickness and air temperature changes during the experimental period.

2.2. Cantilever Beam Test

Full-scale cantilever beam tests were conducted to determine the flexural strength and effective elastic modulus of brackish ice during different ice periods. The tests were conducted from 17 January to 12 March, with corresponding ice thicknesses ranging from 0.28 to 0.45 m. All tests were conducted in areas free of visible cracks. If snow cover was present at the test site, it was removed manually using a shovel. During the first four tests, the ice surface was covered by approximately 3 to 4 cm of snow, which revealed a smooth surface after removal. Between the fourth and fifth tests, the surface snow experienced melting and refreezing due to solar radiation, resulting in a rougher ice surface. In the sixth test, conducted during the melting stage, the ice surface appeared porous with visible meltwater, which included both melted ice and water intruding upward from the underlying lake. At each test site, 4–10 cantilever beams were prepared using chainsaws (Dongcheng Power Tools Co., Ltd., Nantong, China), following the dimensional guidelines recommended by the International Association for Hydraulic Engineering and Research (IAHR). Each beam had a width of twice the ice thickness and a length of eight times the ice thickness (Figure 2a). To reduce stress concentrations at the beam root, a 20 cm diameter semi-circular notch was cut using an ice driller (Dongcheng Power Tools Co., Ltd., Nantong,

China) [25,26]. Additionally, a spacing of approximately 20 cm was maintained between adjacent beams to prevent fracture propagation and mutual interference (Figure 2b).

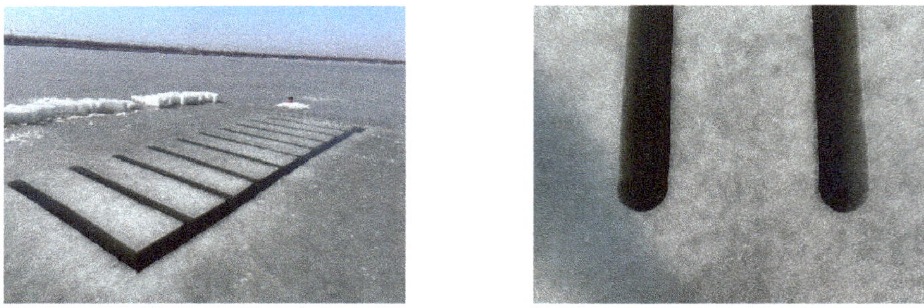

(**a**) (**b**)

Figure 2. Preparation process for cantilever beam samples: (**a**) Array of cantilever beam samples; (**b**) Trimming the roots of the beams to a rounded shape.

The experimental setup integrated sensors, data acquisition devices, and a loading system (Figure 3). A load sensor (10,000 N range, ±5 N accuracy) (Changsha Taihe Electronic Equipment Co., Ltd., Changsha, China) was used to measure the applied force, and a linear variable differential transformer (LVDT)displacement sensor (50 mm range, ±0.05 mm accuracy) (Beijing King Sensor Technology Co., Ltd., Beijing, China) was used to measure the displacement of the cantilever end of ice beams. Both force and displacement signals were synchronously recorded at 1000 Hz. The loading system consisted of a stainless-steel A-frame structure for downward loading and a rake-type frame with four steel bars for upward loading (Dalian Northern Analytical Instruments Co., Ltd., Dalian, China). The steel bars were welded to the A-frame's horizontal beam at a distance exceeding ice thickness. The load was exerted at the beam end. A hinged connection was implemented between the loading lever and A-frame to ensure consistently vertical downward force application. The loading lever was manually actuated upward and downward. The loading system was stabilized by a support base secured to the ice surface with ice cones to prevent slippage during tests. The loading time from applying force to beam failure was no more than 10 s. After the beam failure, the distance between the beam end to the failure crack was measured. Afterwards, a small ice block was taken through the ice thickness at the beam root to measure the ice temperature, salinity, and density (see section below).

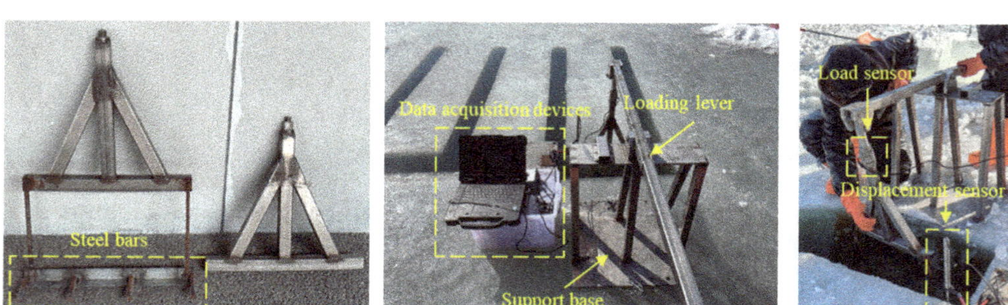

(**a**) (**b**) (**c**)

Figure 3. Cantilever beam testing equipment: (**a**) Two types of loading frames; (**b**) Collecting data and integrating mechanical components; (**c**) Load and displacement sensors.

In summary, the cantilever beam tests in this study were conducted under controlled conditions, with strict regulation of beam preparation, loading configuration, and sensor measurements to ensure data reliability. The six groups of tests covered different ice periods, beam dimensions, and loading configurations. The detailed conditions of each group of tests, including test date, ice period, ice thickness, number of beams, and ice surface characteristics, are summarized in Table 1. For each test set, an equal number of upward- and downward-loaded beams was attempted whenever possible. However, in the fifth test, only one upward-loaded test was completed because the vertical cut at the beam end did not perfectly match the rake-type A-frame, leading to hook detachment during upward loading.

Table 1. Summary of cantilever beam tests conducted on brackish ice.

No.	Date of Test	Ice Period	Ice Thickness (m)	Number of Beams (Upward/Downward)	Ice Surface Condition
1	17 January 2024	Growth	0.38	4 (2/2)	Snow-covered and smooth surface
2	22 January 2024	Growth	0.37	4 (2/2)	Snow-covered and smooth surface
3	26 January 2024	Stable	0.43	6 (3/3)	Snow-covered and smooth surface
4	30 January 2024	Stable	0.40	6 (3/3)	Snow-covered and smooth surface
5	24 February 2024	Stable	0.45	10 (1/9)	Rough surface
6	12 March 2024	Melt	0.28	10 (5/5)	Wet surface

In the cantilever beam tests, the flexural strength and effective modulus were determined from Equations (1) and (2), respectively:

$$\sigma_{\mathrm{f}} = \frac{6PL_c}{bh^2} \tag{1}$$

$$E = \frac{4PL_c^3}{bh^3\delta_c} \tag{2}$$

where σ_{f} and E are the flexural strength and effective modulus derived from cantilever beam tests (in Pa), P is the load at the point of beam failure (in N), δ_{c} is the deflection at the beam cantilever end (in m), L_{c} is the distance between the beam end to the failure crack (in m), b is the beam width (in m), and h is the ice thickness (in m).

2.3. Ice Physical Properties

Following the cantilever beam tests, measurements of ice temperature, salinity, and density were conducted by extracting a 10 × 10 cm cross-section ice block through the full thickness at the beam root. Immediately after lifting the ice block, ice temperature measurements were performed using a fast-response probe thermometer (accuracy ±0.2 °C) (Testo Instruments International Trading (Shanghai) Co., Ltd., Shanghai, China) inserted into 2 mm diameter holes drilled to the block center spaced at 5 cm intervals along the ice thickness. The ice block was then rapidly transported to the lakeside low-temperature laboratory for density measurements. Ice density was determined using the mass/volume method. The ice block was first sectioned along the thickness into 5 cm long segments, which were then machined into 4 cm cubes using a band saw (Hangzhou ABCH Intelligent Technology Co., Ltd., Hangzhou, China). The mass of the cube was weighed using a balance (±0.1 g) (Shenzhen Meifu Electronics Co., Ltd., Shenzhen, China), and the volume was calculated based on the dimensions measured using a caliper (±0.02 mm) (Delixi Instruments Co., Ltd., Shanghai, China). The uncertainty of the estimated ice density can be evaluated with an error propagation analysis [27,28]. Here, the accuracy of dimension measurement was ±0.02 mm and an ice cube typically had a mass of 40–70 g, and thus, the uncertainty due to the measurement error was approximately 5%. Subsequently, the ice cubes were melted for salinity measurements using a salinometer (accuracy ±0.1 psu) (Shanghai Bangwo Instrument Equipment Co., Ltd., Shanghai, China) at a room temperature. The brine and gas volume fractions, as well as porosity of the ice beam, were then determined using ice temperature, salinity, and density according to the method of Cox and Weeks for below −2 °C [29], and for the interval (−2, 0) °C, that by Leppäranta and Manninen [30].

Observations of ice crystal structures were conducted in the low-temperature laboratory during the stable ice period (5 February 2024) when ice thickness reached its maximum. Both vertical and horizontal sections were prepared by first cutting 2 mm thick sections at 5 cm intervals along the ice thickness using a band saw. These thick sections were attached to glass plates and thinned down further to an approximate 0.5 mm thickness using a planer. The thin slices were placed on a universal stage (Dalian Northern Analytical Instruments Co., Ltd., Dalian, China) to observe the crystal structures under crossed polarized light, recorded by photography.

3. Results

3.1. Physical Properties of Ice Layers

3.1.1. Crystal Structure

Figure 4 presents thin-slice observations prepared both parallel and perpendicular to the ice surface. The colored vertical lines on the left indicate the tested ice layer depth ranges of each test. According to Eicken's classification scheme [31], the upper 0–3 cm of the ice consisted of more than 80% polygonal granular crystals, indicating granular ice, while the section from 3 to 48 cm was composed of vertically aligned columnar crystals typical of columnar ice. The vertical thin sections show that granular ice accounted for only a small fraction of the total thickness, and no distinct transition zone was observed between the granular and columnar ice. In the 3–15 cm layer, the columnar crystals were less than 10 cm long along the ice thickness, tightly packed, and evenly distributed. In contrast, crystals between 15 and 48 cm increased significantly in length (15–30 cm), exhibited larger widths, and showed a marked reduction in the number of complete crystals within the observation area. The horizontal thin slices reveal that at depths of D = 5 cm and D = 10 cm, a larger number of intact crystals with elongated shapes were observed. When $D \geq 25$ cm, the crystals became fewer in number and more equant in shape. This stratification indicates a growth transition from an initial rapid freezing stage, with weaker dynamic influences, to a subsequent thermodynamically dominated regime characterized by steady vertical heat flux. The granular-to-columnar interface was sharp, with no visible discontinuities or refrozen boundaries, suggesting uninterrupted in situ growth. The absence of prominent brine channels or sediment inclusions further supports the thermodynamic stability of the ice formation. The tested ice thicknesses and corresponding crystal structure ranges for each experiment are marked by colored vertical lines. It is noteworthy that the locations for different tests were not identical, so the observed ice thickness did not strictly follow the sequence of thickening and thinning. In particular, the depth of tested ice cover during the melt period did not start from the surface (0 cm) because the uppermost layer had already undergone partial melting.

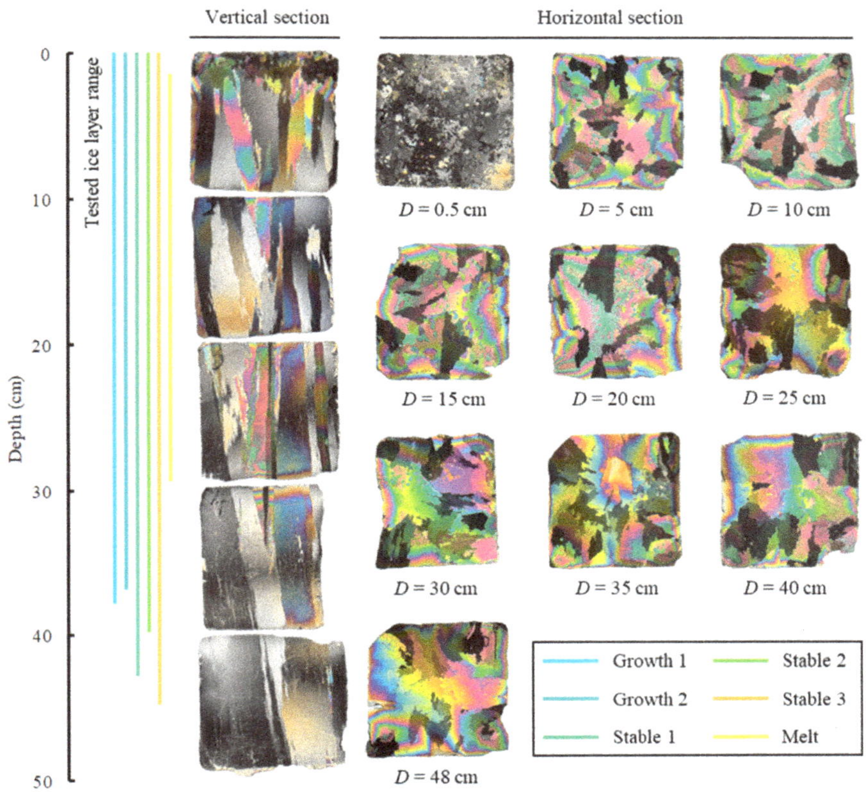

Figure 4. Crystal structures of brackish ice with corresponding tested ice layer depth ranges for each test group. Note: The colors observed in the vertical and horizontal thin sections represent the natural birefringence patterns of ice crystals with different growth orientations under cross-polarized light and are not artificially added.

3.1.2. Physical Parameters

Figure 5 presents the vertical distributions of temperature, salinity, and density for six test groups, where the salinity and density values represent the averages of all beams in each group. During the growth and stable periods, the temperature profiles exhibited a characteristic "C-shaped" distribution: relatively higher temperatures near the ice surface, decreasing with depth, and then increasing again toward the ice bottom, where temperatures approached that of the underlying water. This feature is related to the timing of the experiments. Both growth and stable tests were conducted in the afternoon, following a night without solar radiation, during which the temperature profile was close to a linear or parabolic increase from the cold surface to the warmer bottom. After 8–10 h of solar radiation on the day of testing, the upper ice warmed, resulting in the observed "C-shaped" profile. In contrast, the melt-period test was conducted in the morning, and the temperature profile still displayed a monotonic increase from the surface to the bottom. The salinity profiles generally showed higher values near the ice surface, decreasing with depth, with a slight increase near the bottom due to salt expulsion during freezing. In the Stable 3 test, the surface salinity was reduced because of snowmelt refreezing, but below 10 cm, the profile was consistent with other cases. During the melt period, internal brine channels had become connected and significant brine drainage occurred, producing an overall low-salinity profile. Ice density showed relatively lower values at the surface and bottom, with higher values in the middle section. Similar density magnitudes were observed during the growth and stable periods, whereas the melt-period density was significantly reduced.

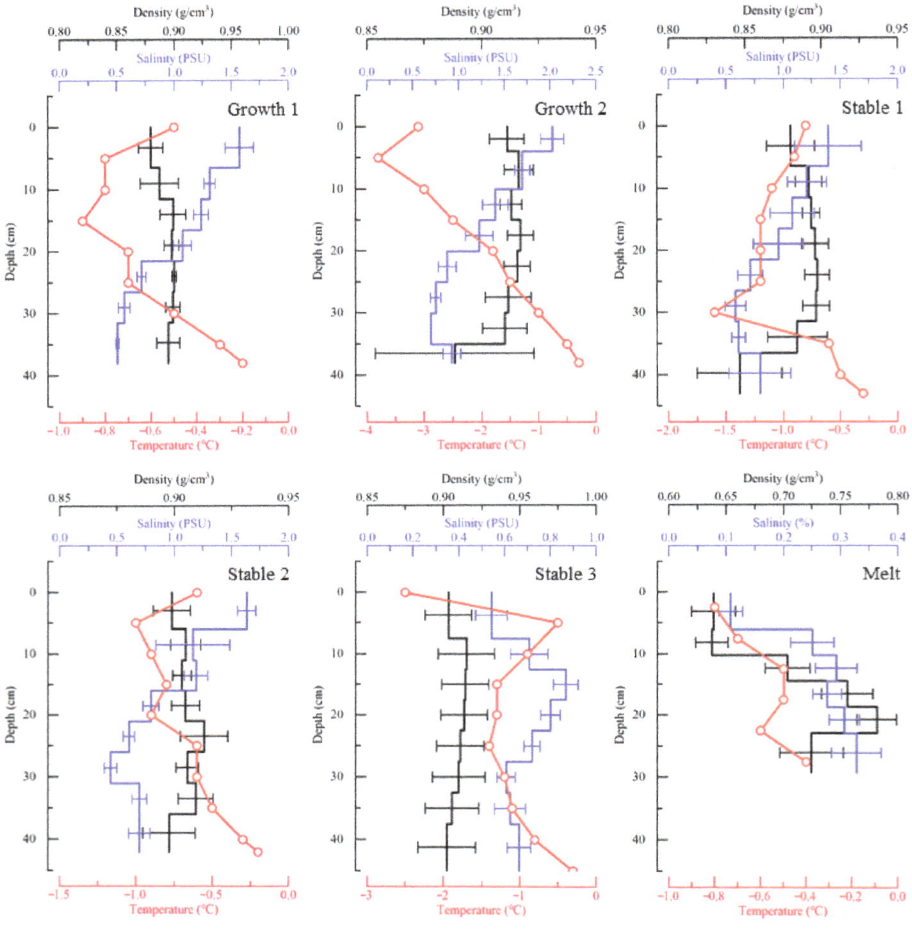

Figure 5. Vertical distributions of temperature, salinity, and density of the ice layer for each test group.

Figure 6 shows the calculated vertical distributions of brine volume fraction, gas volume fraction, and porosity derived from ice temperature, salinity, and density. In the growth and stable periods, these parameters generally decreased with depth in the upper ice and then increased toward the bottom. In the Stable 3 test, snowmelt refreezing led to an overall lower porosity, indicating a more consolidated structure. In contrast, the melt-period ice exhibited lower brine volume fractions and higher gas fractions, as brine loss during melting transformed brine pockets into air bubbles. This effect was most pronounced in the surface layers exposed to strong solar radiation. The porosity profile during the melt stage followed a similar depth-dependent trend to that observed during the growth and stable periods, but with consistently higher values, indicating substantial weakening of the ice matrix.

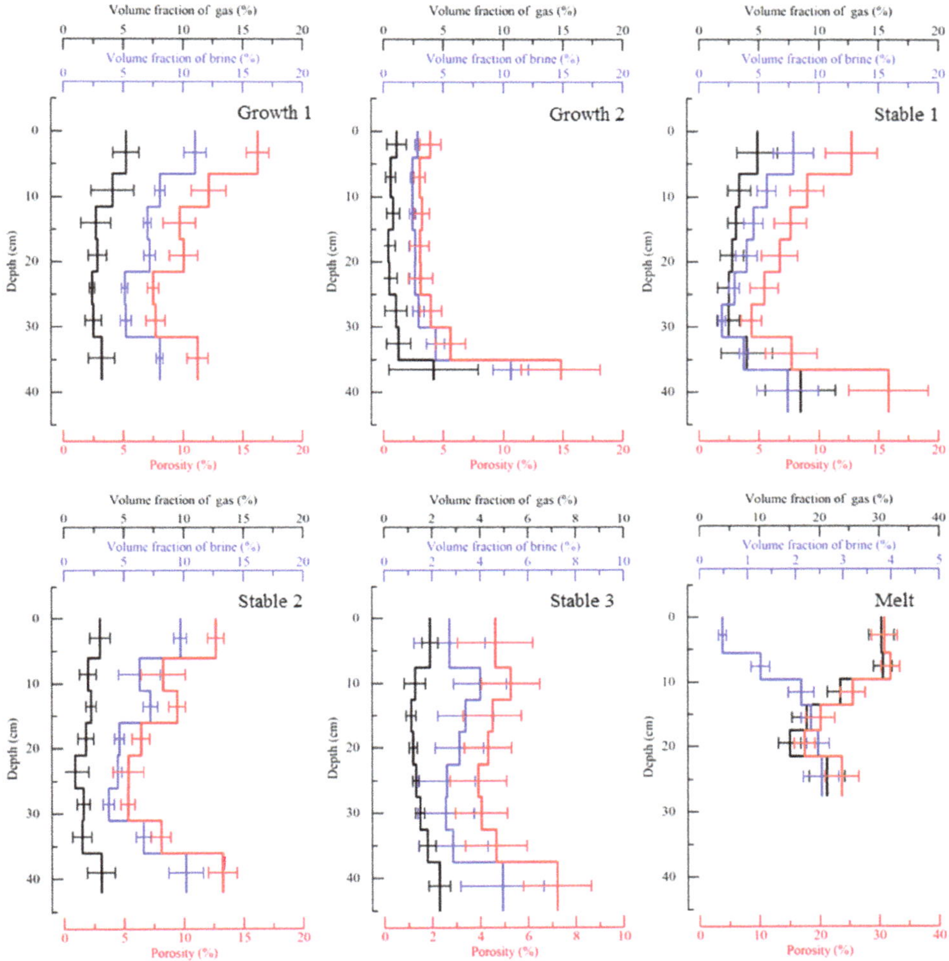

Figure 6. Vertical distributions of volume fraction of gas, volume fraction of brine, and porosity of the ice layer for each test group.

3.2. Flexural Strength and Effective Modulus Derived from Cantilever Beam Tests

A total of 39 valid datasets were obtained from 40 full-scale cantilever beam tests. Table 2 summarizes the experimental results, including beam volume, flexural strength, and effective elastic modulus. The beam volumes ranged from 0.38 to 1.37 m^3, with substantial variations observed among different ice periods, whereas variations within individual groups were negligible. Overall, the flexural strength of upward-loaded beams ranged from 202.69 to 415.52 kPa, with effective elastic moduli between 1.46 and 4.65 GPa. For downward loading, flexural strength varied between 309.61 and 523.78 kPa, and effective elastic moduli ranged from 1.68 to 3.53 GPa. Considerable scatter was observed both within and between groups, reflecting the influence of ice microstructural heterogeneity and testing conditions.

Table 2. Summary of flexural strength and effective elastic modulus of brackish ice beams of each test group. Note: One test conducted during the melt period was excluded due to displacement sensor freezing, which prevented valid data recording.

Group	Beam Volume (m^3)	Loading Direction	Number of Valid Datasets (n)	Flexural Strength (kPa)	Effective Modulus (GPa)	Mean Flexural Strength (MPa)	Mean Effective Modulus (GPa)
Growth 1	0.99 ± 0.005	Upward	2	290.12 ± 35.91	2.86 ± 0.90	346.75 ± 68.9	2.62 ± 0.60
		Downward	2	403.37 ± 11.33	2.39 ± 0.26		
Growth 2	0.99 ± 0.004	Upward	2	415.52 ± 7.97	4.65 ± 0.75	469.65 ± 62.69	3.84 ± 1.04
		Downward	2	523.78 ± 2.34	3.02 ± 0.24		
Stable 1	1.30 ± 0.009	Upward	3	318.96 ± 44.18	3.22 ± 0.74	383.24 ± 76.69	2.81 ± 0.70
		Downward	3	447.53 ± 39.06	2.41 ± 0.42		
Stable 2	1.06 ± 0.011	Upward	3	300.66 ± 28.34	2.79 ± 0.34	396.09 ± 107.82	3.16 ± 0.56
		Downward	3	491.52 ± 30.63	3.53 ± 0.50		
Stable 3	1.37 ± 0.031	Upward	1	370.46	3.22	440.07 ± 35.74	3.23 ± 0.58
		Downward	9	447.81 ± 27.64	3.23 ± 0.61		
Melt	0.38 ± 0.022	Upward	5	202.69 ± 12.88	1.46 ± 0.22	250.21 ± 58.58	1.56 ± 0.28
		Downward	4	309.61 ± 21.47	1.68 ± 0.32		

3.2.1. Effects of Physical Parameters

To identify the physical parameters that primarily govern the flexural mechanical behavior of brackish ice, three parameters were considered: brine volume fraction, gas volume fraction, and porosity. In the downward loading configuration, the upper side of the ice beam is subjected to tension while the lower side experiences compression; since the tensile strength of ice is substantially lower than its compressive strength, flexural failure is essentially governed by tensile failure. Therefore, the physical properties of the tensile side, taken from fractional layers between 1/8 and 8/8 (bulk) of the ice thickness, were selected as potential influencing factors. Their relationships with flexural strength were fitted using logarithmic, power-law, linear, and exponential functions. A total of 96 fitting schemes were evaluated, and the coefficients of determination (R^2) were summarized in heatmaps for comparison (Figure 7).

The heatmaps indicate that, among different fractions of the tensile side, the use of bulk physical parameters as predictors yielded the highest goodness-of-fit. Among the functional forms, exponential fits consistently produced the best performance. Consequently, the square roots of bulk brine volume fraction, gas volume fraction, and porosity were fitted against flexural strength using exponential models. The results show that brine volume fraction exhibited the weakest correlation, and flexural strength even increased with the square root of brine volume fraction, which contradicts previous findings in the literature. In contrast, flexural strength decreased with the square roots of gas volume fraction and porosity, with porosity presenting the strongest correlation among the three parameters. In a similar manner, the influence of physical parameters on the effective elastic modulus was also examined (Figure 8). Consistent with the results for flexural strength, the square root of bulk porosity yielded the best fits, and its R^2 values were even higher than those obtained for flexural strength. The mean flexural strength and effective elastic modulus of each group were fitted against the square root of bulk porosity, yielding the empirical formulas for the full-scale flexural mechanical properties of brackish ice, as shown in Equations (3) and (4).

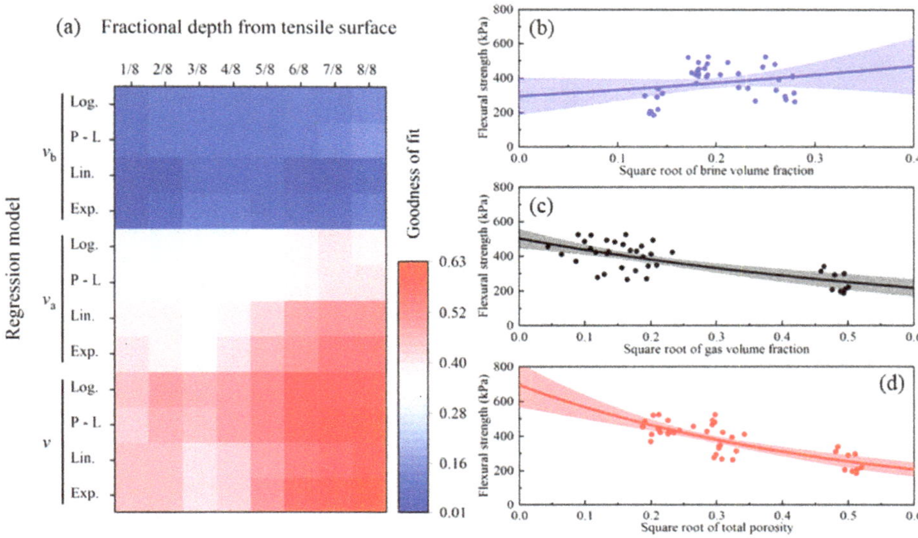

Figure 7. Relationship between ice physical parameters and flexural strength: (**a**) Goodness-of-fit heat map for different influencing factors (brine volume fraction v_b, gas volume fraction v_a, and porosity v) at various fractional depths from the tensile surface, under four regression models: logarithmic (Log.), power-law (P–L), linear (Lin.), and exponential (Exp.); (**b**–**d**) Relationships between flexural strength and the square root of brine volume fraction, gas volume fraction, and porosity, respectively, with exponential fits shown.

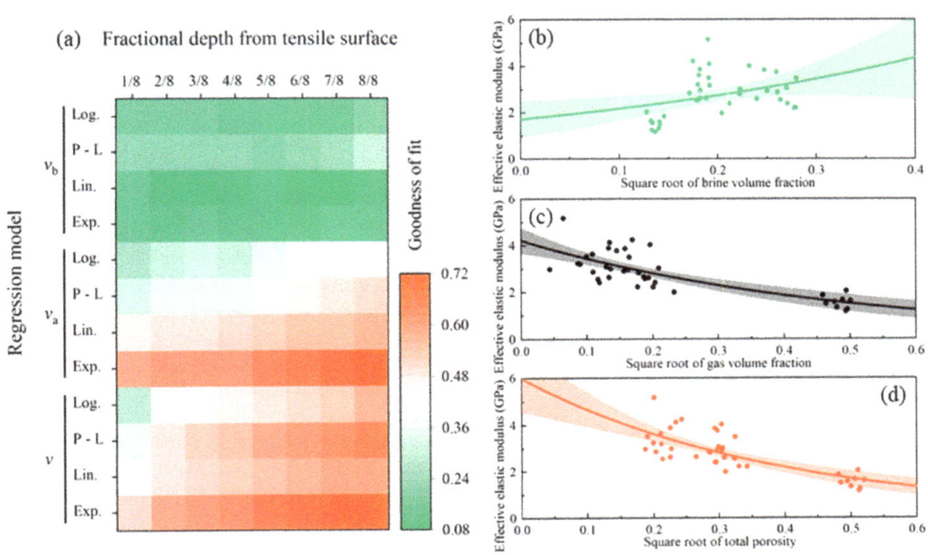

Figure 8. Relationship between ice physical parameters and effective elastic modulus: (**a**) Goodness-of-fit heat map for different influencing factors at various fractional depths from the tensile surface, under four regression models (notations as in Figure 7); (**b**–**d**) Relationships between effective elastic modulus and the square root of brine volume fraction, gas volume fraction, and porosity, respectively, with exponential fits shown.

In the above result analysis, all scatter data were used in the exploratory fitting stage to enable a clear comparison among different parameterizations and functional forms. Using mean values at this stage would result in a uniformly high R^2 across all models, obscuring the distinction between fitting types and influencing factors. In contrast, the final empirical formulas for the full-scale flexural mechanical properties of brackish ice (Equations (3) and (4)) were established based on the mean values of each test group, yielding high coefficients of determination (R^2 = 0.98 and 0.92), which represent validated relationships applicable to engineering practice.

$$\sigma_{\mathrm{f}} = 0.703\exp\left(-2.061\sqrt{v}\right),\ \ R^2 = 0.98 \tag{3}$$

$$E = 6.519\exp\left(-2.738\sqrt{v}\right),\ \ R^2 = 0.92 \tag{4}$$

where v is the bulk porosity.

3.2.2. Effects of Loading Direction

Figure 9 shows the comparison between upward and downward loading for each test group. The bars represent group-mean values of flexural strength and effective elastic modulus, accompanied by scatter points for individual beams and error bars for within-group variability, while the dashed lines indicate the mean values of strength and modulus within each ice period. For flexural strength, a consistent pattern was observed: in all groups, the mean values obtained from downward loaded were higher than those from upward loading. The differences ranged from 17.3% to 38.8%, indicating that downward loading systematically exhibited greater resistance to bending failure. In contrast, the effective elastic modulus did not show a clear or consistent dependence on loading direction. In some groups, downward loading produced slightly higher values, whereas in others, upward loading was higher. Overall, the available data do not provide sufficient evidence to confirm a definitive relationship between loading direction and elastic modulus. These results suggest that flexural strength is more sensitive to loading direction, whereas any directional effect on the elastic modulus remains uncertain.

To examine whether the strength difference between upward and downward loading could be attributed to the use of bulk porosity as the influencing factor, which may neglect vertical variations in porosity within the ice, we further compared different porosity parameterizations. Figure 10 presents the regression of upward and downward-loaded flexural strengths against the square root of porosity. In addition to bulk porosity, porosity values taken from different tensile-side depths (1/8, 4/8 of the ice thickness) were also considered as predictors. The results show that, regardless of the porosity parameter used, the flexural strength obtained under downward loading was consistently higher than that under upward loading at the same porosity level. Moreover, using the square root of bulk porosity as the influencing factor yielded the highest goodness-of-fit for both loading directions, outperforming other parameterization approaches. These findings indicate that bulk porosity remains the most effective predictor, and the systematic strength difference between upward and downward loaded cannot be explained solely by the choice of porosity representation.

A similar analysis was performed for the effective elastic modulus (Figure 11). Unlike flexural strength, the modulus did not exhibit a consistent magnitude relationship between upward and downward loading. At lower porosity values ($\sqrt{\nu} \approx 0.2$), the upward modulus was slightly higher than the downward modulus, but the relationship gradually reversed with increasing porosity, and at $\sqrt{\nu} \approx 0.5$, the upward modulus became lower. Overall, the observed differences between the two loading directions were not systematic, and there is currently insufficient evidence to establish a clear directional dependence for the effective elastic modulus. Nevertheless, consistent with the results for flexural strength, the square root of bulk porosity remained the most effective predictor, yielding the highest goodness-of-fit among the parameterizations.

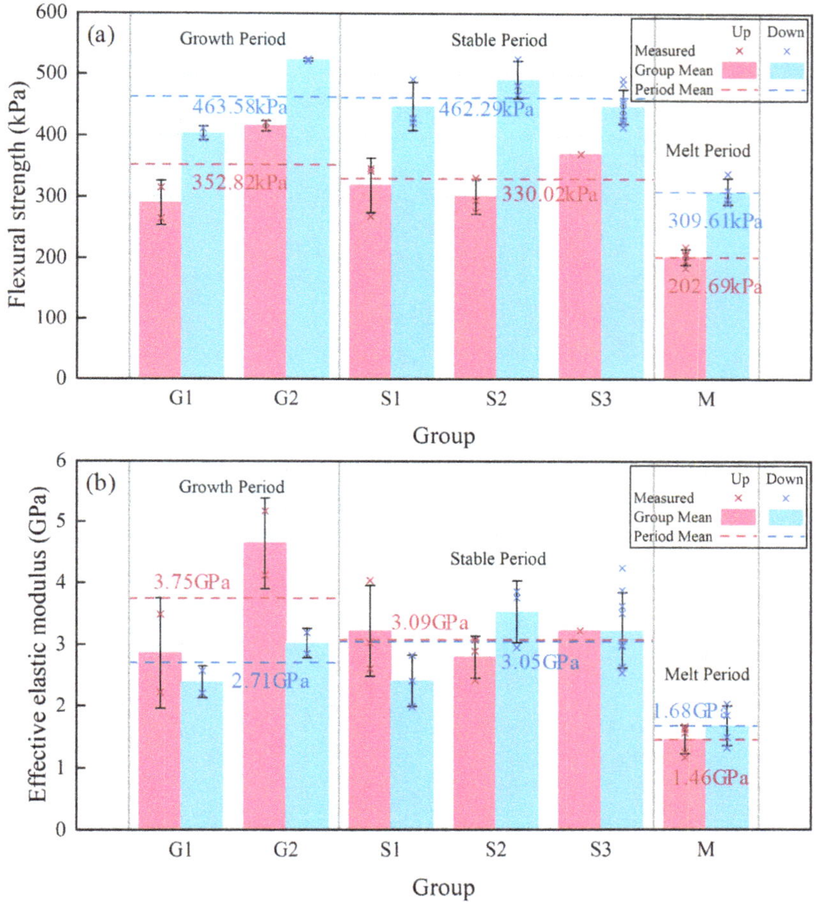

Figure 9. Mean flexural strength (**a**) and effective elastic modulus (**b**) of brackish ice under different groups, loading directions, and ice periods.

The systematic difference in flexural strength between downward and upward loading can be rationalized by the crystallographic stratification of the ice cover. The upper fibers are in tension under downward loading. The top 3 cm consists of granular ice with abundant grain boundaries and inclusions, and the columnar crystals beneath exhibit an increase in diameter with depth. This gradient implies fewer effective bonding interfaces and a reduced resistance to tensile cracking as the tensile side migrates downward, thereby promoting earlier crack initiation and unstable propagation. Although the tensile-side porosity was accounted for in the regressions, porosity alone does not capture microstructural attributes such as grain-boundary orientation, grain-size gradients, and columnar connectivity. These factors plausibly explain why, at comparable porosity levels, downward loading still yields higher strengths than upward loading.

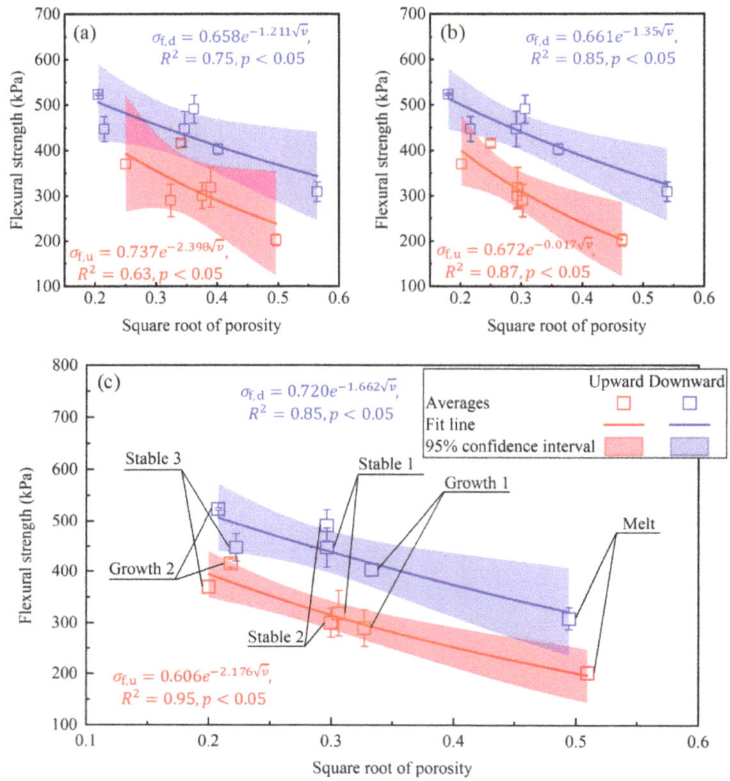

Figure 10. Relationships between flexural strength and the square root of porosity under different loading directions with various porosity averaging methods: Porosity averaged over the (**a**) upper 1/8 tensile side; (**b**) upper 4/8 tensile side; (**c**) bulk. Where d means downward loaded and u means upward loaded.

By contrast, unlike flexural strength, which reflects the threshold at which failure occurs, the modulus describes the rate of increase in load with deflection, and the two are not governed by the same mechanisms. For example, in the case of upward-loaded cantilever beams, the lower ice in tension has a lower tensile strength, making it more prone to failure than downward-loaded cantilever beams. However, this does not imply that the tensile modulus of the lower ice is lower than that of the upper ice. In other words, the magnitude of the failure threshold does not directly determine the slope of the load–deflection curve. Theoretically, the effective elastic modulus is primarily governed by the overall elastic response of the ice matrix rather than local failure initiation, and therefore its sensitivity to loading direction is expected to be weaker than that of flexural strength. Experimentally, there is currently insufficient evidence to establish a clear relationship between loading direction and elastic modulus. However, due to the inherent variability and scatter in full-scale in situ measurements, it is premature to conclude that the modulus is entirely independent of loading direction. Further accumulation of data from multiple ice conditions is needed to draw a more definitive conclusion.

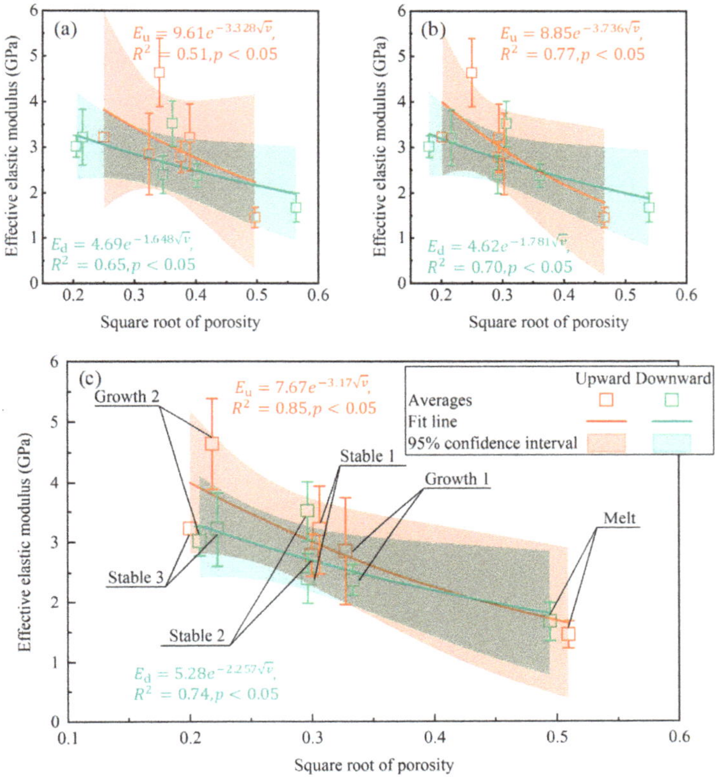

Figure 11. Relationships between effective elastic modulus and the square root of porosity under different loading directions with various porosity averaging methods: Porosity averaged over the (**a**) upper 1/8 tensile side; (**b**) upper 4/8 tensile side; (**c**) bulk.

3.2.3. Effects of Ice Period and Beam Volume

As shown in Figure 9, the mean values for each ice period are indicated by dashed lines, and therefore no additional figures are provided in this section. The flexural mechanical performance during the growth and stable periods was comparable, whereas that of the melt period was markedly lower. Specifically, the average flexural strengths under upward and downward loading were approximately 352.82 kPa and 463.58 kPa in the growth period and 330.02 kPa and 462.29 kPa in the stable period, but decreased to 202.69 kPa and 309.61 kPa in the melt period, representing reductions of about 33–33.2% and 38.6–41.8% relative to the growth and stable periods. The average effective elastic moduli were 3.75 GPa and 2.71 GPa in the growth period, 3.09 GPa and 3.05 GPa in the stable period, and decreased to 1.46 GPa and 1.68 GPa in the melt period, which were approximately 52.8–61.1% and 38.0–44.9% lower than those in the growth and stable periods, respectively. The pronounced weakening of flexural mechanical properties during the melt period is consistent with the physical property results presented in Section 3.1.2: higher gas volume fraction and porosity, together with connected brine channels and surface warming, weakened the effective skeleton of the ice and promoted premature tensile failure. Microstructural degradation, such as grain-boundary wetting, brine drainage, and air-bubble accumulation, further reduced the stiffness, explaining the decrease in strength and modulus compared with the growth and stable periods.

Within each ice period, neither flexural strength nor effective modulus showed a monotonic trend with ice thickness or beam volume. For instance, in the growth period, the two test groups had nearly identical beam volumes (approximately 0.99 m^3), yet their group-mean flexural strengths differed considerably (Growth 1: 346.75 ± 68.9 kPa; Growth 2: 469.65 ± 62.69 kPa), as did their effective moduli (Growth 1: 2.62 ± 0.60 GPa; Growth 2: 3.84 ± 1.04 GPa). During the stable period, beam volumes ranged from 1.06 to 1.37 m^3, but the corresponding strength values varied non-monotonically (383.24 ± 76.69 to 440.07 ± 35.74 kPa), and the effective modulus fluctuated between 2.81 ± 0.70 and 3.23 ± 0.58 GPa, showing no systematic increase with beam volume. Although the beam size was determined by the in situ ice thickness, the variation in beam volume within each ice period was relatively small, and the limited number of tests makes it difficult to conclusively assess the size effect. According to the empirical relationship proposed by Aly et al. [9], the flexural strength of ice tends to decrease with increasing specimen volume due to a higher probability of internal flaws and cracks. However, in the present study, the observed variation in strength across different beam volumes does not follow this theoretical trend, suggesting that the size effect was not a dominant factor. In the melt period, where the beam dimensions varied more noticeably, the substantial changes in ice physical properties, such as temperature, salinity, and porosity, likely overshadowed the geometric influence, making it challenging to isolate the pure contribution of specimen size to the mechanical behavior.

3.2.4. Evaluation of Buoyancy Effects on Test Results

Taking the cantilever beams from the growth and stable periods as examples (which had comparable beam volumes), the influence of buoyancy during loading on the test results was quantitatively evaluated. The beams were in hydrostatic equilibrium prior to loading, with their self-weight balanced by buoyancy. Therefore, only the additional (or reduced) buoyancy generated during loading due to downward (or upward) deflection could influence the test results. Using the measured group means (water density $\rho_w = 1\text{g}/\text{cm}^3$; ice density 0.90 g/cm^3; beam length L_c = 3.34 m, width b = 0.85 m, thickness h = 0.42 m; end displacement δ_c = 3.02 mm; failure load P = 3047.75 N) and assuming a small-deflection parabolic profile $\delta(x) = \delta_c\left(\frac{x}{L_c}\right)^2$, the incremental submerged volume produces an upward distributed buoyancy $\rho_w g b \delta(x)$. The corresponding buoyant bending moment about the beam root can be expressed by Equation (5). The equivalent end load induced by buoyancy is then obtained from Equation (6).

$$M_{\text{buoy}} = \rho_w g b \int_0^{L_c} x\delta(x)dx = \rho_w g b \delta_c \frac{{L_c}^2}{4} = 70.23 \text{ N}\cdot\text{m} \tag{5}$$

$$P_{\text{eq}} = \frac{M_{\text{buoy}}}{L_c} = 21.05 \text{ N} \tag{6}$$

As shown by Equation (6), the equivalent buoyant end load P_{eq} is only 21.05 N. Compared with the measured failure load P = 3047.75 N, which corresponds to 0.69% of the total load. Since both flexural strength and effective elastic modulus are linearly proportional to the applied load, the first-order influence of this buoyancy effect on the derived mechanical parameters is also below 1%. Therefore, the buoyancy-induced correction can be regarded as negligible under the present geometric and loading conditions. In practical conditions, the bending behavior of floating ice sheets is also affected by buoyancy; however, its magnitude is inherently small and therefore was not explicitly corrected for in the analysis.

4. Discussion

4.1. Differences in Mechanical Properties of Brackish Water Ice from Saline Water Ice and Freshwater Ice

Table 3 summarizes the literature datasets selected for comparison with the present brackish ice results. Freshwater ice data were taken from Refs. [16,32,33], all of which reported mean ice temperatures, while sea ice data and brackish ice data were collected from Refs. [16,18,19,34] and Ref. [35], each providing mean brine volume fractions of the ice. The table lists ice type, author, test location, test period, number of data points, and the reported statistics of flexural strength and effective elastic modulus; where available, regression relationships with temperature and the square root of brine (porosity) volume fraction are also included. It should be noted that, except for Ref. [19], all entries were specifically obtained from the full-scale cantilever beam test sections of the original studies. In the referenced datasets, Refs. [18,19,32] did not explicitly specify the loading direction of the cantilever beam tests; in Ref. [34], all 12 tests were conducted under downward loading; and in Ref. [16], all 15 freshwater ice tests were downward loaded, while only 10 of 62 sea ice tests were performed under upward loading. The brackish ice dataset in Ref. [35] represents ice with a mean ice salinity of approximately 1‰ and under-ice water-layer salinity of about 4‰, which is comparable to the physical conditions observed in this study. To facilitate a more straightforward comparison across datasets, loading direction was not distinguished in the present analysis. For consistency, the mean values from the six test groups in this study were fitted against the mean ice temperature and brine volume fraction (porosity).

Table 3. Summary of literature data on flexural strength and effective elastic modulus of freshwater and sea ice, with inclusion of brackish ice results from this study and newly derived regression relationships.

Ice Type	Author	Location	Test Date	Number of Data Points	Flexural Strength (kPa)	Effective Modulus (GPa)	Fit Line
Brackish ice	This study	Lake Hanzhang	2024	39	374.21 ± 99.93	2.77 ± 0.93	$\sigma_f = -0.029T + 0.333$ $E = -0.322T + 2.23$ $\sigma_f = 0.703\exp(-2.061\sqrt{v})$ $E = 6.519\exp(-2.738\sqrt{v})$
	[35]	Gulf of Finland	2006	/	/	/	$\sigma_f = 0.7\left(1 - \sqrt{\frac{v_b}{0.202}}\right)$
Fresh water ice	[16]	Spitsbergen Fjords	2012–2015	15	486.60 ± 149.67	2.91 ± 1.26	/
	[32]	Lake Ulansuhai	2016	9	665.75 ± 93.48	5.32 ± 1.14	$\sigma_f = -0.022T + 0.564$ $E = -0.319T + 3.86$
	[33]	Chassell Bay and Keweenaw Bay	1957–1958	49	438.73 ± 170.36	/	/
Sea ice	[16]	Spitsbergen Fjords and North-West Barents Sea	2010–2018	62	258.58 ± 85.38	1.24 ± 0.62	$\sigma_f = 0.527\exp(-2.804\sqrt{v_b})$ $E = 3.103\exp(-3.385\sqrt{v_b})$
	[18]	South-East Barents Sea and North-East Barents Sea	1996–2006	138	252.00 ± 84.00	/	$\sigma_f = -0.346\sqrt{v_b} + 0.372$
	[19]	Labrador, Greenland, Alaska, Antarctica, Beaufortsea, Baffin, Bothnia, Japan, Spitzbergen	1955–1992	2495	/	/	$\sigma_f = 1.76\exp(-5.88\sqrt{v_b})$
	[34]	North-West Barents Sea	2013	12	225.25 ± 81.15	1.03 ± 0.55	$\sigma_f = 0.656\exp(-4.312\sqrt{v_b})$

Figure 12 illustrates the relationships between flexural strength and mean ice temperature, comparing the present brackish ice results with several freshwater ice datasets. Both the present data and those of Ref. [33] are presented as group-averaged values. It can be observed that, with the exception of Ref. [16], all datasets show decreasing strength with increasing temperature. Freshwater ice generally occupies a higher range, with flexural strengths of 201 to 807 kPa and effective moduli of 1.25 to 6.71 GPa. In contrast, brackish ice in this study exhibits lower values, with flexural strengths of 250.21 to 469.65 kPa and effective moduli of 1.56 to 3.84 GPa. The regression lines further confirm this offset, as the brackish ice fit consistently lies below that of freshwater ice. Nevertheless, the slopes are similar, indicating that thermal weakening is a common mechanism in both ice types. In Ref. [16], the relationship between mechanical properties and temperature is less clear, likely due to inter-annual variability and differences in ice physical properties across testing years. The dataset of Ref. [33], which is concentrated in a higher temperature range, still shows a mean strength of 438.73 ± 170.36 kPa, exceeding the mean value of 374.21 ± 99.93 kPa obtained for brackish ice in this study. The microstructural context helps explain why brackish ice exhibits lower flexural properties than freshwater ice. Zhang et al. [36] conducted laboratory observations on Hanzhang Lake ice and revealed the presence of brine inclusions and drainage features, indicating partial brine connectivity within the ice. Although Liu et al. [37] focused on under-ice primary productivity rather than physical mechanics, their micrographs nonetheless show air bubbles occurring in proximity to, and in some cases apparently coalescing with, brine pockets. In contrast, freshwater ice from Lake Ulansuhai [38] exhibits bubble layers that are more discrete and less connected. Taken together, these observations suggest that the presence of brine channels and the adjacency of bubble-brine inclusions in brackish ice reduce the effective bonding within the ice skeleton, thereby weakening its strength and stiffness. This interpretation is consistent with Timco and O'Brien [19], who attributed the lower strength of sea ice relative to freshwater ice to the presence and connectivity of brine inclusions.

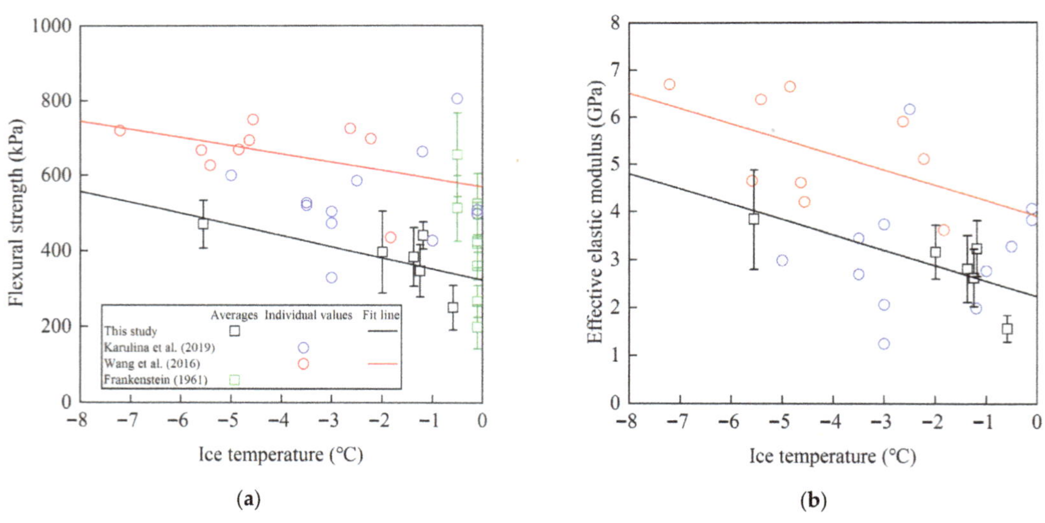

Figure 12. Comparison of brackish ice with freshwater ice from previous studies: (**a**) flexural strength as a function of ice temperature; (**b**) effective elastic modulus as a function of ice temperature. Note: The references mentioned in the figure legend are [16,32,33] (from top to bottom).

Figure 13 presents the relationships between flexural strength and effective elastic modulus with the square root of brine (or porosity) volume fraction, comparing the present results with previously published datasets for brackish and sea ice. In the main figure, the present data are plotted against the square root of porosity, as the correlation between the experimental results and brine volume fraction was relatively weak. To maintain consistency with existing literature, however, the other datasets for sea ice and brackish ice are plotted using the square root of brine volume fraction as the influencing factor. Due to the large number of data points, the results from Refs. [18,19] are shown only as fitted curves. To enable a more direct comparison under similar brackish ice conditions, the inset of Figure 13a re-plots the present data using the square root of brine volume fraction as the independent variable, in accordance with the empirical relationship for brackish ice proposed in Ref. [35]. Overall, both flexural strength and modulus decrease with increasing brine volume fraction (porosity) across all datasets. However, at comparable brine levels, brackish ice consistently exhibits higher values than sea ice. For example, at a $\sqrt{v_b}$ of approximately 0.3, the flexural strength of brackish ice remains 378.82 kPa, whereas sea ice typically ranges between 179.93 and 301.59 kPa. Similarly, the effective elastic modulus of brackish ice falls remains 1.56 to 3.84 GPa, compared to 0.25 to 2.30 GPa for sea ice under the same conditions. The fitted curves further emphasize this offset, with the brackish ice regression consistently lying above those of sea ice. This indicates that, although porosity governs the mechanical weakening of both brackish and sea ice, brackish ice retains higher strength and stiffness due to the absence of extensive brine drainage systems and weaker dynamic forcing during growth, which promote a denser internal structure. The fitted curve for brackish ice reported in Ref. [35] agrees well with the present results during the growth and stable periods, confirming the reliability of the empirical relationship. However, during the melt period, the measured data deviate from this trend due to brine drainage and a concurrent increase in air volume fraction, further demonstrating the limitation of using brine volume fraction alone as the influencing factor.

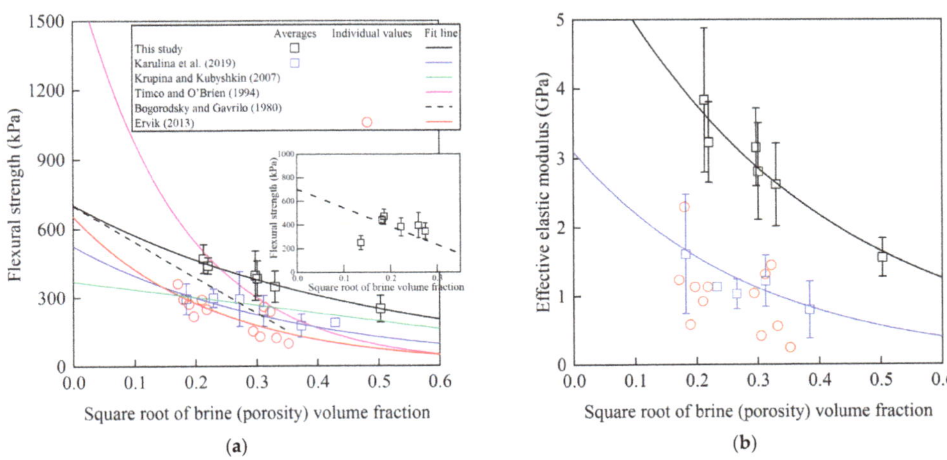

Figure 13. Comparison of brackish ice with sea ice and brackish ice from previous studies: (**a**) flexural strength and (**b**) effective elastic modulus as a function of square root of brine(porosity) volume fraction. Note: In the main figure, the present brackish-ice data are plotted using porosity, whereas the literature datasets are plotted using brine volume fraction. In the inset figure, the present brackish-ice data are re-plotted as a function of brine volume fraction for direct comparison. Note: The references mentioned in the figure legend are [16,18,19,34,35] (from top to bottom).

It should be noted that when the $\sqrt{v_b}$ is below 0.24, the fitted strength curve of Ref. [19] exceeds that of the present study. This discrepancy arises because the regression in Ref. [19] was not solely based on full-scale cantilever beam tests but also incorporated simply supported beams and small-scale specimens. Previous studies have demonstrated that simply supported beams and cantilever beams exhibit fundamentally different mechanical responses, with the discrepancy being particularly pronounced in floating cantilevers. In such cases, loading is not purely governed by self-weight but instead approaches a "pure bending" condition, where shear stresses induced by lateral forces are minimal. Consequently, cantilever deformation is controlled by a combination of tension and compression, whereas free-beam tests involve significant lateral forces that systematically overestimate flexural strength, even when beam dimensions are identical. In the present study, the fitted curve predicts a flexural strength of 703 kPa at zero brine volume fraction, which is in good agreement with the full-scale freshwater cantilever beam tests reported in Ref. [19]. This further demonstrates that the formula of Ref. [19] tends to overestimate sea-ice full-scale flexural strength in the low-salinity regime.

4.2. Fracture of Ice During Melt Period

Figure 14 shows the melting of the ice cover on selected dates during the melt period. Figure 14a–c were captured with a mobile phone, while Figure 14d–f were captured by a fixed surveillance camera from the same angle, providing a clear record of the advance of the melt front. The yellow triangle indicates the location of the ice thickness monitoring device, the red line marks the boundary between melting ice and open water, and the yellow dashed line serves as a reference line. Melting typically initiated from the side near the sea outlet and along the shoreline, and then gradually advanced toward the opposite side of the lake. Rising air temperature and solar radiation increased the surface ice temperature and promoted the interconnection of internal brine channels, resulting in a sharp increase in porosity and a substantial weakening of the load-bearing skeleton. At the macroscopic scale, this was manifested as surface meltwater and initial cracks (Figure 14a). As open-water areas expanded (Figure 14b–e), hydrodynamic forcing from waves and lake currents became concentrated along the ice edges, triggering bending failure in the marginal regions. The newly exposed ice–water interfaces further accelerated brine connectivity and gas accumulation, leading to block detachment of the ice cover under continued wave action. Ultimately, the melt front progressively advanced toward the lake center, and by March 18 (Figure 14f), the ice cover at the monitoring site had completely disintegrated. This process explains the markedly reduced flexural strength and effective elastic modulus during the melt period and highlights the mechanism by which the ice cover can rapidly collapse within a short timescale.

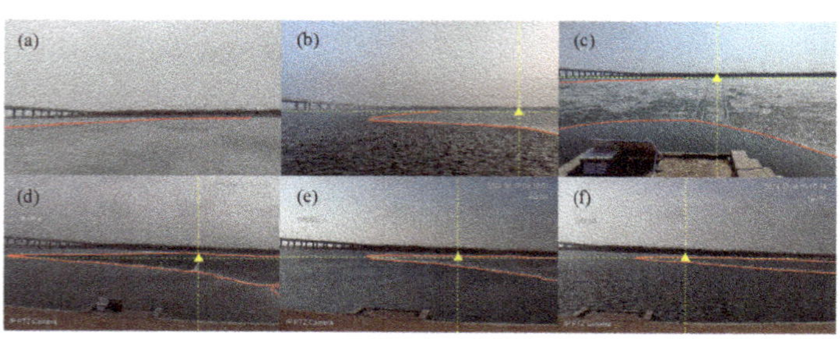

Figure 14. Photographs showing surface melting conditions of the ice cover at different times during the melt season: (**a**) 11 March 2024; (**b**) 14 March 2024; (**c**) 15 March 2024; (**d**) 16 March 2024; (**e**) 17 March 2024; (**f**) 18 March 2024.

5. Conclusions

This study systematically investigated the flexural strength and effective elastic modulus of brackish ice through full-scale in situ tests on ice cover of Hanzhang Lake. The analysis further identified the most effective predictors and regression models, and examined the influences of loading direction and ice period on the mechanical behavior. The main findings are as follows:

(1) Exponential functions were identified as the most effective form for fitting the relationships between flexural strength, effective elastic modulus, and ice physical properties. Although fracture under different loading modes is governed by the physical properties of different parts of the ice layer, comparison among alternative parameterizations revealed that the bulk mean values of physical properties provided the best predictive performance. Both flexural strength and effective elastic modulus decreased with increasing porosity and gas volume fraction, with porosity yielding the highest goodness-of-fit, highlighting its dominant role in controlling ice mechanical behavior. The empirical formulas for the full-scale flexural mechanical properties of brackish ice, as shown in Equations (3) and (4).

(2) Flexural strength obtained under downward loading was systematically higher than that under upward loading, with differences ranging from 17.3 to 38.8%. This behavior can be attributed to the presence of granular crystals in the upper ice layer and the progressive increase in columnar crystal diameters with depth, which weakened intergranular bonding in the tensile zone and reduced tensile resistance. In contrast, the effective elastic modulus reflects the overall stress–strain response rather than the failure threshold. Although no clear directional dependence of the modulus was observed in this study, the inherent variability of full-scale in situ measurements suggests that more data are needed before a definitive conclusion can be drawn regarding the relationship between loading direction and elastic modulus.

(3) No significant differences were observed in flexural strength or effective elastic modulus between the growth and stable periods, as both exhibited comparable mechanical levels. In contrast, the melt period showed a marked reduction: flexural strength decreased by approximately 33–42%, and effective elastic modulus decreased by 49–61% relative to the growth and stable periods. This deterioration is primarily attributed to the interconnection of brine channels and substantial brine drainage during melting, combined with increased gas volume fraction and porosity, which weakened the load-bearing skeleton. Furthermore, grain-boundary wetting and bubble accumulation reduced the overall stiffness, thereby accelerating the degradation of mechanical properties.

(4) Comparisons with existing datasets indicate that the mechanical properties of brackish ice are lower than those of freshwater ice but higher than those of sea ice. At the same temperature, its flexural strength and effective elastic modulus are about 30–50% lower than freshwater ice, due to brine inclusions and bubbles weakening the load-bearing skeleton. At the same porosity, however, brackish ice outperforms sea ice; for example, at a porosity square root of 0.3, its strength (379 kPa) and modulus (1.56–3.84 GPa) exceed those of sea ice (180–302 kPa, 0.25–2.30 GPa). This difference reflects the denser structure of brackish ice, formed under weaker brine drainage and dynamic forcing.

This study systematically analyzed the flexural strength and effective elastic modulus of brackish ice, bridging the gap between freshwater and sea ice and providing valuable in situ data for ice mechanics databases in both polar and mid-latitude regions. Nevertheless, several limitations remain. First, the preparation of cantilever beams was time-consuming, and the cutting process inevitably altered the thermal state of the beams. Although temperatures were measured immediately after testing, they may not have fully reflected the in situ conditions within the ice. Second, while bulk porosity was statistically identified as the optimal predictor of flexural properties, this conclusion was derived empirically and lacks mechanistic interpretation. Finally, the presence of granular ice near the surface contributed to the higher strength observed under downward loading, yet crystal size was not quantitatively incorporated into the predictive formulations for strength and modulus. In future work, efforts should focus on reducing beam preparation time, improving thermal measurements, clarifying the mechanistic role of porosity, and integrating crystal-scale parameters, thereby enabling a more comprehensive understanding of the mechanical behavior of brackish ice.

Author Contributions: Writing—original draft, investigation, data curation, methodology, S.J.; writing—original draft, investigation, formal analysis, Y.L.; conceptualization, writing—review and editing, supervision, formal analysis, resources, Q.W.; conceptualization, writing—review and editing, supervision, resources, P.L.; investigation, data curation, S.Y. All authors have read and agreed to the published version of the manuscript.

Funding: This research was funded by the Open Fund of State Key Laboratory of Coastal and Offshore Engineering (SL2507); the Hi-Tech Ship Project of the Ministry of Industry and Information Technology (CBG2N21-2-3); the National Natural Science Foundation of China (U24A20582).

Data Availability Statement: The original contributions presented in this study are included in the article. Further inquiries can be directed to the corresponding author.

Conflicts of Interest: The authors declare no conflicts of interest.

References

1. Blagrave, K.; Sharma, S. Projecting climate change impacts on ice phenology across Midwestern and Northeastern United States lakes. *Clim. Change* **2023**, *176*, 119. [CrossRef]
2. Ruan, Y.A.; Zhang, X.; Xin, Q.; Qiu, Y.; Sun, Y. Prediction and analysis of lake ice phenology dynamics under future climate scenarios across the inner Tibetan Plateau. *J. Geophys. Res.-Atmos.* **2020**, *125*, e2020JD033082. [CrossRef]
3. Stroeve, J.; Notz, D. Changing state of Arctic sea ice across all seasons. *Environ. Res. Lett.* **2018**, *13*, 103001. [CrossRef]
4. Huo, P.; Peng, L.; Cheng, B.; Yu, M.; Wang, Q.; Li, X.; Li, Z. Reconstructing ice phenology of a lake with complex surface cover: A case study of Lake Ulansu during 1941–2023. *Cryosphere* **2025**, *19*, 849–868. [CrossRef]
5. Wang, X.; Feng, L.; Qi, W.; Cai, X.; Zheng, Y.; Gibson, L.; Tang, J.; Song, X.; Liu, J.; Zheng, C.; et al. Continuous loss of global lake ice across two centuries revealed by satellite observations and numerical modeling. *Geophys. Res. Lett.* **2022**, *49*, e2022GL099022. [CrossRef]
6. Sharma, S.; Richardson, D.C.; Woolway, R.I.; Imrit, M.A.; Bouffard, D.; Blagrave, K.; Daly, J.; Filazzola, A.; Granin, N.; Korhonen, J.; et al. Loss of ice cover, shifting phenology, and more extreme events in northern hemisphere lakes. *J. Geophys. Res-Biogeo.* **2021**, *126*, e2021JG006348. [CrossRef]
7. Zhang, C.; Zhao, X. Theoretical model for predicting the break-up of ice covers due to wave-ice interaction. *Appl. Ocean Res.* **2021**, *112*, 102614. [CrossRef]
8. Masterson, D.M. State of the art of ice bearing capacity and ice construction. *Cold Reg. Sci. Technol.* **2009**, *58*, 99–112. [CrossRef]
9. Aly, M.; Taylor, R.; Dudley, E.B.; Turnbull, I. Scale effect in ice flexural strength. *J. Offshore Mech. Arct.* **2019**, *141*, 051501. [CrossRef]
10. Wei, M.; Dai, F. Laboratory-scale mixed-mode I/II fracture tests on columnar saline ice. *Theor. Appl. Fract. Mech.* **2021**, *114* (Suppl. SC), 102982. [CrossRef]
11. Wei, M.; Prasanna, M.; Cole, D.M.; Polojärvi, A. Response of dry and floating saline ice to cyclic compression. *Geophys. Res. Lett.* **2022**, *49*, e2022GL099457. [CrossRef]
12. Wei, M.; Polojärvi, A.; Cole, D.M.; Prasanna, M. Strain response and energy dissipation of floating saline ice under cyclic compressive stress. *Cryosphere* **2020**, *14*, 2849–2867. [CrossRef]
13. Marchenko, A.; Karulin, E.; Chistyakov, P.; Karulina, M.; Sakharov, A. Meso-scale stress relaxation test on floating sea ice. In Proceedings of the 27th International Conference on Port and Ocean Engineering under Arctic Conditions, Glasgow, UK, 12–16 June 2023.

14. Marchenko, A.; Grue, J.; Karulin, E.; Frederking, R.; Lishiman, B.; Chistyakov, P.; Karulina, M.; Sodhi, D.; Renshaw, C.; Sakharov, A.; et al. Elastic moduli of sea ice and lake ice calculated from in-situ and laboratory experiments. In Proceedings of the 25th IAHR International Symposium on Ice, Trondheim, Norway, 23–25 June 2020.
15. Voermans, J.J.; Rabault, J.; Marchenko, A.; Nose, T.; Waseda, T.; Babanin, A.V. Estimating the elastic modulus of landfast ice from wave observations. *J. Glaciol.* **2023**, *69*, 1823–1833. [CrossRef]
16. Karulina, M.; Marchenko, A.; Karulin, E.; Sodhi, D.; Sakharov, A.; Chistyakov, P. Full-scale flexural strength of sea ice and freshwater ice in Spitsbergen Fjords and North-West Barents Sea. *Appl. Ocean Res.* **2019**, *90*, 101853. [CrossRef]
17. Tian, Y.; Zhao, W.; Yu, C.; Gang, X.; Lu, P.; Yue, Q. Investigations on flexural strength of a columnar saline model ice under circular plate central loading. *Water* **2023**, *15*, 3371. [CrossRef]
18. Krupina, N.A.; Kubyshkin, N.V. Flexural strength of drifting level first-year ice in the Barents Sea. In Proceedings of the 17th International Offshore and Polar Engineering Conference, Lisbon, Portugal, 1–6 July 2007.
19. Timco, G.W.; O'Brien, S. Flexural strength equation for sea ice. *Cold Reg. Sci. Technol.* **1994**, *22*, 285–298. [CrossRef]
20. Lainey, L.; Tinawi, R. The mechanical properties of sea ice — A compilation of available data. *Can. J. Civ. Eng.* **1984**, *11*, 884–923. [CrossRef]
21. Han, H.; Jia, Q.; Huang, W.; Li, Z. Flexural strength and effective modulus of large columnar-grained freshwater ice. *Cold Reg. Sci. Technol.* **2015**, *30*, 04015005. [CrossRef]
22. Murdza, A.; Marchenko, A.; Schulson, E.M.; Renshaw, C.E. Cyclic strengthening of lake ice. *J. Glaciol.* **2021**, *67*, 182–185. [CrossRef]
23. Murdza, A.; Schulson, E.M.; Renshaw, C.E. Behavior of saline ice under cyclic flexural loading. *Cryosphere* **2021**, *15*, 2415–2428. [CrossRef]
24. Wang, Q.; Liu, Y.; Lu, P.; Li, Z. The mechanical properties of gas-enriched sea ice. *Cold Reg. Sci. Technol.* **2025**, *239*, 104576. [CrossRef]
25. Frederking, R.M.W.; Timco, G.W. On measuring flexural properties of ice using cantilever beams. *Ann. Glaciol.* **1983**, *4*, 58–65.
26. Frederking, R.M.W.; Svec, O.J. Stress-relieving techniques for cantilever beam tests in an ice cover. *Cold Reg. Sci. Technol.* **1985**, *11*, 247–253. [CrossRef]
27. Hutchings, J.K.; Heil, P.; Lecomte, O.; Stevens, R.; Steer, A.; Lieser, J.L. Comparing methods of measuring sea-ice density in the East Antarctic. *Ann. Glaciol.* **2015**, *56*, 385–386. [CrossRef]
28. Pustogvar, A.; Kulyakhtin, A. Sea ice density measurements: Methods and uncertainties. *Cold Reg. Sci. Technol.* **2016**, *131*, 46–52. [CrossRef]
29. Cox, G.F.N.; Weeks, W.F. Equations for determining the gas and brine volumes in sea ice samples. *J. Glaciol.* **1983**, *29*, 306–316. [CrossRef] [PubMed]
30. Leppäranta, M.; Manninen, T. *The Brine and Gas Content of Sea Ice with Attention to Low Salinities and High Temperatures*; Finnish Institute of Marine Research: Helsinki, Finland, 1988.
31. Eicken, H.; Dieckmann, G.S.; Thomas, D.N. From the microscopic, to the macroscopic, to the regional scale: Growth, microstructure and properties of sea ice. In *Sea Ice*, 2nd ed.; David, N.T., Gerhard, S.D., Eds.; Blackwell Science Ltd: Oxford, UK, 2003; pp. 22–81. [CrossRef]
32. Wang, J.; Cao, X.; Wang, Q.; Yan, L.; Li, Z. Experimental relationship between flexural strength, elastic modulus of ice sheet and equivalent ice temperature. *J. South-North Water Transf. Water Sci. Technol.* **2016**, *14*, 75–380. [CrossRef]
33. Frankenstein, G.E. Strength data on lake ice. In *U.S. Army Snow Ice and Permafrost Research Establishment*; Corps of Engineers: Wilmette, IL, USA, 1961; p. 20.
34. Ervik, Å. Experimental and Numerical Investigations of Cantilever Beam Tests in Floating Ice Covers. Master's Thesis, Norwegian University of Science and Technology, Lonyearbyen, Norway, 2013.
35. Bogorodsky, V.V.; Gavrilo, V.P. Ice. Physical properties. In *Modern Methods of Glaciology*; Gidrometeoizdat: Leningrad, Russia, 1980; p. 384.
36. Zhang, H.; Yu, M.; Lu, P.; Zhou, J.; Zhou, J.; Xie, F.; Wang, Q.; Li, Z. Experimental investigation of the partitioning of radiation in the melt pond–ice–ocean system. *Cold Reg. Sci. Technol.* **2024**, *219*, 104107. [CrossRef]
37. Liu, T.; Zhang, Y.; Li, Z.; Yu, M.; Xie, F.; Georgiy, K.; Yang, Z. Optical properties of ice in a shallow Chinese lake (Hanzhang) with consequent impacts on primary production. *Sci. Total Environ.* **2023**, *957*, 177512. [CrossRef]
38. Yang, F.; Li, C.; Shi, X.; Zhao, S.; Hao, Y. Impact of seasonal ice structure characteristics on ice cover impurity distributions in Lake Ulansuhai. *J. Lake Sci.* **2016**, *28*, 455–462. [CrossRef] [PubMed]

water

MDPI

Article

Impact of the Reynolds Numbers on the Velocity of Floating Microplastics in Open Channels

Jiachen Li [1], Zhichao Wang [1,2,3,*], Weiping Li [1,2], Shuangyi Jing [1,2], Caio Graco-Roza [3] and Lauri Arvola [3]

1 College of Environment and Energy Resources, Inner Mongolia University of Science and Technology, Baotou 014010, China; w201806@126.com (J.L.); sjlwp@163.com (W.L.); 13500622199@163.com (S.J.)

2 Cooperative Innovation Center of Ecological Protection and Comprehensive Utilization in Inner Mongolia Section of the Yellow River Basin, Baotou 014010, China

3 Lammi Biological Station, University of Helsinki, FI-16900 Lammi, Finland; caio.roza@helsinki.fi (C.G.-R.); lauri.arvola@helsinki.fi (L.A.)

* Correspondence: wzc5658@126.com

Abstract: Quantitatively analyzing the factors influencing the horizontal migration of microplastics (MPs) in water bodies and understanding their movement patterns are crucial for explaining and predicting their transport principles and final destinations. This study used nearly spherical polyethylene (PE), polypropylene (PP), and polystyrene (PS) MPs as experimental subjects. By tracking their motion characteristics through video recording, we established relationships among the Reynolds number (Re), MP density, and floating velocity. The results showed that the Re and MP density jointly affect the horizontal drift of MPs. The horizontal floating velocity of MPs significantly increases with the increase in the Re and shows a power function growth trend. The difference in density of MPs mainly affects their dispersion during the floating process. Moreover, the coefficient of variation (CV) of PP's horizontal floating velocity increased with the Re, suggesting PP's motion is more random and discrete than that of PE and PS. Ultimately, we fitted the horizontal floating velocity of MPs to the equation to comprehensively evaluate the relationship between the floating velocity, Re, and density of MPs. This analysis underscores that the Re predominantly influences the MP velocity in water, while the MP density chiefly impacts the discrete nature of their motion.

Keywords: microplastics (MPs); Reynolds number (Re); velocity; open channels

1. Introduction

As polymer organic compounds, plastics offer advantages such as their light weight, corrosion resistance, and effective insulation properties [1]. They are extensively utilized in human production and daily life. Global plastic production is projected to reach 400.3 million tons in 2023 [2] and is expected to rise to 1.2 billion tons by 2050 [3,4]. Despite the convenience they bring, the disposal of plastics has emerged as a significant issue. Nevertheless, the improper disposal of plastic waste allows some of it to enter the aquatic environment with precipitation, runoff, and sewage discharges, leading to severe water pollution [5]. Upon entering water bodies, microplastics (MPs) can adversely affect the natural environment, including aquatic plants and animals [6]. Upon ingestion by plankton, MPs can enter the human food chain, posing severe risk [7]. Furthermore, MPs are found not only in natural aquatic environments but also in everyday consumables like bottled mineral water, as well as in industrial and domestic pipelines [8]. As a result, MPs gradually enter the human body and damage human organs [9].

Due to differences in the shape and density of MPs, their retention in the water column also differs. Some MPs settle on riverbeds due to gravity, while others move along water currents, overcoming friction [10]. In estuaries that flow into lakes and oceans, MPs undergo rapid diffusion influenced by changes in the flow dynamics. Liu et al. investigated the distribution patterns of the horizontal and vertical transport of MPs in Lake Ulansuhai's water column [11]. Variations in flow conditions at different locations led to varying MP concentrations. The movement of MPs follows a three-dimensional model characterized by vertical settling and horizontal migration [12]. Despite considering flow velocity vectors and MP characteristics (density, volume, shape), hydrodynamic conditions and MP characteristics were not quantitatively analyzed. For example, turbulence significantly impacts MPs' migration [13]. Stride et al. analyzed the migration path of polyethylene (PE) MPs in containers, finding that under overflowing conditions from manholes, most PE followed the path of solutes, with a portion being retained on the surface or in dead zones near manhole inlets [14]. Isachenko and Chubarenko [10] conducted indoor flume experiments and demonstrated how abrupt changes in roughness affect particle movement from stagnation and MP deposition on channel bottoms, influenced by physical bottom properties. Larsen et al. estimated the Lagrangian particle transport velocity by analyzing the average stranding time of MP particles released at various coastal locations [15].

Water temperature is also a key factor influencing the migration and accumulation of MPs. Thermal stratification in lake water and changes in water temperature can affect the water's kinematic viscosity, which in turn alters the Reynolds number (Re) and the behavior of particles [16]. Ahmadi et al. found that the settling of MPs in cold water is relatively slower [17]. When the temperature decreased from 20 °C to 10 °C and 4 °C, the TSV of MPs decreased by an average of 32% to 46%. Nakayama and Osako evaluated the spatiotemporal dynamics of MPs in a watershed by developing an ecological hydrological model [18]. Their research revealed that water temperature significantly influences the transport of MPs. Liu et al. developed a regression model using seawater temperature, salinity, and flow rate as independent variables [19]. Changes in water temperature can influence the formation of a biofilm on the surface of MP particles. Zhang et al. observed that biofilm colonization on MPs occurs more rapidly during the high temperatures of summer, which also affects their movement [20].

In recent years, research on MP pollution in water bodies has grown and advanced significantly. However, the fate of MPs remains poorly understood, and the impact of different hydrodynamic conditions on their migration process is still not fully clear [21–23]. Most existing studies examine the movement of MPs by treating flow velocity, water temperature, and other conditions as separate variables. However, further research is crucial to understand how hydrodynamic conditions influence the horizontal movement of MPs. Such insights are essential for accurately assessing MP pollution levels and safety risks in river ecosystems, as well as for predicting the migration, capture, and removal dynamics of MPs in natural aquatic environments. The horizontal movement of MPs is primarily driven by two factors: external influences and internal attributes. The Re, which combines dynamic viscosity (related to water temperature), flow velocity, and characteristic size (as external factors), is a key variable for quantifying changes in fluid flow patterns. Therefore, this study first uses the Re as a starting point, simplifying complex hydrodynamic conditions into a single parameter. Secondly, this study mainly considers the type (density) of MPs as a typical internal factor (Figure 1). The three types of plastics, polyethylene (PE), polypropylene (PP), and polystyrene (PS), have a wide range of applications in daily life and industrial production, these plastics are prevalent in the marine environment in the form of MPs and are the most representative in the study of MP pollution in water bodies [24,25]. Although PVC and PT are also plastic polymers with significant global production and waste volumes, these MPs have high specific gravity and quickly settle during the experimental process. The results would be hard to observe and measure in this study. The morphology and size of MPs hosted in different water body regions differ somewhat. The migration of spherical MPs is easier to study than other shapes, and most MP sizes are between 100 and 5000 μm in marine and river basins [26,27]. Therefore, PE, PP, and PS particles characterized as shown in Table 1, which are white, nearly spherical, and have a particle size of approximately 950–1090 μm, were used in this study. These MPs are easier to identify during the experimental process.

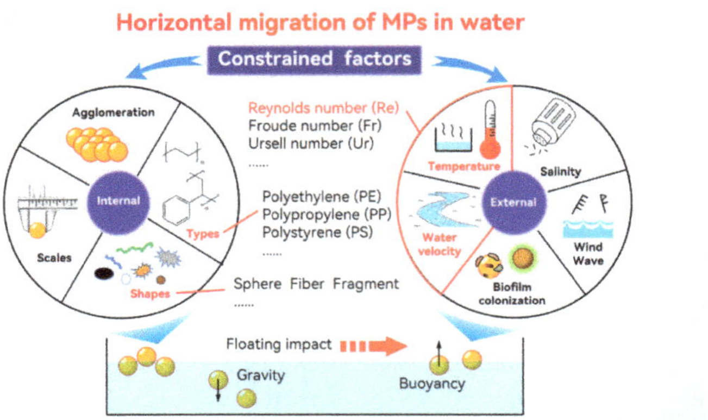

Figure 1. Constrained factors affecting the horizontal migration of MPs in water.

Table 1. MP selection.

MPs	Shapes	Density (g·cm^{-3})
Polyethylene (PE)		0.941
Polypropylene (PP)	Approximate spherical shape	0.908
Polystyrene (PS)		1.010

Given the importance and complexity of MPs in water bodies, this study was conducted using indoor water tank experiments to simulate natural hydrodynamic conditions by varying the Re. The motion dynamics of MPs were tracked using green-light-source reflection and video recording technology. By integrating internal and external factors influencing MPs' horizontal migration, the horizontal floating velocity of MPs was incorporated into an equation to comprehensively analyze the relationship between floating velocity, the Re, and MP density. This study investigates the combined effects of MP density (type) and channel Re changes on the horizontal floating velocity of MPs, aiming to provide a theoretical foundation and methodological framework for understanding MPs' migration behavior in real-world water environments.

2. Materials and Methods

2.1. Experimental Setup

In this study, a rectangular cross-section flume was used for the experiments. The experimental section for MPs' movement was approximately 3.33 m in length, 0.12 m in width, and 0.30 m in height (Figure 2a). Cleaning and a water change were required before the experiment took place and water was transferred to the tank by a submersible pump at ground level to ensure a constant head water supply, after which the flume was returned to form a closed-loop water supply and drainage system. The water on the other side passed through a double freedom diaphragm valve and an effluent master valve, which was used to control the flow rate precisely, and then entered the experimental flume. The upstream flume was equipped with a perforated pipe and deflector to stabilize the water flow, and the downstream flume was equipped with a thermometer. The bottom of the flume was lined with a layer of quartz sand with a slope of i = 0.0015 to simulate the roughness of the channel. A 200-mesh sieve was installed on the floor tank to intercept particles and MPs during the experiment (Figure 2b,c).

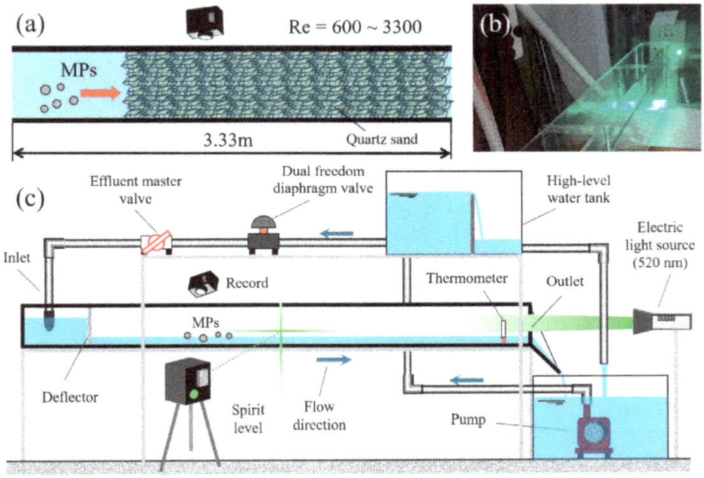

Figure 2. Experimental setup: (**a**) schematic of experimental section, (**b**) concrete drawing, and (**c**) schematic of the experimental flow.

2.2. Hydraulic Characteristics

The flow rate of the flume was determined via the volume flow method. At the beginning of the measurement, the diaphragm valve was adjusted to the appropriate position, and the master valve was opened. At the end of a single measurement, only the master valve was closed, and the diaphragm valve was kept at a fixed value. The above measurements were repeated, and the flow rate in the flume under different working conditions, average flow rate of the cross-section Q, water depth h, average velocity of water flow V_W, and Re were calculated. Finally, the hydraulic characteristics of the flume were statistically determined (Table 2). The relevant calculation was carried out using the following mathematical Equations (1)–(4):

$$Q = \frac{V}{t} \tag{1}$$

$$V_W = \frac{Q}{A} \tag{2}$$

$$\mu = \frac{0.001779}{1 + 0.03368T + 0.000221 \cdot T^2} \tag{3}$$

$$Re = \frac{\rho V_w R}{\mu} \tag{4}$$

where, Q ($m^3 \cdot s^{-1}$) is the flow rate; V (m^3) is the volume of water flow; R (m) is the hydraulic radius; A (m^2) is the cross-sectional area; μ (Pa·s) is the coefficient of viscosity of the movement of water; T (°C) is the temperature of the water; ρ is the density of the water ($kg \cdot m^{-3}$); V_w ($m \cdot s^{-1}$) is the average velocity of the water flow; and the Re (dimensionless) is the Reynolds number.

Table 2. Working condition settings.

No.	Q ($L \cdot s^{-1}$)	H (m)	V_w ($m \cdot s^{-1}$)	Re
1	0.0498	0.0070	0.0593	360
2	0.0717	0.0088	0.0683	509
3	0.0850	0.0095	0.0746	597
4	0.1291	0.0110	0.0978	888
5	0.1604	0.0126	0.1065	1079
6	0.2134	0.0150	0.1186	1389
7	0.2556	0.0165	0.1291	1631
8	0.3063	0.0180	0.1418	1917
9	0.3420	0.0190	0.1500	2114
10	0.3788	0.0200	0.1578	2312
11	0.4404	0.0217	0.1691	2632
12	0.4744	0.0226	0.1749	2804
13	0.5245	0.0240	0.1821	3049
14	0.6141	0.0255	0.2007	3507
15	0.6812	0.0265	0.2142	3845
16	0.7740	0.0289	0.2232	4251
17	0.8604	0.0297	0.2414	4684

According to the working condition measurements illustrated in Table 2, in the range of Re = 600–3300, the width-to-depth ratio of the channel is greater than or equal to 5 ($b/h \geq 5$) in the vast majority of cases, and the wall effect is relatively low.

2.3. Video Tracking

Before starting, we made sure that the water flow had stabilized. About 10 MP particles were placed near the water's surface. For every experimental run, the actual number of effectively recorded MP particles was within the range of three to eight. After each experiment, the flow rate was changed. Based on the hydraulic characteristics and pre-experimental results, the Re was controlled to between 600 and 3300. During the movement of the MPs, video recording was carried out with a camera with a resolution of 1920 × 1080 and a frame rate of 240 fps. With a wavelength of about 520 nm, a green light source was provided downstream to ensure that the MPs could be clearly recorded in the video (Figure 3). The entire process of MP movement was processed frame by frame by the video processing software DaVinci Resolve, from which the entry and exit moments of each MP were deduced to calculate the horizontal velocity of the MPs.

Figure 3. Light source and MP tracking image.

2.4. Data Processing

The movement time of the MPs was recorded through video footage, and because the default frame rate is 30 fps in DaVinci Resolve 18 by Black Magic (http://www.blackmagicdesign.com/products/davinciresolve), it was necessary to convert the frame rate to 240 fps using Equation (5) to obtain the actual movement time of the MPs (in seconds). Statistical analysis was performed using OriginPro, Version 2024b (Learning Edition) (OriginLab Corporation, Northampton, MA, USA). Combining the Re and the density of different types of MPs (ρ) with the horizontal motion of MPs during the experiment velocity (V_p), the results were curve fitted and surface fitted. The Nash–Sutcliffe efficiency coefficient (NSE), mean absolute percentage error (MAPE), and symmetric mean absolute percentage error (sMAPE) were calculated to determine the goodness of fit of the fitted equations. The magnitude of the error and the form of the equation were taken into account to determine the final MP horizontal floating velocity relation V_p (Re, ρ).

$$\textit{Actual total time}\ (s) = \frac{(\textit{Entry time}\ (s) - \textit{Terminal time}\ (s)) \times 30}{240} \tag{5}$$

Since the mean floating velocities of the three MPs varied over the range of Re = 600–3300, it would be inaccurate to use the standard error to characterize the degree of dispersion of the MPs' motion. Therefore, the coefficient of variation (CV) was used to represent the degree of dispersion of the data in Equation (6). The larger the coefficient of variation is, the greater the dispersion of the data.

$$\mathrm{CV} = \frac{\text{Standard Deviation (SD)}}{\text{Arithmatic Mean }(\overline{\mathrm{A}})} = \frac{\sqrt{\sum_{t=1}^{N} \frac{(A_t - \overline{A})^2}{N}}}{\overline{A}} \times 100\% \tag{6}$$

The Nash efficiency coefficient (NSE) is commonly used to verify the accuracy of the simulation results of hydrological models [28]. The NSE ranges from −∞ to 1 Equation (7). The closer the NSE is to 1, the more credible the model is and the better the fit is. The NSE is close to 0, which indicates that the model is generally credible when the process simulation error is large. If the NSE is much less than 0, the model is unrealistic.

$$\mathrm{NSE} = 1 - \frac{\sum_{t=1}^{N}(A_t - F_t)^2}{\sum_{t=1}^{N}(A_t - \overline{A})^2},\ \mathrm{NSE} \in (-\infty \sim 1] \tag{7}$$

The mean absolute percentage error (MAPE) is capable of visually representing the magnitude of prediction error in the form of a percentage, which makes it easy to compare the prediction accuracy of different models or variables. Moreover, the MAPE is less sensitive to outliers and will not have a conspicuous impact on the overall results due to the existence of individual outliers [29]. The MAPE ranges from 0 to +∞ Equation (8). A MAPE closer to 0 indicates that the accuracy of the fit is higher.

$$\text{MAPE} = \frac{1}{N} \cdot \sum_{t=1}^{N} \frac{|A_t - F_t|}{A_t} \times 100\%,\ \text{MAPE} \in [0 \sim +\infty) \quad (8)$$

The MAPE has a potential drawback in that it does not limit the uncertainty of predictions higher than the actual value. Instead, the symmetric mean absolute percentage error (sMAPE) is a relevant metric that attempts to address this problem [30]. The sMAPE is in the range of 0–200% Equation (9). The closer the sMAPE is to 0, the more accurate the fit is.

$$\text{sMAPE} = \frac{1}{N} \cdot \sum_{t=1}^{N} \frac{|A_t - F_t|}{(A_t + F_t)/2} \times 100\%,\ \text{MAPE} \in [0 \sim 200\%] \quad (9)$$

3. Results

3.1. Effect of Different Reynolds Numbers on the Horizontal Floating Velocity of MPs

The experimental results show that the variation in the Re caused by the change in the flow rate of the water tank has a significant effect on the horizontal floating velocity of the floating MPs. The variation in the horizontal floating velocity of different kinds of MPs in the range of Re = 600–3300 is shown in Figure 4. With an increasing Re, the horizontal floating velocities of PE, PP, and PS MPs showed a power-function growth trend. The horizontal floating velocity of PE changed from 0.137 $m \cdot s^{-1}$ at Re = 614 to 0.333 $m \cdot s^{-1}$ at Re = 3248, the velocity increased by 143% with an increase in the Re of 429%, and the velocity increased by 143%; the horizontal floating velocity of PP changed from 0.144 $m \cdot s^{-1}$ at Re = 664 to 0.314 $m \cdot s^{-1}$ at Re = 3173, which increased by 118% with a 378% increase in the Re; and the horizontal floating velocity of PS changed from 0.145 $m \cdot s^{-1}$ at Re = 659 to Re = 3184 0.321 $m \cdot s^{-1}$, a 122% increase in velocity for a 383% increase in the Re.

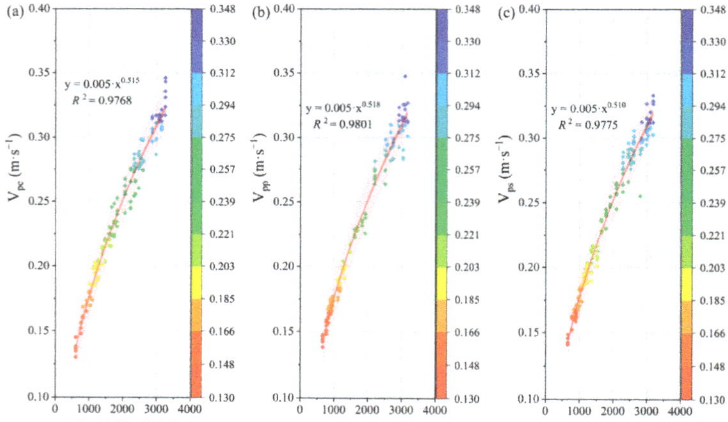

Figure 4. Effect of different Re on the horizontal floating velocity of MPs: (**a**) PE, (**b**) PP, (**c**) PS.

The difference in the density of MPs results in differences in the discreteness and stability of their movement in water. The CV values of the three MPs vary in the range of 0.3–6.9%, but the rates of change are different (Figure 5). With an increasing Re, the rates of change in the CVs of the three MPs, PE, PP, and PS, are 8.67×10^{-7}, 7.52×10^{-6}, and 2.27×10^{-6}, respectively. The regular pattern presented is $k_{PP} > k_{PS} > k_{PE}$. These results imply that the rates of change in the CVs is from PE to PS to PP.

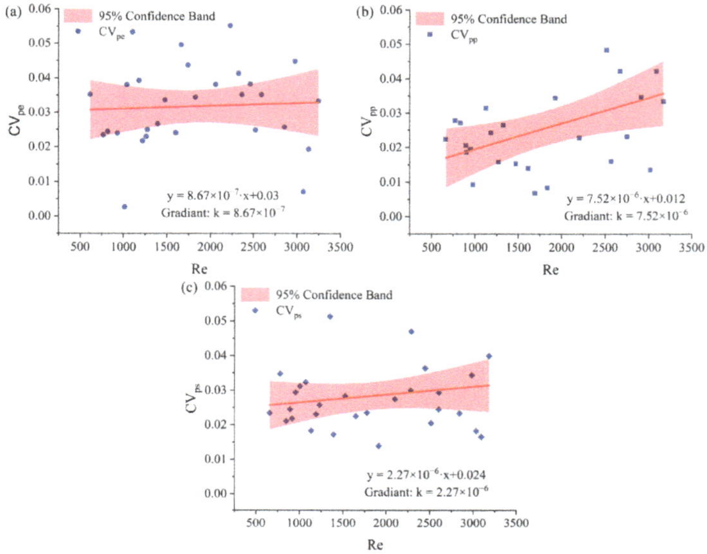

Figure 5. Coefficient of variation (CV) and rate of change (k) of the horizontal floating velocity of different types of MPs: (**a**) PE, (**b**) PP, (**c**) PS.

3.2. Functional Form, Error Analysis, and Formula Fitting

The variation in the Re will cause turbulent disturbances, and the difference in the density of MPs will affect the discrete nature of their motion. For this reason, surface fitting with the Re and MP density as independent variables and the magnitude of MPs horizontal motion velocity as the dependent variable can more realistically reflect the motion of different kinds of MPs in water. In this study, four functional forms, namely, the primary polynomial, quadratic polynomial, power function, and logarithmic function, were chosen for fitting.

Quadratic polynomials, power functions, and logarithmic functions show better fitting results compared to binary primary polynomials (Table 3). Specifically, these three equations exhibit a superior fit for the PE, PP, and PS MPs, with PP > PS > PE in terms of fitting effectiveness. For example, in the case of the binary logarithmic function, the NSE of PE is 0.9769, which is 5.56‰ greater than that of 0.9715 for the binary polynomial fit, and the values of MAPE and sMAPE are 2.54% and 2.53%, which are 20.87% and 21.18% lower than those of 3.21% for the binary logarithmic function fit, respectively; the NSE of PP is 0.9835, which is 5.21‰ greater than that of 0.9784 for the binary logarithmic function fit, the values of MAPE and sMAPE are 2.36% and 2.34%, which are 22.62% and 22.26% lower than those of 3.05% and 3.01%, respectively, for the primary polynomial fit; PS has an NSE of 0.9802, which is an improvement of 9.16‰ compared to 0.9713 for the primary polynomial fit; and the MAPE and sMAPE are 2.75% and 2.77%, which are 14.06% and 12.89% lower than those of 3.20% and 3.18%, respectively. The fitting effects of quadratic polynomials and power functions show a similar pattern, so the fitting method of primary polynomials is rounded off. The equations are complicated due to the presence of higher-order terms (e.g., -1.01×10^{-8}, -4.62×10^{-3}, etc.) in the parameters of quadratic polynomials and power functions, which cannot be overlooked. Despite their complexity, these three types of functions exhibit comparable fitting effects. In contrast, logarithmic fitting shows fewer errors and is less intricate. Hence, the logarithmic function is chosen as the preferred fitting model, providing an approximate expression for the horizontal velocity of MPs motion within the range of Re = 600–3300, considering MP density and the Re as V_p (Re, ρ) = 0.518 lg (0.030×Re + 1.33×ρ + 42.87) − 0.790. Figure 6 depicts the fitted surface based on this function.

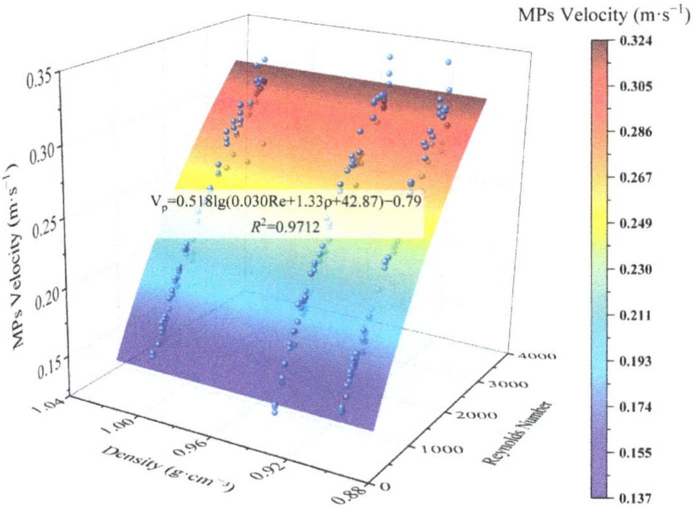

Figure 6. Binary logarithm-fitted surface for the horizontal velocity of MPs motion considering the MP density and Re.

Table 3. Functional form and error analysis.

Functional Form: z (x, y)	Fitting Equation: Vp (Re, ρ)	R^2	MPs	NSE	MAPE	sMAPE
z = Ax + By + C	$V_p = 6.77 \times 10^{-5} \cdot Re + 9.66 \times 10^{-3} \cdot \rho + 0.101$	0.9630	PE	0.9715	3.21%	3.21%
			PP	0.9784	3.05%	3.01%
			PS	0.9713	3.20%	3.18%
z = Ax^2 + By^2 + Cxy + Dx + Ey + F	$V_p = -1.01 \times 10^{-8} \cdot Re^2 + 0.13 \cdot \rho^2 - 3.51 \times 10^{-5} Re \cdot \rho + 1.40 \times 10^{-4} \cdot Re - 0.179 \cdot \rho + 0.135$	0.9714	PE	0.9765	2.59%	2.58%
			PP	0.9841	2.37%	2.35%
			PS	0.9792	2.80%	2.80%
z = A·x$^{B·y+C}$ + D	$V_p = 6.58 \times 10^{-3} \cdot Re^{-4.62 \times 10^{-3} \cdot \rho + 0.490} - 0.011$	0.9714	PE	0.9771	2.53%	2.52%
			PP	0.9832	2.42%	2.40%
			PS	0.9793	2.79%	2.82%
z = A·lg(Bx + Cy + D) + E	$V_p = 0.518 lg(0.030 \times Re + 1.33 \times \rho + 42.87) - 0.790$	0.9712	PE	0.9769	2.54%	2.53%
			PP	0.9835	2.36%	2.34%
			PS	0.9802	2.75%	2.77%

4. Discussion

Both the density and Re of MPs collectively impact their floating velocity in water. As the Re increases, indicating higher turbulence energy in the water body, both the transverse flow velocity and the overall velocity sum increase, intensifying fluid pulsation at the liquid surface [31]. These results are approximate to the research findings of He et al. [32]. They found that within the Re range of 2800 to 5600, the average horizontal velocity of PVC and PS particles increased with the increase in the average cross-sectional flow velocity of subcritical flow. The CV changes of PP are somewhat different from those of PE and PS. In the Re range of 600 to 3300, PP exhibits more random and discrete motion, consistent with the findings of Stride et al. [33]. They studied the settling characteristics of polymers with densities ranging from 0.9 to 1.4 g·cm^{-3} and their relationship with the critical horizontal flow velocity in channels. The results indicate that regardless of the horizontal flow velocity of the channel, PP tends to remain trapped on the water's surface and undergoes horizontal migration without settling or suspending. PP has a lower density than PE and PS, making its motion more significantly influenced by pulsation compared to the higher-density polymers. Moreover, the activity of PP after treatment with surfactant sodium dodecylbenzenesulfonate (CTAB) is higher than that of PS, and its flowability is further increased [34]. PP in CTAB solution exhibits higher stability and a stronger migration ability than other MPs. These studies have shown that PP exhibits the highest horizontal migration rate, likely due to its unique properties, which allow it to move faster horizontally in water compared to other MPs. Notably, in this study, PE and PP, with densities lower than water, floated on the water's surface, while PS, with a density close to that of water, also drifted on the surface until it travelled downstream. No significant settling or suspending behavior was observed during their migration. Previous studies suggest that the density of PS is nearly identical to that of water, and when combined with its low mass and hydrophobic properties, it is harder to settle. For example, Ahmadi et al. calculated the final settling velocity (TSV) of PS MPs through computational fluid dynamics (CFD) simulations, indicating that hydrophobic PS exhibited a phenomenon of harder to settle [17]. In addition, using the MPs static water settling formula fitted by Kaiser et al., in Equation (10) [35], the calculated static water settling velocity of PS is around 100 μm per second, which is relatively smaller than the channel water depth used in this study.

$$w_s = 11.68 + 0.1991 \times ESD + 0.0004 \times ESD^2 - 0.0993 \times \Delta\rho + 0.0002 \times \Delta\rho^2 \qquad (10)$$

where w_s ($m{\cdot}d^{-1}$) is the sedimentation velocity; ESD (m) is used to calculate particle size expressed as the equivalent spherical diameter; and $\Delta\rho$ ($kg{\cdot}m^{-3}$) is the particle density minus fluid density.

Moreover, in this study, the environment where MPs are located is not static water; instead, turbulence generates pulsating velocities in the x, y, and z directions [36,37]. The low settling velocity of MPs during brief intervals is likely offset by the vertical turbulent velocity of the aquatic environments. Moreover, due to the hydrophobic properties of MPs, PS can achieve a certain migration rate in a short period. The conclusions drawn by Stride et al. [33] regarding the buoyancy characteristics of PS are not entirely consistent with the results of this study. They assume that in rivers with a flow velocity of $\geq$0.101 $m{\cdot}s^{-1}$, the majority of spherical polymers with a density of 0.94–1.32 $g{\cdot}cm^{-3}$ will eventually settle in the marine ecosystem. Although the density of PS falls within the aforementioned threshold range, in this study, PS did not show significant sedimentation in water. This can be attributed to two main reasons. Firstly, PS MPs do not settle significantly in a short period. In natural water bodies, the factors influencing PS deposition are more complex. Long-term external influences and the accumulation of vertical velocity ultimately result in PS deposition [33,38]; secondly, Stride et al. conducted experiments using PMMA and PEEK as polymers with a density threshold in the range of 0.94–1.32 $g{\cdot}cm^{-3}$, without considering PS MPs. Therefore, the buoyancy characteristics of PS may be related to its unique molecular structure and require further research and discussion. Hence, for MPs with properties similar to PS, a comprehensive management strategy should be established. Prioritizing the proactive capture and removal of PS MPs before their complete sedimentation is crucial. This approach can prevent their accumulation in benthic sediments as well as potential resuspension. Such sedimentation and resuspension could exacerbate environmental toxicity via complex ecological interactions.

In this study, as a feature number, the Re was combined with the water flow velocity (flow velocity), channel characteristic size, and water temperature as external environmental variables, and MP density as the internal factor to fit a binary equation between these two and the horizontal movement velocity of MPs. Based on previous evidence, changes in water flow patterns caused by environmental factors such as waves and wind also affect the migration of MPs and other light polymers [39]. The quantification of these flow pattern changes will introduce dimensionless numbers other than the Re, such as the Ursell number (Ur) [40]. In addition, the salinity parameters play a crucial role in influencing the migration of MPs. Salinity affects not only the density of water but also its conductivity, which, in turn, impacts the behavior of MPs in water [41]. These environmental factors will increase the variables in the equation, especially when there are more than three independent variables. The visualization level of the equation in the traditional Cartesian coordinate system will decrease. In future studies, it is necessary to consider the internal and external factors affecting MPs' migration separately. For instance, the wave height and wavelength, which vary uniformly under wind and wave conditions, along with the salinity of the water body, can be considered as external independent variables, while the density changes of MPs following biofilm colonization serve as internal independent variables [42,43]. This will allow for exploring their relationship with the movement speed of MPs, and fitting the actual hydrodynamic conditions of rivers and lakes from shallow to deep.

5. Conclusions

The horizontal velocity of three types of MPs—PE, PP, and PS—exhibited an increasing trend with the Re in the range of 600 to 3300, following a power function. Particularly noteworthy is the increased randomness and dispersion in the motion of PP MPs as the Re increases, whereas the dispersion in the motion of PE and PS MPs shows less dependence on the Re relatively.

Based on the horizontal floating velocity and the Re of MPs, alongside the density data for MPs, this study proposes an approximate expression, Vp (Re, ρ), to describe MPs' floating velocity in water, which is finally expressed as V_p (Re, ρ) = 0.518 lg (0.030 × Re + 1.33 × ρ + 42.87) − 0.790.

The results showed that the Re influences the movement velocity of MPs in water. The difference in density of MPs mainly affects their dispersion during the floating process. These findings enhance our understanding of MPs' motion behavior in aquatic environments, offering a valuable scientific foundation for assessing the environmental implications of MP pollution.

Author Contributions: Conceptualization, J.L.; methodology, Z.W. and S.J.; formal analysis, J.L.; investigation, Z.W.; writing—the original draft, J.L.; writing—review and editing, Z.W., C.G.-R. and L.A.; supervision, Z.W. and W.L.; project administration, W.L.; funding acquisition, Z.W. All authors have read and agreed to the published version of the manuscript.

Funding: This research was funded by the Natural Science Foundation of Inner Mongolia Autonomous Region (2023MS05016), the Science and Technology Plan Project of Inner Mongolia Autonomous Region (2023YFHH0041; 2023KJHZ0026), and the China Scholarship Council (202308150174).

Data Availability Statement: The datasets generated and analyzed in this current study are available from the corresponding author upon reasonable request.

Acknowledgments: Thanks to the Inner Mongolia University of Science and Technology, Cooperative Innovation Center of Ecological Protection and Comprehensive Utilization in the Inner Mongolia Section of the Yellow River Basin, and the University of Helsinki for providing technical support for this study. We are grateful to the editors and reviewers for their comments and suggestions.

Conflicts of Interest: The authors declare no conflicts of interest.

References

1. Geyer, R.; Jambeck, J.R.; Law, K.L. Production, use, and fate of all plastics ever made. *Sci. Adv.* **2017**, *3*, e1700782. [CrossRef]
2. Plastics Europe. Plastics—The Fast Facts. 2023. Available online: https://plasticseurope.org/knowledge-hub/plastics-the-fast-facts-2023/ (accessed on 16 February 2024).
3. Dai, L.; Zhou, N.; Lv, Y.; Cobb, K.; Chen, P.; Wang, Y.; Liu, Y.; Zou, R.; Lei, H.; Mohamed, B.A. Catalytic reforming of polyethylene pyrolysis vapors to naphtha range hydrocarbons with low aromatic content over a high silica ZSM-5 zeolite. *Sci. Total Environ.* **2022**, *847*, 157658. [CrossRef] [PubMed]
4. Huang, H.; Mohamed, B.A.; Li, L.Y. Accumulation and fate of microplastics in soils after application of biosolids on land: A review. *Environ. Chem. Lett.* **2023**, *21*, 1745–1759. [CrossRef]
5. Lin, L.; Pan, X.; Zhang, S.; Li, D.; Zhai, W.; Wang, Z.; Tao, J.; Mi, C.; Li, Q.; Crittenden, J.C. Distribution and source of microplastics in China's second largest reservoir—Danjiangkou Reservoir. *J. Environ. Sci.* **2021**, *102*, 74–84. [CrossRef] [PubMed]
6. Dias, B.C.G.; Amarathunga, A.A.D.; De Silva, D.S.M.; Bakir, A.; McGoran, A.R.; Athukorala, A.; Sivyer, D.B.; Reeve, C.; Maddumage, M.D.S.R. The Ecological Implication of Microplastic in Crabs from a Tropical Lagoon: Ingested Microplastic in Mud Crab Scylla serrata. *Water* **2024**, *16*, 3534. [CrossRef]
7. Riaz, H.H.; Lodhi, A.H.; Munir, A.; Zhao, M.; Farooq, U.; Qadri, M.N.M.; Islam, M.S. Breathing in danger: Mapping microplastic migration in the human respiratory system. *Phys. Fluids* **2024**, *36*, 043338. [CrossRef]
8. Arbabi, A.; Gholami, M.; Farzadkia, M.; Djalalinia, S. Microplastics removal technologies from aqueous environments: A systematic review. *J. Environ. Health Sci. Eng.* **2023**, *21*, 463–473. [CrossRef] [PubMed]
9. Winiarska, E.; Jutel, M.; Zemelka-Wiacek, M. The potential impact of nano- and microplastics on human health: Understanding human health risks. *Environ. Res.* **2024**, *251*, 18535. [CrossRef]
10. Isachenko, I.; Chubarenko, I. Transport and accumulation of plastic particles on the varying sediment bed cover: Open-channel flow experiment. *Mar. Pollut. Bull.* **2022**, *183*, 114079. [CrossRef] [PubMed]
11. Liu, Y.; Hao, R.; Shi, X.; Zhang, S.; Sun, B.; Zhao, S.; Huotari, J. Application of a microplastic trap to the determination of the factors controlling the lakebed deposition of microplastics. *Sci. Total Environ.* **2022**, *843*, 156883. [CrossRef]
12. Dittmar, S.; Ruhl, A.S.; Altmann, K.; Jekel, M. Settling velocities of small microplastic fragments and fibers. *Environ. Sci. Tech.* **2024**, *58*, 6359–6369. [CrossRef] [PubMed]
13. Mendrik, F.; Houseago, R.C.; Hackney, C.R.; Parsons, D.R. Microplastic trapping efficiency and hydrodynamics in model coral reefs: A physical experimental investigation. *Environ. Pollut.* **2024**, *342*, 123094. [CrossRef] [PubMed]
14. Stride, B.; Dykes, C.; Abolfathi, S.; Jimoh, M.; Bending, G.D.; Pearson, J. Microplastic transport dynamics in surcharging and overflowing manholes. *Sci. Total Environ.* **2023**, *899*, 165683. [CrossRef]
15. Larsen, B.E.; Al-Obaidi, M.A.A.; Guler, H.G.; Carstensen, S.; Goral, K.D.; Christensen, E.D.; Kerpen, N.B.; Schlurmann, T.; Fuhrman, D.R. Experimental investigation on the nearshore transport of buoyant microplastic particles. *Mar. Pollut. Bull.* **2023**, *187*, 114610. [CrossRef] [PubMed]
16. Singh, P.; Bagrania, J.; Haritash, A.K. Seasonal behaviour of thermal stratification and trophic status in a sub-tropical Palustrine water body. *Appl. Water. Sci.* **2019**, *9*, 139. [CrossRef]
17. Ahmadi, P.; Elagami, H.; Dichgans, F.; Schmidt, C.; Gilfedder, B.S.; Frei, S.; Peiffer, S.; Fleckenstein, J.H. Systematic evaluation of physical parameters affecting the terminal settling velocity of microplastic particles in lakes using CFD. *Front. Environ. Sci.* **2022**, *10*, 875220. [CrossRef]
18. Nakayama, T.; Osako, M. Development of a process-based eco-hydrology model for evaluating the spatio-temporal dynamics of macro- and micro-plastics for the whole of Japan. *Ecol. Model.* **2023**, *476*, 110243. [CrossRef]
19. Liu, J.; Liu, H.; He, D.; Zhang, T.; Qu, J.; Lv, Y.; Zhang, Y. Comprehensive Effects of Temperature, Salinity, and Current Velocity on the Microplastic Abundance in Offshore Area. *Pol. J. Environ. Stud.* **2022**, *31*, 1727–1736. [CrossRef]
20. Zhang, B.; Yang, X.; Liu, L.; Chen, L.; Teng, J.; Zhu, X.; Zhao, J.; Wang, Q. Spatial and seasonal variations in biofilm formation on microplastics in coastal waters. *Sci. Total Environ.* **2021**, *770*, 145303. [CrossRef]
21. Akdogan, Z.; Guven, B. Microplastics in the environment: A critical review of current understanding and identification of future research needs. *Environ. Pollut.* **2019**, *254*, 113011. [CrossRef] [PubMed]
22. Moodley, T.; Abunama, T.; Kumari, S.; Amoah, D.; Seyam, M. Applications of mathematical modelling for assessing microplastic transport and fate in water environments: A comparative review. *Environ. Monit. Assess* **2024**, *196*, 667. [CrossRef] [PubMed]
23. Wu, P.; Huang, J.; Zheng, Y.; Yang, Y.; Zhang, Y.; He, F.; Chen, H.; Quan, G.; Yan, J.; Li, T.; et al. Environmental occurrences, fate, and impacts of microplastics. *Ecotox. Environ. Safe.* **2019**, *184*, 109612. [CrossRef] [PubMed]
24. Nitzberg, E.J.; Parmar, S.; Arbuckle-Keil, G.; Saba, G.K.; Chant, R.J.; Fahrenfeld, N.L. Microplastic concentration, characterization, and size distribution in the Delaware Bay. *Chemosphere* **2024**, *361*, 142523. [CrossRef] [PubMed]

25. Campillo, A.; Almeda, R.; Vianello, A.; Gómez, M.; Martínez, I.; Navarro, A.; Herrera, A. Searching for hotspots of neustonic microplastics in the Canary Islands. *Mar. Pollut. Bull.* **2023**, *192*, 115057. [CrossRef] [PubMed]
26. Gedik, K.; Eryaşar, A.R.; Öztürk, R.Ç.; Mutlu, E.; Karaoğlu, K.; Şahin, A.; Özvarol, Y. The broad-scale microplastic distribution in surface water and sediments along Northeastern Mediterranean shoreline. *Sci. Total Environ.* **2022**, *843*, 157038. [CrossRef]
27. dos Santos Silva, J.; Cidade, M.J.A.; Panero, F.S.; Ribeiro, L.B.; Campos da Rocha, F.O. Microplastic pollution in the Amazon Basin: Current scenario, advances and perspectives. *Sci. Total Environ.* **2024**, *946*, 174150. [CrossRef] [PubMed]
28. Yu, S.; Qin, H. Modeling the effects of plant uptake dynamics on nitrogen removal of a bioretention system. *Water Res.* **2023**, *247*, 120763. [CrossRef]
29. Zeng, S.; Lei, S.; Li, Y.; Lyu, H.; Dong, X.; Li, J.; Cai, X. Remote monitoring of total dissolved phosphorus in eutrophic Lake Taihu based on a novel algorithm: Implications for contributing factors and lake management. *Environ. Pollut.* **2022**, *296*, 118740. [CrossRef] [PubMed]
30. Johannesen, N.J.; Kolhe, M.L.; Goodwin, M. Smart load prediction analysis for distributed power network of Holiday Cabins in Norwegian rural area. *J. Clean. Prod.* **2020**, *266*, 121423. [CrossRef]
31. Kumar, P.; Sharma, A. Reynolds number effect on the parameters of turbulent flows over open channels. *AQUA-UK* **2024**, *73*, 1030–1047. [CrossRef]
32. He, L.; Jiang, J.; Zheng, S.; Yu, N.; Zhu, Y.; Han, W. Numerical simulation and experimental study of microplastic transport under open channel shear flow: Roles of particle physical properties and flow velocities. *Ecotox. Environ. Safe* **2024**, *287*, 117308. [CrossRef]
33. Stride, B.; Abolfathi, S.; Bending, G.D.; Pearson, J. Quantifying microplastic dispersion due to density effects. *J. Hazard. Mater.* **2024**, *466*, 133440. [CrossRef]
34. Jiang, Y.; Zhou, S.; Fei, J.; Qin, Z.; Yin, X.; Sun, H.; Sun, Y. Transport of different microplastics in porous media: Effect of the adhesion of surfactants on microplastics. *Water Res.* **2022**, *215*, 118262. [CrossRef]
35. Kaiser, D.; Estelmann, A.; Kowalski, N.; Glockzin, M.; Waniek, J.J. Sinking velocity of sub-millimeter microplastic. *Mar. Pollut. Bull.* **2019**, *139*, 214–220. [CrossRef]
36. Duraisamy, K.; Iaccarino, G.; Xiao, H. Turbulence Modeling in the Age of Data. *Annu. Rev. Fluid. Mech.* **2019**, *51*, 357–377. [CrossRef]
37. Huang, H.; Zeng, S.; Luo, C.; Long, T. Separate effect of turbulent pulsation on internal mass transfer in porous biofilms. *Environ. Res.* **2023**, *217*, 114972. [CrossRef] [PubMed]
38. Boos, J.P.; Gilfedder, B.S.; Frei, S. Tracking microplastics across the streambed interface: Using laser-induced-fluorescence to quantitatively analyze microplastic transport in an experimental flume. *Water. Resour. Res.* **2021**, *57*, e2021WR031064. [CrossRef]
39. Lee, J.; Choi, J.M. Measurement of near-surface current shear using a Lagrangian platform and its implication on microplastic dispersion. *J. Mar. Sci. Eng.* **2023**, *11*, 1716. [CrossRef]
40. Sherwood, C.R.; Dongeren, A.; Doyle, J.; Hegermiller, C.A.; Hsu, T.J.; Kalra, T.S.; Olabarrieta, M.; Penko, A.M.; Rafati, Y.; Roelvink, D.; et al. Modeling the Morphodynamics of Coastal Responses to Extreme Events: What Shape Are We In? *Annu. Rev. Mar. Sci.* **2022**, *14*, 457–492. [CrossRef] [PubMed]
41. Rahman, A.M.N.A.A.; Rusli, A.; Abdullah, M.K.; Shuib, R.K.; Hamid, Z.A.A.; Ishak, K.M.K.; Makhtar, M.M.Z.; Jaafar, M.; Shafiq, M.D. A review of microplastic surface interactions in water and potential capturing methods. *Water. Sci. Eng.* **2023**. [CrossRef]
42. Alsina, J.M.; Jongedijk, C.E.; van Sebille, E. Laboratory measurements of the wave-induced motion of plastic particles: Influence of wave period, plastic size and plastic density. *J. Geophys. Res. Oceans* **2020**, *125*, e2020JC016294. [CrossRef] [PubMed]
43. Mendrik, F.; Fernández, R.; Hackney, C.R.; Waller, C.; Parsons, D.R. Non-buoyant microplastic settling velocity varies with biofilm growth and ambient water salinity. *Commun. Earth Environ.* **2023**, *4*, 30. [CrossRef]

water

MDPI

Article

Mathematical Description of the Immunochemical Response of the Manila Clam (*Ruditapes philippinarum*) to Extreme Temperature Reductions

Runling Li [1,2], Jianjun Wang [1], Wei Han [3], Jianying Gong [1] and Jun Ding [4,*]

[1] Department of Municipal and Environmental Engineering, Hebei University of Architecture, Zhangjiakou 075000, China; lirunlings@163.com (R.L.); w1wjj@sina.com (J.W.); janlin1987@163.com (J.G.)
[2] Hebei Key Laboratory of Water Quality Engineering and Comprehensive Utilization of Water Resources, Zhangjiakou 075000, China
[3] Zhangjiakou Shangqing Environmental Testing Technology Co., Ltd., Zhangjiakou 075000, China; zjksqhjjc@163.com
[4] Key Laboratory of Mariculture & Stock Enhancement in North China's Sea, Ministry of Agriculture, Dalian Ocean University, Dalian 116023, China
* Correspondence: dingjun1119@dlou.edu.cn; Tel.: +86-13322257023

Abstract: The activity levels of superoxide dismutase (SOD), catalase (CAT), lysozyme (LZM), acid phosphatase (ACP), and alkaline phosphatase (AKP) can reflect the immune status of an organism. The immune status may be affected by extreme changes in the weather, especially rapid declines in temperature. In this study, the SOD, CAT, LZM, ACP, and AKP activity levels of the Manila clam (*Ruditapes philippinarum*) were measured for 24 h while the seawater temperature rapidly decreased to the freezing point from 8 °C, 4 °C, and 2 °C to analyze its immunochemical response to temperature decline. The results showed that the enzyme activity levels fluctuated with time as the temperature declined. By fitting the data, a model was obtained to describe the variation in immune enzyme activity within a short time period as temperature declined. The mathematical description included the stress response and the direct temperature response. The enzyme activity was adjusted rapidly as a stress response in the short term as the temperature declined, before it tended to stabilize. The direct temperature response also caused the enzyme activity to change as the temperature declined to the freezing point. The correlation coefficient between the fitted model and the actual enzymatic activity levels exceeded 0.87, which demonstrated that the mathematical description was adequate.

Keywords: Manila clam; enzyme activity; water temperature; stress response; process formula

1. Introduction

Decades of intense study have been devoted to biological immunity, but the ecological impacts of immunity on disease outbreaks in the ocean remain unknown, even when these diseases affect aquaculture. Thus, the lack of baseline information on the normal immune levels in the ocean affects our ability to assess the novelty of recent disease outbreaks. The total energy in a body is constant, but energy must be expended in response to environmental stress, thus the energy used by immune responses should decrease. Therefore, the weakened immune system will allow pathogens to invade and reproduce more easily [1,2]. Temperature is the main limiting factor in marine environments; low or high environmental temperatures can affect the immune regulatory ability of fish [3], which means that it is essential to determine the impact of water temperature on the immune activity of marine organisms before we can assess the influence of climate.

The innate immune system is simple in invertebrates, but it is important because the innate immune system is highly sensitive to changes in the environment. Shellfish are representative examples of marine invertebrates [4]. In particular, the Manila clam (*Ruditapes philippinarum*) is a type of eurythermal and euryhaline bivalve mollusk. The Manila clam is an important aquaculture product because of its adaptable and rapid growth.

The innate immune system is the only weapon to defend against diseases in invertebrates. The innate immune system comprises cellular and humoral components. The humoral system includes various enzymes and components of the complement pathways, as well as agglutinins, precipitins, natural antibodies, cytokines, chemokines, and antibacterial peptides. Several external and internal factors can influence the activity of innate immune parameters. A better understanding of environmental stress factors, including environmental temperature, salinity, and photoperiods, may provide insights into immune suppression in aquaculture systems.

It has been widely accepted that lysozyme functions as a crucial effector molecule in innate immunity [5]. The effect of water temperature on the activity of lysozyme (LZM) has been examined in previous studies of fish and shellfish. A correlation between the LZM activity and water temperature has been reported in many species [6–12]. Many results demonstrate that the acid phosphatase (ACP) and alkaline phosphatase (AKP) levels are constrained by temperature. The ACP and AKP activity levels of some aquatic animals decreased when the water temperature was excessively high or low, where the immune function decreased, or death even occurred [13–17]. The superoxide dismutase (SOD) and catalase (CAT) activity levels are related to water temperature [18–23], although it was reported that the SOD activity remained stable when the temperature changed slightly [12]. The SOD activity is weaker at low water temperatures [14,24]. The CAT activity level in plants is suppressed by low temperature and it has been shown that the seasonal change in the CAT activity level can be represented by a parabolic curve [25,26].

In the present study, aiming to determine the effects of water temperature on immunity in the Manila clam, the activity levels of LZM, ACP, AKP, SOD, and CAT are measured to assess the innate immunity of the Manila clam to determine their trends as the water temperature declined.

Current research results on the law of change in ecological immunity, whether changes in natural environmental factors or response to environmental pollution [27], are predominantly presented through bar graphs corresponding to test points [6,9,12]. While this conventional approach can reflect the changes in enzyme activity over time or in response to environmental factors, it fails to fully capture the complete process of immune response. Consequently, it hinders the integration of other research outcomes with biological immune alterations resulting from climate change. So, this study employs ecological immunology theory and laboratory test data to deconstruct the enzymatic activity changes in Manila clams into three components: (1) reference values for enzyme activity at specific water temperatures; (2) fluctuations in enzyme activity as a stress response caused by temperature differences during sudden drops in water temperature; and (3) variations in enzyme activity as a cold stress response induced by low-temperature environments. Mathematical expressions that align with their respective response characteristics are then selected accordingly. The sum of these mathematical expressions is utilized to fit the test data and obtain a comprehensive understanding of how enzyme activity levels evolve over time. The results not only reveal ecological immunological patterns depicted within histograms but also provide insights into peak and trough levels as well as their temporal occurrence during fluctuating enzyme activities. By employing this method when quantifying immune-related enzymatic activities across various shellfish species, a predictive model can be established for anticipating immune changes under temperature stress conditions among shellfish populations. If we can predict the impact of temperature on immunity, it is feasible to enhance immunity by artificial means in the case that shellfish immunity is suppressed.

2. Materials and Methods

2.1. Materials

The clams used in this experiment were bought from the same farm. After 1 week of artificial feeding at 10 °C, the water temperature was decreased by 1 °C/day and then maintained at 8 °C for 1 week before the experiment. Oxygen was supplied continually during the experiment and the minimum air temperature in the cold room was −18 °C.

2.2. Experimental Methods

There were three groups in the experiment. The water temperature was decreased rapidly from 7.50 °C (A), 4.20 °C (B), or 1.95 °C (C) to the freezing point in each separate group. The experiment started when the temperature began to decrease. The water temperatures for the three control groups were set at 8 °C, 4 °C, and 2 °C, respectively, and remained stable throughout the entire experiment. All the groups were sampled when the experiment commenced and at 0, 1, 3, 6, 9, and 24 h after the water temperature reached freezing point. The cooling rates were 0.88 °C/h, 0.89 °C/h, 0.88 °C/h. Thirty samples were selected randomly from every group at each time point. Each sample was divided into three subsamples, which were treated as shown in Table 1.

Table 1. Cooling rates for the three groups.

Groups	A	B	C
Initial temperature (°C)	7.50	4.20	1.95
Cooling rate (°C/h)	0.88	0.89	0.88

We randomly selected 50 Manila clams and used the number of dead ones to calculate the mortality rate. Then, we randomly selected 30 live samples from them as a set of experimental samples. To reduce the impact of individual differences, we grouped them into sets of 10, took mixed samples, and analyzed the results by averaging the three parallel groups. The adductor muscle was chosen as the sample tissue.

After the treatment, the adductor muscle was cut from each Manila clam sample. Since the sample was small, to reduce the possible effect of room temperature on the sample, equal weights of the adductor muscle samples were frozen in liquid nitrogen and stored at −80 °C. The activity levels of SOD, CAT, LZM, ACP, and AKP, and the protein concentration in the muscle were measured subsequently in all the samples.

The protein concentrations in the Manila clams were also measured in our study to determine whether there was a correlation between the changes in the protein content and the cold temperature stress. The protein content can be replaced by the protein concentration because the weight of each sample was the same. The metabolism of protein is slow at low temperatures, so the protein content will decrease if protein synthesis does not occur. In our study, the protein concentration tended to decrease. This change agreed with the changes in all the enzyme activities that we measured (Figure 1). The results showed that the changes in the enzyme activity measurements were modulated by physiological changes, and they were not random changes, thereby supporting the credibility of our results.

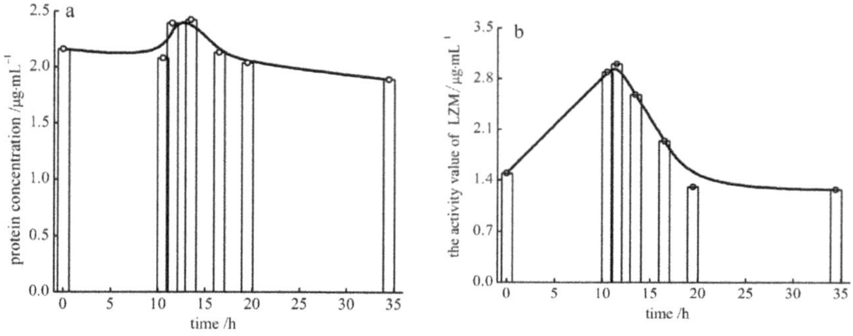

Figure 1. The comparison between response processes of protein concentration (**a**) and enzymatic activity (**b**).

2.3. *Mathematical Description of the Effects of Water Temperature on the Immune Activity Levels*

The physiological changes that occur under stress are complex and a specific function that expresses these changes is not available. Thus, we described the effects of water temperature on the immune activity levels using a process decomposition method, where the overall process (y) was decomposed into three components: the baseline enzyme activity value (y_0), the stress response in terms of the enzyme activity caused by an abrupt decline in temperature (y_1), and the stress response in terms of the enzyme activity caused by low temperature (y_2). Each of these components can be described by a simple mathematical expression, thereby describing the effects of the change in water temperature on the immune activity level (Equation (1)).

$$y = y_0 + y_1 + y_2 \tag{1}$$

2.3.1. Baseline Enzyme Activity Value at Different Temperatures (y_0)

Organisms have the ability to adapt to different temperatures. The thermal acclimation process allows organisms to respond to changes in the environmental temperature by adjusting their physiological responses or behaviors [28]. The activities of enzymes involved in metabolism are maintained at high expression levels to allow an organism to adapt to low temperatures [29].

The changes in the activities of LZM, ACP, AKP, SOD, and CAT are related to temperature. The correlations indicate that the enzymatic activity level is constrained when the temperature is below a critical temperature for immunity, whereas the activity increases when the temperature exceeds the critical temperature, before decreasing when the temperature exceeds the upper bound of the optimal range. Thus, we assumed that there was a specific enzymatic activity value at a corresponding temperature called the baseline value, which was represented by y_0. This relationship is expressed by Equation (2):

$$y_0 = a_0' + (a_0 - a_0') / \left(1 + ((\theta - \theta') / (v \cdot c_0))^{d_0}\right) \tag{2}$$

where θ is water temperature, θ' is freezing temperature, v is cooling rate, a_0 is the measured baseline value of enzyme activity before water temperatures rapidly decreased, a_0' is the measured baseline value of enzyme activity corresponding to the freezing temperature, and c_0 and d_0 are coefficients of curve variation fitted values based on the measured data.

Using the y_0 value of LZM as an example, the change in y_0 (temperature at 7.5 °C to the freezing point and keeping 24 h) with temperature is shown in Figure 2.

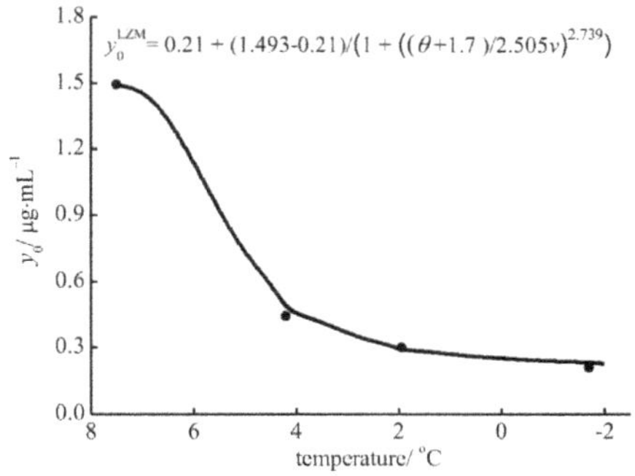

Figure 2. Relationship between y_0 and the measured values.

2.3.2. Mathematical Description of the Stress Response Caused by the Temperature Declines (y_1) and Low Temperature (y_2)

It is important for all organisms to maintain homeostasis. The stress response will be induced to resist changes in the external environment to maintain homeostasis. In shellfish, when a nonspecific stimulus is received, the neuroendocrine axis regulates and controls the stress hormones [30]. These semiochemicals circulate in the blood where they act on the immune system directly or indirectly to stimulate or inhibit the innate immune response.

Shellfish are poikilothermic, so their body temperature matches the external temperature, and no extra energy is needed to maintain a constant body temperature. However, when the outside temperature changes suddenly, a stress response occurs to prevent the body from physical damage, which is caused by the body temperature declining rapidly. Rapid temperature changes lead to a stress response but unsuitable environmental temperatures, such as high and low temperatures, also elicit a stress response. An environmental temperature of −1.5 °C is a stressful environment for most organisms and it will elicit a stress response. The stress response is caused by the temperature declines (y_1) and low temperature (y_2). The changes in enzyme activities in response to stress (y_1 and y_2) with time can be described by Gaussian. This stress response can be described by Equation (3):

$$y_i = a_i \cdot e^{-(t-t_i)^2/c_i^2} \; i = 1,2 \quad (3)$$

where t is time, t_i is peak time, and a_i, c_i are constants, which are determined by curves fitted to the measurements. The results are shown in Figure 3.

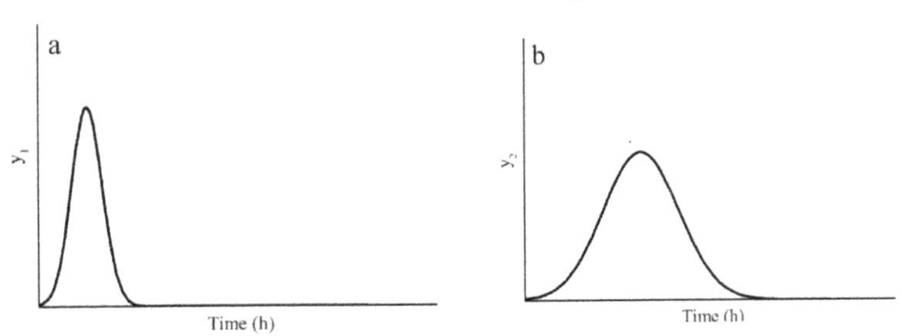

Figure 3. Stress response caused by an abrupt decline in temperature (y_1) (**a**) and by low temperature (y_2) (**b**).

Therefore, according to the general law of this study, after the mutation of environmental factors, the stress response of animals to environmental mutation and the stress response of adapting to new environments have a sequence, and the response to environmental change is the first and the level is higher than that of the latter. That is, the peak time of y_1 is earlier than that of y_2, and the peak of y_1 is greater than y_2.

Taking the LZM activity test data (group C) as an example, the test data are plotted in Figure 4 with a bar chart. The blue solid line in Figure 4 can be obtained by the superposition combination of the three component curves (red solid line, orange solid line, and green solid line). The level of enzyme activity at final stabilization (a_0') expressed by the solid blue line is lower than the level at initiation (a_0). Before it tends to be stable, the response process of LZM shows peaks 1 and 2, whose corresponding activity values are represented by a_1 and a_2, respectively, and their occurrence times are represented by t_1 and t_2. This can be seen in Figure 4, in which a_1 is larger than a_2 and t_1 is earlier than t_2. This is consistent with the stress response process of shellfish immune ecology. It can be seen that the combination of the three components to analyze the changes in shellfish enzyme activity is consistent with the biological principle and the measured LZM data. Trough 1 is caused by the superposition of y_0 rapidly decreasing with time but y_1 and y_2 with lag [30], and trough 2 is formed by the superposition of y_1 and y_2.

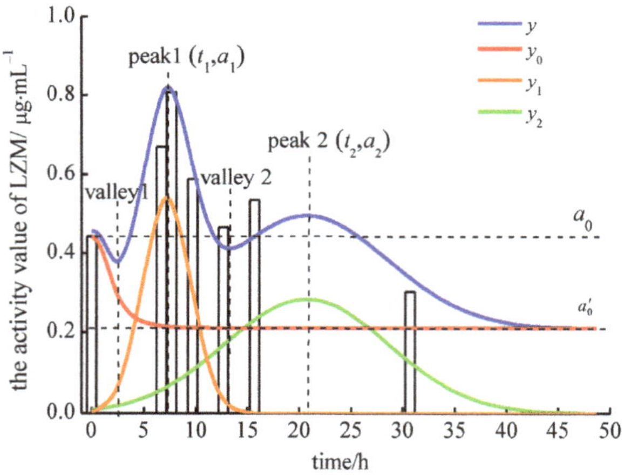

Figure 4. The interpretation of syntagmatic relation between decomposition courses and total course describing the enzyme activity changes on stress responses (taking the example of LZM).

3. Results

According to Figure 4, the response process of enzyme activity to water temperature change can be superimposed by decomposed independent processes. Therefore, the test data should conform to the superimposed combination of Equations (2) and (3), and the whole response process is the sum of three components, that is, Equation (4).

$$y = a_0' + (a_0 - a_0') / \left(1 + \left((T - T') / (\theta \cdot c_0)\right)^{d_0}\right) + \sum_{i=1,2} a_i \cdot e^{-(t-t_i)^2 / c_i^2} \quad (4)$$

3.1. Changes in the Enzyme Activities as the Temperature Declines (y)

The change in the enzyme activity as the temperature declines (y) is equal to the sum of y_0, y_1, and y_2. The changes in the LZM, ACP, AKP, SOD, and CAT activity levels were similar to the temperature range between 8 °C, 4 °C, and 2 °C to the freezing point. The responses of the LZM, ACP, AKP, SOD, and CAT activity levels to the change in temperature are shown in Figure 5a–e.

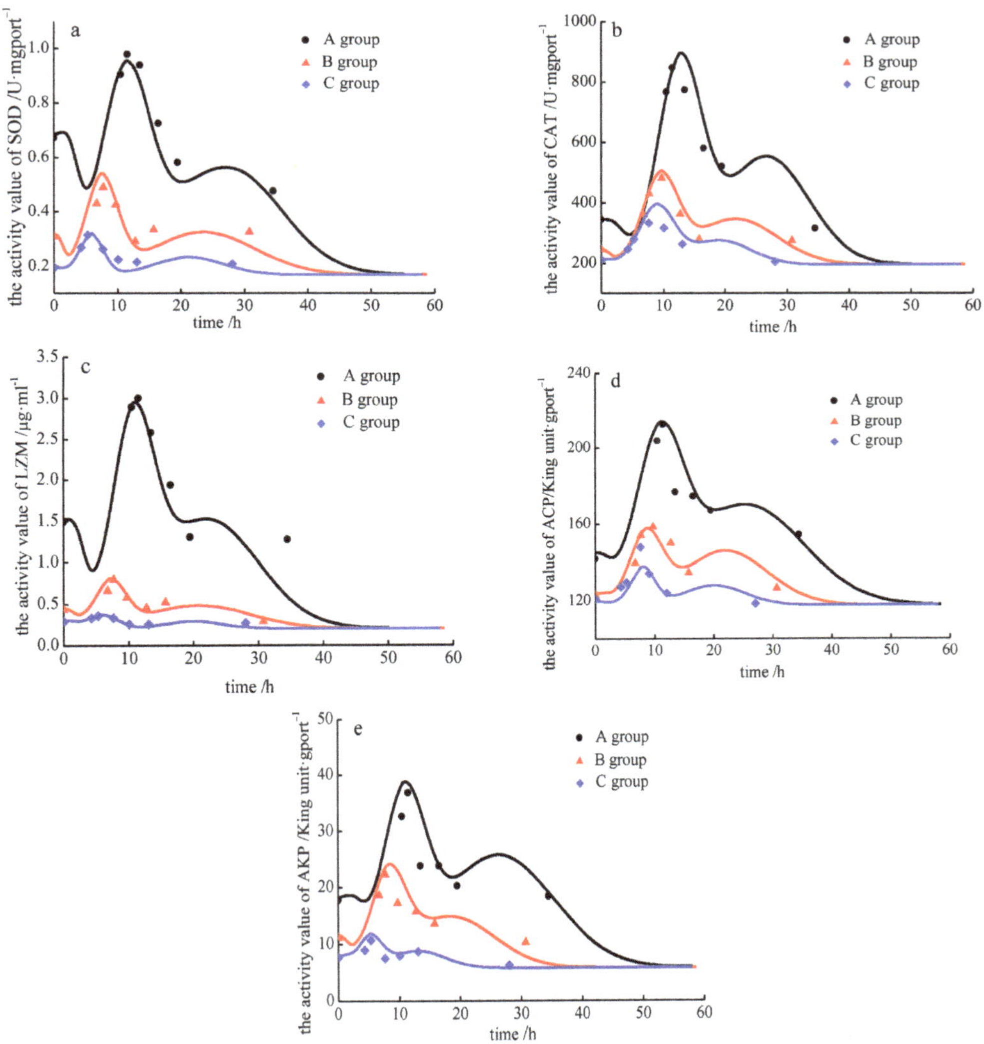

Figure 5. Fitted results for the measured enzyme activity values for the five enzymes. (**a**) SOD; (**b**) CAT; (**c**) LZM; (**d**) ACP; (**e**) AKP.

The relationships between the changes in the enzyme activities were similar (Figure 5), where the enzyme activities varied with the temperature under stress. The peak time of the change in the enzyme activity levels was affected by the temperature difference, where the time when the peak occurred was later when the temperature difference was greater.

The correlation coefficients between the fitted curves and the measurements exceeded 0.87 (Table 2), thereby demonstrating that this new method is suitable for modeling the immune system response in shellfish.

Table 2. Coefficients and correlations of the fitted curves of enzyme activity levels of *Ruditapes philippinarum*.

Enzyme Activity	Group	y_0				y_1			y_2			r
		a_0	a_0'	c_0	d_0	a_1	t_1	c_1	a_2	t_2	c_2	
SOD	A	0.672	0.17	3.893	5.054	0.710	11.265	5.345	0.393	26.469	11.695	0.980
	B	0.312	0.17	1.582	3.931	0.352	7.501	3.490	0.158	23.467	11.520	0.952
	C	0.198	0.17	1.756	2.241	0.150	5.879	2.511	0.065	21.290	7.480	0.955
CAT	A	344.08	196	3.853	4.049	658.297	12.720	5.380	359.352	26.669	9.511	0.968
	B	246.575	196	1.580	2552	280.893	9.497	3.958	152.167	21.665	9.458	0.978
	C	211.962	196	2.100	3.563	192.725	8.922	3.881	81.863	19.096	6.958	0.959
LZM	A	1.493	0.21	3.291	4.656	2.292	10.745	4.402	1.315	22.165	11.015	0.948
	B	0.442	0.21	2.002	2.950	0.542	7.218	3.248	0.283	20.707	11.169	0.874
	C	0.301	0.21	1.978	3.555	0.170	5.942	3.310	0.089	19.906	6.107	0.956
ACP	A	141.88	118	2.884	2.564	78.531	9.914	5.396	52.429	25.468	13.704	0.938
	B	123.363	118	2.643	4.532	35.358	8.579	3.811	28.283	22.021	10.090	0.909
	C	121.157	118	1.414	2.032	19.423	8.005	2.726	10.249	18.069	7.351	0.893
AKP	A	17.782	5.8	4.438	4.045	28.195	10.815	4.791	19.972	26.312	12.780	0.920
	B	11.234	5.8	1.907	2.333	15.186	8.252	3.855	9.093	18.475	9.938	0.939
	C	7.811	5.8	2.182	4.775	5.453	5.193	2.620	3.099	13.004	6.329	0.898

3.2. Changes in Death Rates

The death rates were calculated at each time point in every group and the results are shown in Table 3.

Table 3. Death rates in the three groups (%).

Groups	0	1	3	6	9	24
A	3.0303	6.0606	9.0909	12.8205	11.7647	20.3390
B	3.2258	8.8235	10.0000	13.5128	18.9831	21.6087
C	13.3333	25.0000	28.2609	26.6667	30.0000	36.3636

4. Discussion

Based on the analysis of Figure 5 and Table 2, a clear correlation emerges between the activity of the five immune-related enzymes (SOD, CAT, LZM, ACP, AKP) and the temperature differential. Specifically, a larger cooling temperature difference elicits a more pronounced enzymatic response, as evidenced by the order of peak enzyme activity: Group A (7.50 °C) > Group B (4.20 °C) > Group C (1.95 °C). Furthermore, the magnitude of the peak–valley values suggests that larger temperature differences correspond to increased peak–valley amplitudes.

The timing of peak enzyme activity exhibits a clear correlation with temperature differentials: as the temperature disparities intensify, the duration required for enzyme activity to attain both its zenith and nadir elongates. The alignment between the fitted curve and experimental data underscores that more significant temperature differentials lead to accelerated rates of change in enzyme activity within a confined range. Notably, across all three experimental groups, a consistent ranking of fitting coefficients (a_1 and a_2) was observed for the five immune-related enzyme activities in Manila clam, with CAT ranking highest, followed by ACP, AKP, LZM, and SOD in descending order. All five enzyme activities display an initial peak (t_1) preceding the corresponding second peak (t_2), suggesting that the peak timing of the stress response triggered by an abrupt temperature decline precedes that elicited by prolonged exposure to low-temperature conditions.

The physiological responses of the Manila clam to temperature changes aim to achieve a steady state. However, rapid declines in temperature and ongoing low temperatures trigger a stress response, resulting in initial increases in enzyme activity levels before they decline. Table 3 reveals a significant positive correlation between the mortality rate of the Manila clam and the duration of the experimental period, with an extended reaction time corresponding to an increased mortality rate.

Comparative analysis of the three experiments indicates that the initial temperature plays a crucial role in determining the enzymatic stress response and mortality rate. Our findings suggest that temperature is a major limiting factor in aquaculture, and a predictive model is essential to understand and anticipate the suppression of shellfish immunity under varying temperature conditions.

In conclusion, this study provides a basis for the development of a better predictive model to describe the changes in immune enzyme activity levels in mollusks subjected to sudden temperature drops. By refining our understanding of these responses, we can inform strategies to mitigate the adverse effects of temperature stress on shellfish populations in aquaculture.

5. Conclusions

During the immune-related enzyme activity stress response of mollusks subjected to a sudden temperature drop to freezing point, maintained for 24 h, the enzymatic stress response can be delineated into three distinct phases: (1) the specific benchmark value of enzyme activity at a given temperature; (2) fluctuations in enzyme activity induced by thermal gradients during the cooling process; and (3) variations in enzyme activity resulting from cold-induced stress at low temperatures. The comprehensive response can be mathematically represented as the summation of these three decomposed processes. Notably, the correlation coefficients among experimental data for five immune-related enzyme activities in Philippine scallops exceed 0.87, indicating that this stacking combination method possesses universality and is applicable in practical scenarios.

The relationships between the immune enzyme activity levels and temperature are highly variable because the optimal temperature ranges of organisms are different. The limiting factors of immune enzymes are also diverse and the water temperature differs among areas. Thus, analyzing the effects of temperature on immune enzyme activity levels will be important in future studies. The aim is to describe the changes in the activity levels of immune enzymes using mathematical models, which have the capacity to obtain accurate predictions. Temperature is the main limiting factor in aquaculture; thus, a predictive model may help us to understand the suppression of shellfish immunity. Our study provides a basis for the development of a better predictive model.

Author Contributions: Conceptualization, J.D. and R.L.; methodology, J.D. and R.L.; investigation, R.L.; data curation, R.L.; formal analysis, R.L.; writing—original draft preparation, R.L. and J.D.; writing—review and editing, R.L., J.W., W.H. and J.G.; project administration, J.D. and R.L.; funding acquisition, J.D. All authors have read and agreed to the published version of the manuscript.

Funding: This research was funded by the National Nature Science Foundation of China (51221961), the National Marine Public Welfare Project of China (201105016-3), foundation of State Key Laboratory of Coastal and Offshore Engineering, Dalian University of Technology, China (SL14-5-3), and University-level Research Fund of Hebei University of Architecture (B-201608).

Data Availability Statement: The data are available in the case that it is required.

Conflicts of Interest: Author Wei Han was employed by the company Zhangjiakou Shangqing Environmental Testing Technology Co., Ltd. The remaining authors declare that the research was conducted in the absence of any commercial or financial relationships that could be construed as a potential conflict of interest.

References

1. Paillard, C.; Allam, B.; Oubella, R. Effect of temperature on defense parameters in Manila clam *Ruditapes philippinarum* challenged with *Vibrio tapetis*. *Dis. Aquat. Org.* **2004**, *59*, 249–262. [CrossRef] [PubMed]
2. Cheng, W.; Hsiao, I.S.; Hsu, C.H.; Chen, J.C. Change in water temperature on the immune response of Taiwan abalone *Haliotis diversicolor supertexta* and its susceptibility to *Vibrio parahaemolyticus*. *Fish. Shellfish. Immunol.* **2004**, *17*, 235–243. [CrossRef] [PubMed]
3. Nikoskelainen, S.; Bylund, G.; Lilius, E.M. Effect of environmental temperature on rainbow trout (*Oncorhynchus mykiss*) innate immunity. *Dev. Comp. Immunol.* **2004**, *28*, 581–592. [CrossRef] [PubMed]
4. Chen, M.Y.; Yang, H.S. Molluscan ecological immunology: Review and prospect. *Stud. Mar. Sin.* **2007**, *48*, 140–152.
5. Söderhäll, K. *Invertebrate Immunity*; Springer Science + Business Media, LLC: New York, NY, USA, 2010.
6. Hutchinson, T.H.; Manning, M.J. Seasonal trends in serum lysozyme activity and total protein concentration in dab (*Limanda limanda* L.) sampled from Lyme Bay, U.K. *Fish. Shellfish. Immunol.* **1996**, *6*, 473–482. [CrossRef]
7. Morgan, A.L.; Thompson, K.D.; Auchinachie, N.A.; Migaud, H. The effect of seasonality on normal haematological and innate immune parameters of rainbow trout *Oncorhynchus mykiss* L. *Fish. Shellfish Immunol.* **2008**, *25*, 791–799. [CrossRef] [PubMed]
8. Langston, A.L.; Hoare, R.; Stefansson, M.; Fitzgerald, R.; Wergeland, H.; Mulcahy, M. The effect of temperature on non-specific defence parameters of three strains of juvenile Atlantic halibut (*Hippoglossus hippoglossus* L.). *Fish. Shellfish Immunol.* **2002**, *12*, 61–76. [CrossRef] [PubMed]
9. Bowden, T.J.; Butler, R.; Bricknell, I.R. Seasonal variation of serum lysozyme levels in Atlantic halibut (*Hippoglossus hippoglossus* L.). *Fish. Shellfish Immunol.* **2004**, *17*, 129–135. [CrossRef]
10. Dominguez, M.; Takemura, A.; Tsuchiya, M. Effects of changes in environmental factors on the non-specific immune response of Nile tilapia, *Oreochromis niloticus* L. *Aquac. Res.* **2005**, *36*, 391–397. [CrossRef]
11. Watts, M.; Munday, B.L.; Burke, C.M. Investigation of humoral immune factors from selected groups of southern bluefin tuna, *Thunnus maccoyii* (Castelnau): Implications for aquaculture. *J. Fish Dis.* **2002**, *25*, 191–200. [CrossRef]
12. Yu, J.H.; Song, J.H.; Choi, M.C.; Park, S.W. Effects of water temperature change on immune function in surf clams, *Mactra veneriformis* (Bivalvia: Mactridae). *J. Invertebr. Pathol.* **2009**, *102*, 30–35. [CrossRef]
13. Liu, S.L.; Jiang, X.L.; Hu, X.K.; Gong, J.; Hwang, H.M.; Mai, K.S. Effects of temperature on non-specific immune parameters in two scallop species: Argopec ten irradians (Lamarck 1819) and Chlamys farreri (Jones & Preston 1904). *Aquac. Res.* **2004**, *35*, 678–682.
14. Wang, T.S.; Zhou, X.; Zhao, C.Y.; Xu, Z.H.; Shui, Y. The change about immunoenzyme activity of Procambarus clarkia in different temperature. *Jiangsu Agric. Sci.* **2012**, *40*, 239–241.
15. Zhang, J.P.; Lin, J.C.; Xie, J.J. Purification and Some Characterization of Alkaline Phosphatase from Grass Carps (*Ctenopharyngodon idellus*). *J. Xiamen Univ. (Nat. Sci.)* **2005**, *44*, 684–687.
16. Fang, M.J.; Liu, K.; Li, Z.Q. The Temperature tolerance of immunoenzyme in the Serum of *Hemibarbus maculatus Bleeker*. *J. Zhejiang Ocean Univ. (Nat. Sci.)* **2010**, *29*, 336–339.
17. Castellano, M.; Silva-Álvarez, V.; Fernández-López, E.; Mauris, V.; Conijeski, D.; Villarino, A.; Ferreira, A.M. Russian sturgeon cultured in a subtropical climate shows weaken innate defences and a chronic stress response. *Fish Shellfish Immunol.* **2017**, *68*, 443–451. [CrossRef] [PubMed]
18. Monari, M.; Matozzo, V.; Foschi, J.; Cattani, O.; Serrazanetti, G.P.; Marin, M.G. Effects of high temperature on functional responses of haemocytes in the clam *Chamelea gallina*. *Fish Shellfish Immunol.* **2007**, *22*, 98–114. [CrossRef] [PubMed]
19. Sun, H.S.; Li, G.Y. Activities and properties of superoxide dismutase and catalase in the haemolymph of *Chlamys farreri*. *Oceanol. Limnol. Sin.* **2000**, *31*, 259–265.
20. Kim, M.J.; Kim, J.A.; Lee, D.W.; Park, Y.S.; Kim, J.H.; Choi, C. Oxidative stress and apoptosis in disk abalone (*haliotis discus hannai*) caused by water temperature and pH changes. *Antioxidants* **2023**, *12*, 1003. [CrossRef]
21. Kim, J.A.; Park, Y.S.; Kim, J.H.; Choi, C.Y. Impact of water temperature on oxidative stress and intestinal microbiota in pearl-spot chromis, *Chromis notata* (Temminck & Schlegel, 1843). *Comp. Biochem. Phys. B* **2024**, *275*, 111029.
22. Yu, J.; Yin, Z.H.; Zhang, Y.M.; Bi, J.H.; Yan, X.W.; Nie, H.T. Effects of high water temperature on physiology, survival, and resistance to high temperature air-exposure in the Manila clam *Ruditapes philippinarum*. *Comp. Biochem. Phys. C* **2022**, *262*, 109469. [CrossRef]
23. Karayakar, F.; Yurt, Ö.; Cicik, B.; Canli, M. Accumulation and Elimination of Cadmium by the Nile Tilapia (*Oreochromis niloticus*) in differing Temperatures and Responses of Oxidative Stress Biomarkers. *Bull. Environ. Contam. Toxicol.* **2022**, *109*, 1126–1134. [CrossRef] [PubMed]

24. Song, L.S.; Ji, Y.B.; Cai, Z.H.; Fang, M.; Su, J.G.; Cui, Z.X.; Li, T.W. The immunochemical variation of mitten hand crab Eriocheir sinensis after the increment of temperature. *Oceanol. Limnol. Sin.* **2004**, *35*, 74–77.
25. MacEae, E.A.; Ferguson, L.B. Changes in catalase activity and hydrogen peroxide concentration in plants in response to low temperature. *Physiol. Plant.* **1985**, *65*, 51–56. [CrossRef]
26. Lu, P.; Guo, J.X.; Zhu, L. Soil catalase activity of main plant communities in Leymus chinensis grassland in northeast China. *Chin. J. Appl. Ecol.* **2002**, *13*, 675–679.
27. Rohr, J.R.; Raffel, T.R. Linking global climate and temperature variability to widespread amphibian declines putatively caused by disease. *Proc. Natl. Acad. Sci. USA* **2010**, *107*, 8269–8274. [CrossRef] [PubMed]
28. Wakeling, J.M.; Cole, N.J.; Kemp, K.M.; Johnston, I.A. The biomechanics and evolutionary significance of thermal acclimation in the common carp *Cyprinus carpio*. *Am. J. Physiol.-Regul. Integr. Comp. Physiol.* **2000**, *279*, 657–665. [CrossRef]
29. Lu, X.B.; Wei, Y.L. Progress in the study of cold-adapted enzyme and their cold-adaptation mechanisms. *Biotechnol. Bull.* **2006**, *4*, 51–58.
30. Ottaviani, E.; Caselgrandi, E.; Kletsas, D. Effect of PDGF and TGT-β on the release of biogenic amines from invertebrate immunocytes and their possible role in the stress response. *FEBS Lett.* **1997**, *403*, 236–238. [CrossRef] [PubMed]

MDPI AG
Grosspeteranlage 5
4052 Basel
Switzerland
Tel.: +41 61 683 77 34

Water Editorial Office
E-mail: water@mdpi.com
www.mdpi.com/journal/water

www.ingramcontent.com/pod-product-compliance
Lightning Source LLC
Chambersburg PA
CBHW040316270926
51850CB00007BC/1498
9783725869220